UFOs,
CHEMTRAILS, AND ALIENS

Donald R. Prothero and Timothy D. Callahan
Foreword by Michael Shermer

UFOs,
CHEMTRAILS, AND ALIENS

What Science Says

INDIANA UNIVERSITY PRESS

This book is a publication of

Indiana University Press
Office of Scholarly Publishing
Herman B Wells Library 350
1320 East 10th Street
Bloomington, Indiana 47405 USA

iupress.indiana.edu

Manufactured in the
United States of America

Library of Congress Cataloging-in-
Publication Data

Names: Prothero, Donald R., author.
Title: UFOs, chemtrails, and aliens : what
    science says / Donald R. Prothero
    and Timothy D. Callahan ; foreword by
    Michael Shermer.
Description: 1st [edition]. | Bloomington,
    Indiana : Indiana University Press, 2017. |
    Includes bibliographical references
    and index.
Identifiers: LCCN 2017019981 (print) |
    LCCN 2017006388 (ebook) | ISBN
    9780253027061 (e-book) | ISBN
    9780253026927 (cloth : alk. paper)
Subjects: LCSH: Human-alien encounters. |
    Extraterrestrial beings. | Unidentified
    flying objects. | Curiosities and wonders. |
    Pseudoscience. | Parapsychology. |
    Occultism.
Classification: LCC BF2050 (print) | LCC
    BF2050 .P76 2017 (ebook) | DDC
    001.94—dc23
LC record available at https://lccn.loc.
    gov/2017019981

1 2 3 4 5   22 21 20 19 18 17

We dedicate this book to our wonderful wives,
Teresa LeVelle and Bonnie Callahan,
for all their love and support while writing it.

"Extraordinary claims require extraordinary evidence."

—CARL SAGAN

CONTENTS

11   Cloud's Illusions   335

12   Are They Out There?   349

13   Why Do People Believe in UFOs
      and Aliens?   389

14   The Verdict   408

      Notes   *415*

      Bibliography   *433*

      Index   *449*

FOREWORD

UFOs, UAPs, and CRAPs: Unidentified Aerial Phenomena Offer a Lesson on the Residue Problem in Science

BY MICHAEL SHERMER

One early morning several years ago, a black, triangular object flew over my home in the San Gabriel Mountains of Southern California. It was almost completely silent, made rapid turns and accelerations, and was so nonreflective that it looked like a hole in the sky. It seemed almost otherworldly. It was, in fact, the B-2 Spirit, or "stealth bomber," looping around to make another run over the Pasadena Rose Bowl Parade on January 1—an annual tradition. But had I not known what it was, and had I been seeing it for the first time—say, out in the desert at dusk—I might easily have thought it a UFO. Such is the nature of human perception and belief when it comes to UFOs and aliens, which Donald Prothero and Tim Callahan have so effectively demonstrated in this book—one of the most thorough and effective skeptical analyses of the topic I have read. As they show in case after case, the likelihood that UFOs and aliens are real is exceptionally small, whereas the probability that people make mistakes in their perceptions, have preconceived notions that influence their belief, and cannot seem to say "I don't know" when it comes to an unidentified flying object is extremely high.

In this sense, it is good to remember what the U in UFO stands for: *unidentified. Unidentified* does not mean extraterrestrial, or government conspiracy, or anything else asserted by any other pet theory. The U just means *unidentified*—nothing more. And in science it is always acceptable to say "I don't know." In fact, doing so is a virtue. A case in point is a best-selling book published a few years ago by investigative journalist Leslie Kean, *UFOs: Generals, Pilots, and Government Officials Go on the Record* (Crown, 2010). It is a compilation of anecdotes and stories by military, aviation, and political observers about what she calls Unidentified Aerial Phenomena (UAP). Kean asks her readers to consider "with an open and truly skeptical mind" that such sightings represent "a solid, physical phenomenon that appears to be under intelligent control and is capable of speeds, maneuverability, and luminosity beyond current known technology," that the "U.S. government routinely ignores UFOs and, when pressed, issues false explanations," and that the "hypothesis that UFOs are of extraterrestrial or interdimensional origin is a rational one and must be taken into account, given the data we have."

How much data do we have? And can it help us distinguish between UAPs and what I call Completely Ridiculous Alien Piffle (CRAP) such as crop circles and cattle mutilations, alien abductions and anal probes, and genetic experiments and human–alien hybrids? According to Kean, "roughly 90 to 95 percent of UFO sightings *can* be explained" as "weather balloons, flares, sky lanterns, planes flying in formation, secret military aircraft, birds reflecting the sun, planes reflecting the sun, blimps, helicopters, planes in formation, the planets Venus or Mars, meteors or meteorites, space junk, satellites, swamp gas, spinning eddies, sundogs, ball lightning, ice crystals, reflected light off clouds, lights on the ground or lights reflected on a cockpit window" and more. So the entire extraterrestrial hypothesis is based on the residue of data left over after the foregoing list has been exhausted. What's left? Not much, I'm afraid.

For example, Kean opens her exploration "on very solid ground, with a Major General's firsthand chronicle of one of the most vivid and well-documented UFO cases ever"—the UFO wave over Belgium in 1989–90. Here is Belgian Major General Wilfried De Brouwer's recounting of the first night of sightings: "Hundreds of people saw a majestic triangular craft with a span of approximately a hundred and twenty feet and powerful beaming spotlights, moving very slowly without making any significant noise but, in several cases, accelerating to very high speeds." But even seemingly unexplainable sightings such as De Brouwer's can have simple explanations. It could simply have been an early experimental model of a stealth bomber (U.S., Soviet, or otherwise) that secret-keeping military agencies were understandably loath to reveal.

In any case, compare De Brouwer's narrative to Kean's summary of the same incident: "Common sense tells us that if a government had developed huge craft that can hover motionless only a few hundred feet up, and then speed off in the blink of an eye—all without making a sound—such technology would have revolutionized both air travel and modern warfare, and probably physics as well." Note how a 120-foot craft becomes "huge," how "moving very slowly" changes to "can hover motionless," how "without making any significant noise" shifts to "without making a sound," and how "accelerating to very high speeds" transforms into "speed off in the blink of an eye." This language transmutation—probably unintentional—is common in UFO narratives, making it harder for scientists to provide natural explanations.

In all fields of science, we find a residue of anomalies unexplained by the dominant theory. That does not mean that the prevailing theory is wrong or that alternative theories are right. It just means that we need to do more work to bring those anomalies into the accepted paradigm. In the meantime, it is okay to live with the uncertainty of knowing that not everything has an explanation.

What Prothero and Callahan have accomplished in this book is a model of scientific reasoning, rational analysis, and elegant prose that reveals a phenomenon every bit as interesting as the possibility of alien life's landing here on Earth or existing somewhere out there in the cosmos—the fact that we can conceive of such a concept, study it scientifically, and understand that we are part of the universe as evolved sentient beings capable of such sublime thought. In that sense, science is our myth—the story of who we are and where we came from that ultimately answers the question *are we alone?* The answer at present, however, is *we don't know.*

Michael Shermer is publisher of *Skeptic* magazine (www.skeptic.com), a monthly columnist for *Scientific American*, and a presidential fellow at Chapman University. He is author of *Why People Believe Weird Things*, *Why Darwin Matters*, *The Believing Brain*, and *The Moral Arc*, among other titles.

PREFACE: CONSIDER THE FOLLOWING . . .

Strange lights in the sky. Aircraft flying higher and faster than jet planes at high altitude. Odd-looking objects that appear to be from some alien source. People with stories of seeing alien beings or even being abducted by them and experimented on.

These and many other strange stories are the stuff of modern folklore and all over the media. A large percentage of Americans believe that visitors from space have come to this planet. What do we make of all this?

Perhaps a story from a century ago can give us some insight. It all started when Wallace Tillinghast (a notorious con man) proclaimed in an interview in the *Boston Herald* on December 13, 1909, that he had perfected an airplane that carried three passengers for 30 miles in the air. This was a few years after the Wright Brothers' first flights in 1903. A few other pioneers, such as Glenn Curtis, had made further progress, so the claim seemed plausible—but it was still far beyond the known capability of any aircraft builder of the day. Tillinghast made the hoax even more plausible by claiming that he was planning to win an upcoming aerial competition but that he needed to test his craft at night so that his design couldn't be copied by some other inventor—a reasonable precaution. As soon as the papers began publicizing his claims, people from all over New England and New York claimed to have heard its engines flying overhead or to have seen it fly over them in the light of the moon. People claimed they had seen it fly around the Statue of Liberty, across Boston Harbor, and in many other places—sometimes in multiple places across hundreds of miles in the same night! On December 22 alone, 2,000 people saw it "circling" Boston several times and "flying" for more than three hours, some claiming it had been more than 1,000 feet up in the sky (all impossibilities for any airplane back then). By 7:00, it was reported all over the suburbs of Boston as well, an area spanning many hundreds of kilometers. On December 23, more than 50,000 people claimed to have seen its lights flying over Worcester. The sightings peaked on Christmas Eve 1909 and then began to tail off rapidly. But the day after Christmas, skeptical newspaper accounts began to occur, and on December 27, cries began for Tillinghast to produce proof and fly his plane in broad daylight. Within a few weeks, the hoax had been exposed, and the story soon dropped out of public attention—and public memory.

The New England Airship Hoax is particularly relevant to our examination of the evidence of UFOs, for it is a very similar case to many of the UFO "sightings" we will talk about in this book. Just like many UFO sightings, at its peak thousands of people "saw" lights and "heard" sounds of something that did not exist. Yes, 50,000 observers can be wrong about what they "see"! All it took was one smart hoaxer who made a claim that was plausible in an age of rapid improvements in aeronautics (so that his claim would not seem so outrageous as to be disbelieved), plus a willing media to stoke the frenzy—and thousands of "eyewitnesses" began seeing and hearing something that wasn't there.

ACKNOWLEDGMENTS

This book was the product of many years of activity in the fields of science and skepticism. Both of us have written extensively on these topics for *Skeptic Magazine* and in a number of blog posts for several years. What's more, both of us have read widely in the paranormal and UFO/alien literature and have kept track of many different developments in the paranormal and pseudoscientific world.

Then a series of events led us to write the book. The publication of Loxton and Prothero's *Abominable Science* showed us that there was a demand for a comprehensive, scientific, complete, exhaustively researched book on a paranormal topic—in that instance, cryptozoology. This led us to think that a similar approach, especially one offering extensive discussion of scientific methods, standards of evidence, and the psychology of paranormal beliefs, would be appropriate for the UFO culture as well. In addition, Prothero began to teach meteorology at the college level and found just how much of the UFO literature is simply a product of common weather phenomena.

Accordingly, we approached Indiana University Press with the idea, and this book was born. We thank acquisition and editorial staff at the press for their careful reviews of the manuscript and their help in obtaining images and permissions.

Finally, Callahan thanks his wife, Bonnie Callahan, for her help and support. Prothero thanks his wonderful wife, Dr. Teresa LeVelle, and his sons, Erik, Zachary, and Gabriel, for putting up with his long hours of writing in seclusion and for tolerating the desert heat on our field excursions to Area 51.

We would like to acknowledge the following people for their help in creating this book:

Bonnie Callahan for her part in creating the cover image.

The following artists, editors, companies, and organizations for letting us use their images as illustrations in the book: the Kobal Collection; Phil Langdon; Pat Linse; Daniel Loxtron; Danny Sorocco and Chris Choin of Dimensional Designs; David Sutton, editor of the *Fortean Times*; and Wikipedia and Wikimedia Commons and their contributors, including Diego Delso, the National Museum of E-World Cultures, Rodfico, smn 121, Olaf Tausch, and Napoleon Vier.

The following authors for their writings, websites, and blogs: Jason Colavito, Brian Dunning, and Aaron Sakulich.

UFOs, CHEMTRAILS, AND ALIENS

1

Science and the Paranormal

UFO?

Science is nothing but developed perception, interpreted intent, common sense rounded out and minutely articulated.

—*George Santayana, philosopher*

September 3, 2013: It's the sixth inning of a minor-league baseball game between the Vancouver Canadians and the Everett AquaSox. The game is being played at ScotiaBank Field's Nat Bailey Stadium in Vancouver. A fan is videotaping the game (available on *YouTube*[1]), and you can hear the crowd cheering and clapping and urging the team on the field to play well. The video clip pans from right to left over a mere 26 seconds, and it zooms in on something in the distance, beyond the trees outside the stadium. For those few brief seconds, it appears that there is some sort of flying saucer, complete with a ring of bright lights flashing in all directions, flying off in the distance. Strangely, however, the videographer doesn't hold the zoom on the mysterious UFO but returns to a wide-angle view of the game, then pans farther to the right. Whatever the videographer sees when he or she zooms in on the UFO, it isn't impressive or startling enough to keep him or her focused on it, because he or she goes right back to filming the game.

Nevertheless, this few seconds of footage is soon all over the Internet, mentioned in the news in Vancouver and elsewhere. The *Vancouver Sun* newspaper jokingly calls the object "divine intervention" that helped the local team win the game.[2] Everyone else seems to think that this

1

startling image is proof that UFOs are alien spacecraft, even though few people in Vancouver seem to have noticed it. The other fans in the stands with the videographer who might have reacted much more strongly if it seemed like an alien spacecraft instead were tweeting as if it were nothing unusual. One tweet mentioned[3] that "it hovered for a while, going up and down then gone . . . . weird, but lucky, C's have tacked on 4 runs since it made appearance." The British Columbia news site *The Province* suggested[4] that the "levitating . . . . shiny blue something" may have been a kite or a remote-controlled helicopter.

But this is not enough to stop the huge community of UFO believers from trumpeting the few seconds of footage as "proof" that UFOs are real—without doing any investigation or digging about what was happening in Vancouver that night. They don't even think about the fact that the videographer didn't find the image startling enough to keep filming it and instead went back to the minor-league baseball game. Marc Dantonio, the chief video analyst for the Mutual UFO Network (MUFON), looks at the footage and testifies[5] that the image is not a camera trick or a computer-generated fake inserted into the video after it was filmed. Still, he finds it suspicious that "there was no reaction at all from anyone, nor the videographer." Rather, he says, "I suspect much more strongly, based on the way it was sideslipping to our right a bit while leaning slightly to that side, that this was likely either a lit up kite or a small drone-type object like we created for a National Geographic show. This object's behavior matches either one of those possibilities in the short video snip of it here that we can see. The stability makes this more likely a flying small hobby-type drone."

After the story has spread around the world through the internet, the culprit finally confesses.[6] The "UFO" was indeed a drone. It is eventually revealed that the H. R. MacMillan Space Centre built the drone, shaped like its new planetarium, as a form of gonzo advertising. Working with a local advertising agent, the drone is part of an "extreme tease campaign" to generate excitement and mystery—and lots of free publicity and increased attendance at its new planetarium.

After the hoax is revealed, it continues circulating on the internet as a legitimate UFO with no explanation, even though the fakery has been exposed. Even sadder, an internet poll on *Huffington Post*[7] shows that almost 34% of the people who clicked on the polling buttons *still* believe it is proof of a Canadian UFO sighting!

As we shall discuss in later chapters, this story is typical of most UFO sightings: something is spotted that the observer can't identify, so it becomes an "unidentified flying object," or "UFO." Then the account snowballs into a more exaggerated version as people immediately jump to the conclusion that it is an alien craft. The "unidentified" flying object is then exposed as a hoax or as some other more prosaic natural phenomenon. But even after a satisfactory explanation is given (and even when the hoaxer confesses), people *still* refuse to believe the truth and insist that the object was an alien spacecraft.

We will look at all these elements of typical UFO stories in later chapters. But first: how does a scientist examine the claims about UFOs and aliens?

SCIENCE AND PSEUDOSCIENCE

There are many hypotheses in science which are wrong. That's perfectly all right; they're the aperture to finding out what's right. Science is a self-correcting process. To be accepted, new ideas must survive the most rigorous standards of evidence and scrutiny.

—*Carl Sagan*

Stories such as this one raise important questions: How do we evaluate the claims? How do we decide whether they're credible? As we shall see in this book, there have been a huge number of hoaxes whenever the topic of UFOs and aliens is involved. How can we tell whether we're being conned and fooled?

In our modern society, critical thinking and science have proven to be the most consistent and effective methods of distinguishing reality from illusion. As Carl Sagan put it, "skeptical scrutiny is the means, in both science and religion, by which deep thoughts can be winnowed from deep nonsense." Even though many of us like to imagine that we can have a "lucky streak," and though we might adhere to superstitions such as avoiding walking on sidewalk cracks, when important issues such as money are involved, we all try to be skeptics. As mature adults, we have learned not to be naïve about the world. By hard experience, we are all equipped with a certain degree of healthy skepticism. We have learned that politicians and salesmen are often dishonest, deceptive, and untruthful. We see exaggerated advertising claims everywhere, but deep inside we are experienced enough to recognize that they are often false or misleading. We try not to buy products based on a whim but rather look for the best price and the best quality. We try to live by the famous Latin motto *Caveat emptor*: "Let the buyer beware." When we are dealing with important matters, science and critical thinking are the only techniques we can rely on to avoid being fooled. But what are the principles of science and critical thinking? What do they tell us about UFOs and aliens?

A big problem with our conception of science is that it is based on the classic "mad scientist" stereotypes that are so prevalent in the movies, on television, and in other media. Indeed, there are almost no depictions of scientists in the media that don't include the stereotypical white lab coats and bubbling beakers and sparking Van de Graaff generators—and the "scientist" is usually a nerdy old white guy with glasses and wild hair. But that's strictly Hollywood stereotyping, not reality. Unless a scientist is a chemist or biologist working on material that might spill on your clothes, there is no reason to wear a white lab coat. Even though one of us (Prothero) is a professional scientist, I *never* use a lab coat. I haven't needed one since my days in college chemistry class. Most scientists don't need the fancy glassware or sparking apparatuses—or even have a lab!

What makes someone a scientist is not a white lab coat or lab equipment but rather *how he or she asks questions about nature* and what thought processes he or she employs to solve problems. Science is about suggesting an explanation (a hypothesis) to understand some phenomenon, then *testing* that explanation by examining evidence that might show us whether the hypothesis is right or wrong. Contrary to popular myth, most scientists don't try to prove their hypotheses right. As British philosopher of science Karl Popper pointed out long ago, it's almost impossible to prove statements true, but it's much easier to prove them false. For example, you could hypothesize that "all swans are white," but no matter how many white swans you find, you'll never prove that statement true. But if you find just one nonwhite swan (such as the Australian black swan, Fig. 1.1), you've shot down the hypothesis. It's

FIGURE 1.1. Australian Black Swan. Taken in Victoria, Australia, in January 2009. (Taken by fir0002 | flagstaffotos.com.au Canon 20D + Canon 400mm f/5.6 L (Own work) [GFDL 1.2 (http://www.gnu.org/licenses/old-licenses/fdl-1.2.html)], via Wikimedia Commons.)

finished—over—kaput! Time to toss it on the scrap heap and create a new hypothesis, then try to falsify it as well.

Thus science is not about proving things true—it's about proving them false! This is the exact opposite of the popular myths that scientists are looking for "final truth" or that we can prove something "absolutely true." Scientific ideas must *always* remain open to testing, tentative, and capable of being rejected. If they are held up as "truth" and no longer subject to testing or scrutiny, then they are no longer science—they are dogma. This is the feature that distinguishes science from many other beliefs, such as religion or Marxism or any widely accepted belief system. In dogma, you are told what is true, and you must accept it on faith. In science, no one has the right to dictate what is true, and scientists are constantly testing and checking and reexamining ideas to weed out the ones that don't stand up to scrutiny.

Since Popper's time, not all philosophers of science have agreed with the strict criterion of falsifiability, because there are good ideas in science that don't fit this criterion yet that are clearly scientific. Pigliucci[8] proposed a broader definition of science that encompasses scientific topics that might not fit the strict criterion of falsifiability. All science is characterized by the following: (1) *Naturalism:* We can examine only phenomena that happen in the natural world, because we cannot test supernatural hypotheses scientifically. We might want to say about something that "God did it," but there is no way to test that hypothesis. (2) *Empiricism:* Science studies only things that can be observed by our senses—things that are objectively real to not only ourselves but also any other observer. Science does not deal with internal feelings, mystic experiences, or anything else that is in the mind of one person and that no one else can experience. (3) *Theory:* Science works with a set of theories that are well established ideas about the universe and that have survived many tests.

What is a theory? Some ideas in science have been tested over and over again, and instead of being falsified, they are corroborated by more and more evidence. These hypotheses then reach the status of some idea that is well supported and thus widely accepted. In science, that is what is meant by the word *theory*. Sadly, the word *theory* has completely different meanings in general usage. In the pop culture world, a theory is a wild guess, such as the "theories" about why JFK was assassinated. But as we just explained, in science, the word *theory* means something completely different: an extremely well supported and highly tested idea that scientists accept as provisionally true. For example, gravity is "just a theory," and we still don't understand every aspect of how it works—but even so, objects don't float up to the ceiling. The germ theory of disease was controversial about 100 years ago, when doctors tried to cure people by bleeding them with leeches—but now people who get sick due to a virus or bacterium will follow modern medical practices if they want to get well. Nevertheless, there are people who don't like what science tells them (such as creationists who reject evolution), and they will deliberately confuse these two different uses of the word *theory* to convey the idea that somehow evolution is not one of the best-tested explanations of the world that we have. Yet these same people do not reject the theory of gravity or the germ theory of disease.

Scientists aren't inherently sourpusses or killjoys who want to rain on everyone else's parades. They are just cautious about and skeptical of any idea until it has survived the gauntlet of repeated testing and possible falsification, then risen to the level of something that is established or acceptable. They have good reason to be skeptical. As discussed herein, humans are capable of all sorts of mistakes and false ideas and self-deception. Scientists cannot afford to blindly accept the ideas of one person, or even a group of people, making a significant claim. They are obligated to criticize and carefully evaluate and test it before accepting it as a scientific idea.

But scientists are human, and we are subject to the same foibles as all mortals. We love to see our ideas confirmed and to believe that we are right. And there are all sorts of ways we can misinterpret or overinterpret data to fit our biases. As the Nobel Prize–winning physicist Richard Feynman put it, "the first principle is that you must not fool yourself and you are the easiest person to fool." That is why many scientific experiments are run by the double-blind method: not only do the subjects of the experiments not know what is in sample A or sample B, but neither do the investigators. Samples are coded so that no one knows what is in each, and only after the experiment is run do they open the key to the code to find out whether the results agree with their expectations.

So if scientists are human and can make mistakes, then why does science work so well? The answer is testability and *peer review*. Individuals might be blinded by their own biases, but once they put their ideas forth in a presentation or publications, their work is subject to intense scrutiny by the scientific community. If the results cannot be replicated by another group of scientists, then they have failed the test. As Feynman put it, "It doesn't matter how beautiful your theory is, it doesn't matter how smart you are. If it doesn't agree with experiment, it's wrong."

The mad scientist stereotype that prevails in nearly all media is completely wrong not only because of the stereotypical dress and behavior and apparatuses that are shown but also because the "mad scientist" is not testing hypotheses about nature or experimenting to find out what is really true. In a famous cartoon widely circulated on the internet, someone interrogating a "mad scientist" asks, "Why did you build a death ray?" The mad scientist says, "To take over the world." "No, I mean what hypothesis are you testing? Are you just making mad observations?" The "mad scientist" responds, "Look, I'm just trying to take over the world. That's all." The interrogator continues, "You at least are going to have some of the world as a mad control group, right?" As the cartoon says, he's really not a scientist at all—he's just a "mad engineer." (Although engineers might understand some science, their goal is not to discover truths about nature but rather to apply science to make inventions or practical devices.)

SHAM SCIENCE

The public is happy to admire science as long as they don't have to understand it deeply. Sham inquiry plays to the admiration of science by the public. A lack of familiarity with how science is supposed to work is a major reason why the public has trouble recognizing counterfeit science. Add an '–ology' to the end of whatever you study and it acts like a toupee of credibility—to hide the lack of substance. The public is vulnerable to pseudoscience that resembles real inquiry and genuine knowledge.

—*Sharon Hill*

Because of the prestige and trust that we attach to science, there are lots of con men and zealots out there who try to peddle stuff that looks and even sounds like science but that doesn't actually pass muster through testing, falsification, experimentation, and peer review. Yet it often sounds "sciencey" to most people or imitates the trappings of science, becoming what geologist and skeptic Sharon Hill called "sham science" or "sham inquiry."

A classic case of mistaking the trappings for the real thing are the famous "cargo cults" of the South Pacific islands. During World War II, many of these islands hosted U.S. military bases, and their native peoples came in contact with the advantages of western civilization for the first time. Then the war ended, and the military left. But the natives wanted the airplanes to return and bring their goodies, so they used local materials to build wooden "radio masts," "control towers," and "airplanes" and other replicas of the real things, hoping that they could summon the planes.

A good example of "sham science" is the many paranormal television shows about "ghost hunters" who poke around dark houses with fancy equipment pretending to be "scientific." In other television shows about Bigfoot or other mythical creatures, we see amateur "Bigfoot hunters" blundering around in the bushes in the night with military-style night-vision goggles, completely mystified by each animal noise they hear (which just shows that they are not trained biologists). They set out "camera traps" and other expensive pieces of equipment—which never photograph anything but the common animals of the area. Sure, they are using the trappings of science (expensive machines that look and sound impressive), but are they following the scientific method? No! In these cases, they are violating one of the most important principles that separates science from pseudoscience: *the unexplained is not necessarily unexplainable.* There are many phenomena in nature for which science doesn't yet have an explanation. But scientists know that eventually we'll probably find one. In the meantime, an unexplained mystery is just that: not *yet* explained. Scientists don't jump to the conclusion that it's a ghost or Bigfoot or some other paranormal idea that has never been established to be real by scientific evidence. A UFO is just "unidentified" until further research is done to rule out simple natural causes; it does not automatically become an alien craft.

One common ploy of UFO believers is to bring up a UFO incident that has yet to be explained, assuming that anything not yet explained utterly confounds both science and any possible explanation of the given incident other than its being the work of extraterrestrial aliens. However, just because something hasn't been explained does not mean that we must invoke paranormal causes as a solution for the mystery. In fact, sometimes the most rational approach to a mystery is to accept that it isn't yet, and might never be, solved. A case in point is that of the *Mary Celeste*, an American merchant brigantine that was found adrift and deserted in the Atlantic Ocean, off the Azores Islands, on December 4, 1872, by the Canadian brigantine *Dei Gratia*. She was in a disheveled but seaworthy condition, under partial sail, with no one on board and her lifeboat missing. The last log entry was dated 10 days earlier. She had left New York City for Genoa on November 7 and on discovery was still amply provisioned. Her cargo of denatured alcohol was intact, and the captain and crew's personal belongings were undisturbed. None of those who had been on board—the captain and his wife, their two-year-old daughter, the crew of seven—was ever seen or heard from again. Although there are several theories about what happened to the crew, a number of them quite viable, the simple fact is that we will probably never know what happened to the crew of the *Mary Celeste*. For all that, we needn't invoke either space aliens or the paranormal as the cause of their disappearance. It's quite all right to let a mystery be a mystery. In any case, evidence of extraterrestrial visitation must be positive and testable to be of any worth. The unexplained remains simply the unexplained.

Accordingly, when investigating paranormal claims, we cannot just practice sham science with expensive toys and claim that we're using the "scientific method." No, the first step is to *think* like a scientist, which means testing hypotheses about what the currently mysterious phenomenon might be, then ruling out explanations one by one. If we hear a strange noise on a "ghost hunt" or a "Bigfoot hunt," instead of jumping to the conclusion that it is a ghost or a Bigfoot, first we should rule out the idea that the noise is some common phenomenon, such as the wind blowing through the boards of the "haunted house" or some animal call that we don't happen to recognize. Even if we rule out *every* possible natural explanation for a strange phenomenon, it *still* doesn't give us the right to jump to the conclusion that it must be a paranormal entity. We must still follow the principle that *the unexplained is not necessarily unexplainable*. As scientists, we put the unexplained phenomenon on the back burner, withholding judgment about whether the phenomenon is real until we actually have firm evidence one way or the other. It is not acceptable to jump to a paranormal conclusion without giving science the time to rule out all the normal explanations. Some day we might find out what is really happening, so the paranormal "solution" just gets in the way of doing science properly and distinguishing reality from baloney.

BALONEY DETECTION

Extraordinary claims require extraordinary evidence.

—*Carl Sagan*

So what are the general principles of science and critical thinking that we need to follow if we wish to separate fact from fiction? How can "deep thoughts . . . be winnowed from deep nonsense"? Many of these were outlined in Carl Sagan's 1996 book *The Demon-Haunted World* and Michael Shermer's 1997 book *Why People Believe Weird Things*. Some of the most important principles we must use if we are to decipher fact from fiction in UFO claims include the following.

Extraordinary claims require extraordinary evidence: This famous statement by Carl Sagan (or the similar "Extraordinary claims require extraordinary proof" by Marcello Truzzi) is highly relevant to separating garbage from truth in claims about UFOs. As Sagan pointed out, there are hundreds of routine claims and discoveries made by scientists nearly every day, but most are just small extensions of what was already known and don't require extensive testing by the scientific community. By contrast, crackpots, fringe scientists, and pseudoscientists make revolutionary claims about the world and argue strenuously that they are right. For such claims, it is not sufficient to have just one or two suggestive pieces of evidence, such as blurry photographs or ambiguous eyewitness accounts, when most of the evidence goes against their cherished hypothesis. In these cases, we need extraordinary evidence, such as the actual remains of the alien craft or an alien corpse, to overcome the high probability that these things do not exist.

Burden of proof: In a criminal court, the prosecution has the burden of proof. It must prove its case beyond a reasonable doubt, and the defense need do nothing if the prosecution fails to prove its case. In a civil case, the plaintiff needs to prove his or her case based on a preponderance of the evidence, and the respondent need do nothing. In science, extraordinary claims have a higher burden of proof than do routine scientific advances, because they are claiming to overthrow a larger body of knowledge. When evolution by natural selection was first proposed almost 160 years ago, it had the burden of proof, because it overturned the established body of creationist biology. Since then, so much evidence has accumulated to show that evolution has occurred that the burden of proof is now on the shoulders of the creationists who would seek to overthrow evolutionary biology. Similarly, there is so much evidence to show that the Holocaust occurred (not only the accounts of survivors and eyewitnesses but also detailed records kept by the Nazis themselves) that the burden of proof is on the Holocaust deniers to refute this immense body of evidence. Similarly, most of the extraordinary claims about UFOs and aliens discussed in this book require a much higher degree of proof, because so much of what is claimed about them goes against everything we know from biology, astronomy, geology, and other sciences.

Authority, credentials, and expertise: One of the main strategies of pseudoscientists is to cite the authority and credentials of their leading proponents as proof that their claims are credible. But a PhD degree or advanced training is not enough: a true expert has advanced training *in the relevant field*. It's common for people trying to push an argument to point to their advanced degree (usually a PhD) as they make their case. This is a slick strategy to intimidate the audience into believing that the expert's having a PhD makes him or her smarter than the members of the audience—and an expert in everything. But those of us who have earned a PhD know that it qualifies you to talk only about the field of your training—and, moreover, that during the long, hard slog to get your dissertation project finished and written up, you might actually *lose* some of your breadth of training in other subjects. Most scientists know that anyone who is flaunting his or her PhD in making arguments is "credentialmongering." If a book says "PhD" on the cover, be wary—the arguments between the covers might not be able to stand on their own merits.

Here's a good way to approach it: if you run into someone who flaunts his or her PhD, make sure he or she has some training in a relevant field. Not only did my own training include extensive background in biology and geology (which are relevant to claims about life on other planets), but I have also taught astronomy, meteorology, planetary science, and geophysics at the college level—so I'm familiar with the astronomical issues bearing on the likelihood that life is on other planets (discussed later in this book). As we shall see, most of the "experts" on aliens and UFOs have no training in biology or astronomy, so their "expert opinions" should be taken with a grain of salt. They have no more advanced training in the relevant field than you or I, so their PhD makes no real difference. You wouldn't trust an astronomer to know how to fix your car or write a symphony simply by virtue of his or her having a PhD in astronomy. So why would you trust his or her opinion on biology or some other field in which he or she has no advanced training?

Special pleading and *ad hoc* hypotheses: One of the marks of pseudoscientists is that when the evidence is strongly against them, they do not abandon their cherished hypothesis. Instead, they resort to special pleading to salvage their original idea rather than admitting that it is wrong. These attempts to salvage an idea are known as *ad hoc* (Latin, "for this purpose") hypotheses and are universally regarded as signs of a failed idea. When a psychic conducts a séance and fails to contact the dead, he or she might plead that the skeptic "just didn't believe hard enough" or that the "room wasn't dark enough" or that the "spirits didn't feel like coming out this time." When you demonstrate to a creationist that Noah's ark couldn't possibly have contained the tens of millions of species of animals known, he or she uses evasions such as "only the created kinds were on board" or "fish and insects don't count" or "it was a miracle." If you show a UFO believer the scientific implausibility of his or her claims, he or she usually retreats to some claim that effectively makes the aliens supernatural beings, incapable of being evaluated by any laws of nature or science.

Any time you encounter such special pleading, it is a sure sign that the hypothesis has failed and that the person doing the pleading has abandoned the scientific method and is trying to salvage a favored idea despite the evidence. As the great Victorian naturalist Thomas Henry Huxley said, it is "the great tragedy of science—the slaying of a beautiful hypothesis by an ugly fact."

Ockham's Razor: There is a well-known principle in the philosophy of science known as Ockham's (also spelled "Occam's") Razor, or the Principle of Parsimony. Named after a famous medieval scholar, William of Ockham (1287–1347), who discussed it many times, it basically says that when there are two or more explanations for something that equally explain the facts, the simplest explanation is likely to be the best. In another formulation, it says that we don't need to create overly complex explanations when simpler ones might do. The metaphorical "razor" shaves away the unnecessarily complicated ideas until only the simplest ones are left. Scientists use this as a basic guide when choosing among hypotheses, although we all know that in the real world, once in a while, the complicated explanation is indeed the real one.

Nonetheless, Ockham's Razor is highly relevant when evaluating two versions or explanations of an event. For example, which seems more likely? Is it more likely that aliens traveled many hundreds of light-years just to make a few crop circles in a farmer's cornfield (and do nothing else, including reveal themselves) or that some prankster made the crop circle in the middle of the night using simple tools such as a long board on a rope staked to the ground? As we shall discuss in later chapters, all the difficulties of imagining aliens traveling all that distance just to do something stupid seem positively ridiculous when it's much easier to imagine that it's a prank or a hoax.

In statistics, there is a similar principal: the Null Hypothesis, or the Hypothesis of No Significant Differences. If we want to statistically evaluate whether two things are truly different, we start with the simplifying assumption that there are *no significant differences* ("Null Hypothesis"). Then we must demonstrate that there is a statistically significant difference between the two things before we can make the assertion that they are truly not the same. We can apply this to many other fields in which scientific reasoning is needed. For example, our Null Hypothesis might be that the strange noise in the room is the wind moving the shutters. To prove that it is not something simple and natural such as this but rather a supernatural ghost requires enough evidence to show that the wind could not possibly have caused the noise. Likewise, crop circles or strange lines drawn in the sand can be evaluated this way as well. Our Null Hypothesis is that some simple common phenomenon, such as ordinary pranksters drawing lines in the sand, or dragging a board across the grain field, explains these features. We must prove that these explanations *cannot* explain the feature before we are allowed to embrace the much less parsimonious explanation that aliens came all the way to this planet just to draw lines in the sand or create flattened crops.

SUMMARY

If we wish to scientifically evaluate the claims that UFOs are flying through our skies or aliens landing on this planet, we need to follow the rules of science. In short, we must follow these principles:

· Do not assume that just because we don't have an explanation right now, it cannot be explained by science eventually. The unexplained is not necessarily unexplainable!

· Don't give credence to people who do not have the proper expert training in a field relevant to the claim being examined.

· Don't fall back on special pleading or ad hoc explanations when your favorite explanation falls apart.

· Don't assume a more complicated scenario when a simpler one will do.

· Recognize that for extraordinary claims, the burden of proof lies with the person making that claim to give extraordinary evidence that will overthrow the mountain of evidence against it—not just blurry photos of supposed UFOs or footage of "aliens" that could be hoaxed or just another "eyewitness account" (which are problematic, as we shall discuss in the next chapter).

· No scientific explanation can veer into the paranormal or the supernatural. If you start talking about aliens and spacecraft that violate the laws of science, you're no longer doing science.

· Most important, be prepared to subject your ideas to critical scrutiny and peer review. If you scorn or ignore the criticisms and corrections of others but persist in your beliefs because they are important to you—despite their rejection by science—then you are no longer acting as a scientist but rather are acting like a true believer or a zealot.

The problem boils down to evidence: how good is it? As with the "evidence" for Bigfoot or the Loch Ness monster or other cryptids, photos and videos and footprints and eyewitness accounts are not enough, because all these are easily faked, altered, or hoaxed. A scientist tells the cryptozoologist to "show me the body" (or at least its bones or other convincing piece of tissue). Likewise, a true scientist expects the UFO advocate to "show me the body" of the alien or "show me the spacecraft" before taking these claims seriously and considering them to be valid scientific evidence.

2

The Believing Brain

In the previous chapter, we briefly outlined some of the rules of science, as well as the problems with people's approaching the topics of UFOs and aliens while claiming to be "scientific" but not actually following the rules of science and critical thinking. The issue boils down to the quality of the evidence: how strong is it, and can it pass muster compared to the strict rules of scientific criticism and peer review? Not all evidence is created equal. For example, most of us are not convinced by crudely done drawings or silly stories told by crazy people. But where do we draw the line?

The book *Abominable Science*, by Daniel Loxton and Donald Prothero, discusses the poor quality of the evidence that has been used to argue that Bigfoot, the Loch Ness monster, and other bizarre beasts ("cryptids") are real. A lot of the "evidence" was stuff that was easily hoaxed, such as footprints—and, indeed, most of the original Bigfoot footprints have been shown to be hoaxes. Not much better are the various photographic and video images of different monsters, which either are so fuzzy or blurry or far away that you can't tell what they are or are well-known hoaxes (as confessed by the hoaxers or their descendants). Nevertheless, all this crummy evidence, which even the dedicated cryptozoologists admit is useless, is still touted by these people as somehow supporting their case that these imaginary creatures are real. The cryptozoologists must weed their "lists of evidence" down enormously to get rid of all the garbage, admitted hoaxes, and vague or inconsistent accounts that don't contribute to their cause before anyone can take their field seriously.

The only evidence worthy of attention in the scientific community is undisputable physical evidence: a live specimen, a carcass, or even part of a carcass that can be clearly distinguished from something already known to be real. Even DNA would be a bit more helpful, although the recent claims of "Bigfoot DNA" made by Melba Ketchum and her colleagues have proven to just be incompetent analyses of ordinary human DNA mixed with that of opossums and other wild creatures known to science.[1] Even if we *did* find DNA that didn't match any known species, that still wouldn't make it Bigfoot DNA. It's just DNA from an unknown source until someone actually captures Bigfoot and can determine what its DNA actually looks like.

But the biggest problem with the evidence for paranormal events is that most of this "proof" comes from "eyewitness" accounts and personal testimony. Most of the stories about ghosts, Bigfoot, and Nessie are "eyewitness" accounts, devoid of physical evidence. This is especially true for nearly all the "proofs" used to support claims that UFOs and aliens have visited Earth. They come from human observation, and that kind of evidence is nowhere near as conclusive or strong as most people think.

"I SAW IT WITH MY OWN EYES!"

The human brain is a "belief engine." We form our beliefs for a variety of subjective, personal, emotional, and psychological reasons in the context of environments created by family, friends, colleagues, culture, and society at large; after forming our beliefs we then defend, justify, and rationalize them with a host of intellectual reasons, cogent arguments, and rational explanations. Beliefs come first, explanations for beliefs follow.

—*Michael Shermer, The Believing Brain*[2]

Humans, being storytelling animals, are easily persuaded by the testimony of other individuals. Telemarketers and other advertisers know that if they get a popular celebrity to endorse their product, it will sell well—even if there are no careful scientific studies or FDA approvals to back up their claims. Even the endorsement of your next-door neighbor might be enough to induce you to make simple decisions—but in science, anecdotal evidence counts for very little. As Frank Sulloway put it,[3] "Anecdotes do not make a science. Ten anecdotes are no better than one, and a hundred anecdotes are no better than ten."

Most scientific studies require dozens to hundreds of experiments or cases, along with detailed statistical analysis, before we can accept the conclusion that event A caused event B. In medicine and drug testing, there must be a "control" group that doesn't receive a given treatment, or that instead receives a placebo, so that we can rule out possible effects of the power of suggestion as well as random effects. Only after such rigorous testing, which can rule out the biases of the subjects and the observers, random noise, and all other uncontrolled variables, can scientists make the statement that event A *probably caused* event B. Even then, scientists do not speak in finalistic terms of "cause and effect" but only in probabilistic terms, saying that event A has, for example, a 95% probability of having caused event B.

The same goes for eyewitness testimony, which might have some value in a court of law but is regarded as highly suspect in most scientific studies. Hundreds of studies have shown that eyewitnesses are easily fooled by distractions such as a weapon and can be confused by stress or otherwise misled into confidently "remembering" things that did not happen.[4] This is vividly demonstrated by a video of a stunt with six people passing around a basketball[5] that instructs viewers to count the number of times that players dressed in white shirts pass the ball. Many viewers who focus on counting players' passes completely miss the man in a gorilla suit who walks right through the shot—because their attention is focused on something else.

As Levin and Kramer put it,

> Eyewitness testimony is, at best, evidence of what the witness believes to have occurred. It may or may not tell what actually happened. The familiar problems of perception, of gauging time, speed, height, weight, of accurate identification of persons accused of crime all contribute to making honest testimony something less than completely credible.[6]

Consequently, court systems around the world are undergoing reform as DNA evidence has revealed case after case in which eyewitness testimony resulted in a wrongful conviction. One famous study involved a young woman who conclusively identified the face of her rapist and sent him to prison only to find out, when his DNA was tested, that he was not the culprit. She had seen his face on television and subconsciously associated his face with her (unseen) rapist's face. As psychologist Elizabeth Loftus has shown, eyewitness accounts of events and their memory of them are notoriously unreliable. In her 1980 book *Memory: Surprising New Insights into How We Remember and Why We Forget*, Loftus writes:

> Memory is imperfect. This is because we often do not see things accurately in the first place. But even if we take in a reasonably accurate picture of some experience, it does not necessarily stay perfectly intact in memory. Another force is at work. The memory traces can actually undergo distortion. With the passage of time, with proper motivation, with the introduction of special kinds of interfering facts, the memory traces seem sometimes to change or become transformed. These distortions can be quite frightening, for they can cause us to have memories of things that never happened. Even in the most intelligent among us is memory thus malleable.[7]

Benjamin Radford and Joe Nickell[8] recount several examples that make this point vividly. In 2004, Dennis Plucknett and his 14-year-old son Alex were out hunting in north Florida. Alex was in a ditch about 225 yards away from his father when someone yelled "Hog!" Dennis grabbed his gun, pointed it at a distant moving object and fired. Instead of shooting a hog, he killed his son with a single shot to the head. Alex had been wearing a black stocking cap, not a hog costume or anything else that would have made him look remotely hog-like. Yet at that distance, and primed by the suggestion that there was a wild hog nearby, his father mistook a stocking cap for a wild boar—to tragic effect. Stories of hunting accidents such as these are common, because many hunters shoot first and ask questions later and are often confused by distant objects that are moving—as well as by the propensity of their own imaginations to "see" what they are looking for instead of what's really there.

Take, for example, the Washington, DC, sniper panic of 2002. Early eyewitness descriptions had authorities looking for a white or light-colored box truck or van fitted with a roof rack. Police wasted weeks of time and caused huge traffic jams stopping every vehicle that remotely matched the description. But when the suspects, John Lee Malvo and John Allen Muhammad, were finally caught, the "white van" was actually a dark blue 1990 Chevrolet Caprice. District of Columbia Chief of

Police Charles H. Ramsey said,[9] "We were looking for a white van with white people, and we ended up with a blue car with black people." All that energy and time was wasted because eyewitness accounts were wrong. The *Washington Post* reported[10] that the suspects' Caprice had actually been stopped 10 times during the period of the panic for being near the site of one of the attacks, but each time it was released—because it didn't match the eyewitness descriptions.

Many of these "memories" of strange experiences or eyewitness accounts of bizarre beasts can be attributed to sleep deprivation, dreaming, and hallucination. Michael Shermer[11] describes how he once had an experience of being abducted by aliens. At the time, he was undergoing the stress of an ultra-marathon cross-country bicycle race, and his mind hallucinated the entire experience. As discussed later in this book, most of the famous accounts of alien abductions and out-of-body experiences can clearly be attributed to dreams or hallucinations caused by stress. Despite the "eyewitness" accounts of bizarre aliens, UFOs, or monsters such as Bigfoot or Nessie, there is overwhelming evidence that individuals can hallucinate or imagine things that aren't really there or can be fooled by a stressful, panicked glimpse of something real (such as a bear instead of Bigfoot)—and then their memory and imagination fills in the details later.

What is the proper scientific approach to such evidence? As we shall see, most scientists discount "eyewitness evidence" outright unless there is strong physical evidence to support it. Even the most reliable eyewitness account does not meet the standard of "extraordinary evidence" that we would need to substantiate an "extraordinary claim."

PATTERNICITY

For example, believing that the rustle in the grass is a dangerous predator when it is only the wind does not cost much, but believing that a dangerous predator is the wind may cost an animal its life. The problem is that we are very poor at estimating such probabilities, so the cost of believing that the rustle in the grass is a dangerous predator when it is just the wind is relatively low compared with the opposite. Thus, there would have been a beneficial selection for believing that most patterns are real.

—*Michael Shermer, The Believing Brain, 2011*[12]

Michael Shermer points out that much of the problem with our seeing things that aren't really there is that our brains are programmed to spot patterns even when no pattern exists.[13] He calls this tendency "patternicity." For our early hominid relatives, spotting the pattern of a predator in the grass could save a life, but mistakenly "seeing" a pattern of a predator when there was none costs us nothing, so we are hardwired to spot patterns whether they are there or not.

We all can look at clouds and "see" things even when we know that they are just random patterns in water vapor trapped in the upper atmosphere. The famous Rorschach tests are just random blots of ink on pieces of cardboard, but psychologists still use them today to understand an individual's state of mind. Our brains are so heavily programmed this way that people will see the Virgin Mary in a grilled cheese sandwich or the face of Jesus in a whole range of objects—especially in Catholic parts of the world, where human brains are so heavily bombarded by images of Jesus and Mary that almost any random pattern of light and dark can be shoehorned into appearing like some familiar image.

One familiar form of patternicity is the tendency to force a random set of visual cues into something recognizable, from clouds to Rorschach tests to Jesus in a grilled cheese sandwich. This is also known as *pareidolia*. It is a common phenomenon in the paranormal and pseudoscientific world to "see" things in random noise. A classic case is that of the work of fringe "scientist" Richard Hoaglund, who got huge media attention when he pointed to an image from the 1976 Viking spacecraft as it flew over Mars. In that early, low-resolution image, one particular hill looks somewhat like a face because of how the shadows lie (Fig. 2.1).

Hoaglund not only claimed that this was evidence of Martians' attempts to show their presence but also looked at other shapes on the images and claimed to see huge "pyramids" and other structures that resembled buildings on Earth (except that the ones he claimed to see on Mars would be bigger than our tallest mountains). Even at the time when he made these claims, people who knew something about photography of planetary surfaces found them laughable,[14] but he managed to convince huge numbers of people that a sophisticated Martian civilization had carved a whole mountain into a face that resembled a human one, then built great pyramids and other structures.

FIGURE 2.1. The early low-resolution 1976 Viking image of the Martian surface. At the top is the hill that looked something like a face. (By Viking 1, NASA [Public domain], via Wikimedia Commons.)

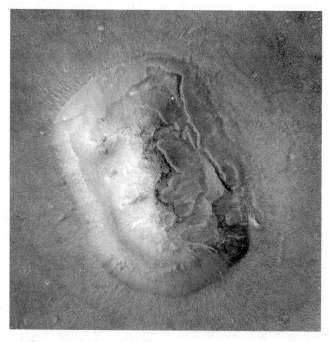

FIGURE 2.2. The same hill, shot on April 8, 2001, with different lighting and at higher resolution, showed that the "face" was just a trick of light and shadow and underscored the human tendency to see patterns when there are none. (NASA/JPL/Malin Space Science Systems.)

But with a few more passes over Mars at a different time of day, the highly suggestive shadows—and thus the "face"—disappeared. Each time the same hill has been looked at again and again with higher resolution and under different lighting conditions, the region appears even less facelike (Fig. 2.2). Astronomer Phil Plait[15] has exposed the bilge that Hoaglund has been spewing—all over one suggestive image—and has shown that Hoaglund has bilked a lot of people out of money, as well as time, with this silly story about "Martian faces." What's more, now that we have rovers wandering Mars, we are getting to know what is *really* there, and it's nothing like the old sci-fi ideas of Mars. At best we find evidence of the presence of liquid water on its surface millions of years ago, but today Mars is a frozen, waterless wasteland, with not the slightest sign that any advanced Martians existed, let alone that they modified the landscape. So far, there are no signs of life—not even the simplest microbes. Forget humanoids carving faces, as so many old sci-fi stories imagined!

Another type of "patternicity" error is *false correlation*. We often make the mistake of assuming that if two events happen together, one must have caused the other. An example is the urban myth of "earthquake weather"—the idea that earthquakes happen on particularly hot days. There is no link between the two, of course, and none would make sense, because earthquakes are generated many kilometers down in the Earth's crust. At that depth, fault lines cannot "feel" daily changes in weather, whose effects penetrate only a few centimeters into the ground. This, like many other examples of two phenomena that seem to be connected but that are not, is a good example of the *post hoc* fallacy (from the Latin *post hoc ergo propter hoc*: "After this, therefore because of this"). The famous Scottish philosopher David Hume put it in a more modern context by pointing out that correlation does not signify causation. Just because two events occur together does not mean that they are causally connected or otherwise related.

Another example of the post hoc fallacy is the old myth that sleeping while wearing your shoes causes a headache. A closer examination of the facts shows that the shoes did not cause a headache, but instead the correlation is due to the fact that drunks who fall asleep without taking off their shoes almost always have a hangover headache the next day.

Athletes and their fans form superstitions in the same way: if they wear a certain shirt and their team wins, they will keep on wearing the "lucky shirt" no matter how many more times the team wins or loses.

A particularly dangerous form of false correlation is the current scare that has (mostly) American and British parents believing in a link between autism and vaccinations. As this "anti-vaxx" movement has become dangerously popular in otherwise well-educated parts of our society, deadly childhood diseases that had been nearly wiped out by vaccination, such as measles, mumps, chickenpox, and whooping cough, are now returning with a vengeance. The parents who buy into the anti-vaccination nonsense are from a generation too young to remember the very real and devastating effects of measles, mumps, whooping cough, tuberculosis, and many other once prevalent diseases—thanks to several decades of effective vaccination programs in our society.

The false correlation came from a fraudulent study (now repudiated and withdrawn) conducted in 1998 by a British doctor, Andrew Wakefield, who is now prohibited from practicing medicine for his fraud and misconduct. He cooked up the phony study as a ploy to generate lawsuits for a lawyer who was secretly working with him.[16] Nevertheless, the myth of the vaccination-autism link swept through the parenting community through internet sites despite its lack of medical credibility. Parents wanted to believe it because autism spectrum disorders (ASD) are heartbreaking and are devastating to a family. By sheer coincidence, it just happens that the first symptoms of autism occur at the age when the MMR (measles, mumps, rubella) shot is given. Thousands of studies later, the possible links have been disproven decisively. More important, recent studies have shown that ASD is largely genetic, being most often passed by older fathers to their sons, so *no* environmental effects (not shots, not food, not siblings, not parenting, nothing) could have changed it.

Anti-vaxxers point to the increase in diagnosed cases of ASD since the mid-1980s, as if that correlation proves causation by shots. Rather, it reflects changes in diagnostic practice: Asperger's syndrome and other variants of ASD were not even defined before then. As these conditions became better known and understood in the medical community, they slowly began to be recognized more and more often in schools and diagnosed more and more often by doctors, the symptoms having become widely known. A number of people have pointed out that correlations could be drawn between cases of ASD and *anything* that has increased since the 1980s. For example, the sales of Macintosh computers have increased since 1984. Does that mean that Macs caused ASD to increase? Or, for that matter, the rise in the fame of Jenny McCarthy (the former *Playboy* centerfold who is the biggest celebrity advocate for anti-vaxxers) corresponds exactly to the rise in number of cases of ASD. Did *she* cause ASD to spread?

The tendency to make false connections between two unrelated events is extremely common in human society. For vulnerable primates on the African savanna, every cue from nature that would save a member of the group from danger was worth remembering, even if the connections were often false. In regions where many snakes are venomous, most animals and people learn to fear all snakes, venomous or not; but in regions that have only nonvenomous snakes, this fear of snakes never develops.[17] All animals learn from mistakes and pain and try not to repeat the behaviors associated with negative consequences. But many of those associations are simply not real.

As Shermer points out, we seek patterns in nature because at one time they were essential for survival.[18] When we were small, vulnerable hominids on the African savannah a few million years ago, it was a matter of life or death whether we spotted the pattern of a lion or leopard in the undergrowth or recognized the rustle of the underbrush as the sound of a predator approaching. Natural selection favored the survival of our ancestors who made Type I errors (seeing a pattern when there was none) and thus were skittish and quick to overreact, because there is no survival penalty for running away from the wind blowing through the bushes. But there *is* a big survival penalty for Type II errors (not seeing a pattern when there is one), because if you fail to recognize the pattern of a predator in the grass, you're lunch! Thus we are heavily programmed to overreact to patterns that aren't there rather than seeing random splotches of light and dark, or random associations of two events in time, as they really are.

AGENTICITY

Souls, spirits, ghosts, gods, demons, angels, aliens, intelligent designers, govern-ment conspirators, and all manner of invisible agents with power and intention are believed to haunt our world and control our lives. Why?

—*Michael Shermer,* The Believing Brain, *2011*[19]

Related to patternicity is another trick of the human brain that Shermer calls "agenticity."[20] In a nutshell, human brains are prepro-grammed to see an invisible, intentional "agent"—be it deity, power, or conspiracy—behind events. As Shermer describes it, our brains cannot easily accept the idea that most things happen by chance, without any plan or design behind them:

As large-brained hominids with a developed cortex and a theory of mind—the capacity to be aware of such mental states as desires and intentions in both ourselves and others—we infer agency behind the patterns we observe in a practice I call "agenticity": the tendency to believe that the world is controlled by invisible intentional agents. We believe that these intentional agents control the world, sometimes invisibly from the top down (as opposed to bottom-up causal randomness). Together patternicity and agenticity form the cognitive basis of shamanism, paganism, animism, polytheism, monotheism, and all modes of Old and New Age spiritualisms. Agenticity carries us far beyond the spirit world. The Intelligent Designer is said to be an invisible agent who created life from the top down. Aliens are often portrayed as powerful beings coming down from on high to warn us of our impending self-destruction. Conspiracy theories predictably include hidden agents at work behind the scenes, puppet-masters pulling political and economic strings as we dance to the tune of the Bilderbergs, the Rothschilds, the Rockefellers or the Illuminati.[21]

As we shall see in later chapters, much of what motivates the believ-ers in aliens is the idea that extraterrestrials are powerful and can affect things on Earth. Do the pyramids seem too well designed for the crude technology of early Egypt? Then an alien intelligence must have done it! Does this string of lights moving in a mysterious fashion at night seems to act too strangely to be a normal airplane? Then it must be an alien spacecraft! Has the evidence for UFOs and aliens from Area 51 or Roswell never been released? Then there must be a government con-spiracy to prevent its release!

THE TRUTH IS OUT THERE!

Modern political religions may reject Christianity, but they cannot do without demonology. The Jacobins, the Bolsheviks and the Nazis all believed in vast conspiracies against them, as do radical Islamists today. It is never the flaws of human nature that stand in the way of Utopia. It is the workings of evil forces.

—*John Gray, political philosopher*

Conspiracy theories: They're just fairy tales adults tell each other on YouTube.

—*John Oliver, television host and comedian*

Conspiracy thinking, in particular, is a huge part of the UFO community's way of thinking. Conspiracy theories are everywhere in our culture, and lots of people indulge in them. Ever since President Kennedy was shot in 1963, there have been dozens of different conspiracy theories about who shot him and why. Conspiracies have been hatched raising doubts about the nature of Princess Diana's death, or claiming that the moon landing was a hoax, or asserting that climate change science is a hoax perpetrated by the entire scientific community in an attempt to destroy capitalism. Just days after the 9/11 terror attacks, dozens of 9/11 "Truthers" emerged claiming that it was all a conspiracy—an "inside job" by powerful forces (insert your favorite conspirator here) for unstated motives. Of course, the 9/11 attacks *were* a conspiracy—by 19 Muslim men who hijacked planes according to a plan hatched by al-Qaeda. But this is not what the "9/11 Truthers" want to accept. No, it must be something bigger and more sinister—usually something planned by the Bush administration.

There are conspiracies about everything now, including the crazy idea that an ordinary military training exercise was an Obama "black helicopter" plot to take over west Texas (even the governor of Texas himself bought into this garbage). Saddest and most sickening are the nutjobs who claim the Newtown massacre of innocent Sandy Hook school kids was a "false flag" operation by the Obama administration that aimed to provide a pretext for confiscating citizens' firearms. Some of these crazies have even harassed the bereaved parents, claiming that the death of their child never happened!

High percentages of Americans (on the order of 25–40%) believe in at least one or more conspiracy theory. What's more, studies have shown that those who believe one conspiracy theory tend to accept many others—sometimes even inconsistent ones. As William Saletan wrote:

> The appeal of these theories—the simplification of complex events to human agency and evil—overrides not just their cumulative implausibility (which, perversely, becomes cumulative plausibility as you buy into the premise) but also, in many cases, their incompatibility. Consider the 2003 survey in which Gallup asked 471 Americans about JFK's death. Thirty-seven percent said the Mafia was involved, 34 percent said the CIA was involved, 18 percent blamed Vice President Johnson, 15 percent blamed the Soviets, and 15 percent blamed the Cubans. If you're doing the math, you've figured out by now that many respondents named more than one culprit. In fact, 21 percent blamed two conspiring groups or individuals, and 12 percent blamed three. The CIA, the Mafia, the Cubans—somehow, they were all in on the plot.
>
> Two years ago, psychologists at the University of Kent led by Michael Wood (who blogs at a delightful website on conspiracy psychology) escalated the challenge. They offered UK college students five conspiracy theories about Princess Diana: four in which she was deliberately killed and one in which she faked her death. In a second experiment, they brought up two more theories: that Osama Bin Laden was still alive (contrary to reports of his death in a U.S. raid earlier that year) and that, alternatively, he was already dead before the raid. Sure enough, "The more participants believed that Princess Diana faked her own death, the more they believed that she was murdered." And "the more participants believed that Osama Bin Laden was already dead when U.S. special forces raided his compound in Pakistan, the more they believed he is still alive."[22]

Conspiracy thinking is strongly self-reinforcing. Polls[23] show that those who accepted the JFK assassination conspiracy were twice as likely to believe that a UFO crashed at Roswell (32% believed, versus 16% among those who didn't accept a different conspiracy). The people who believed in Roswell UFO stories, in turn, were far more likely to believe that the CIA had distributed crack cocaine, that the government had "knowingly allowed" the 9/11 attacks, and that the government adds fluoride to our water for sinister reasons.

Psychological studies have shown that conspiracy thinking is all about the need for control and certainty in a random, frightening world in which everything seems out of control. Conspiracies are nice, simple explanations for scary phenomena that we don't want to believe are simply the result of random events. "Conspiracy nuts" tend to be people who have high levels of anxiety about their lives, their jobs, and their futures and who thus need someone to blame for their troubles and failures. Various psychological surveys have shown that they have a very low level of trust in their fellow human beings or human institutions and tend to have a high degree of political cynicism. They are disposed to believe the worst about other humans. In broader terms, they are people who focus on intention and agency rather than randomness and complexity.

At one time, conspiracy nuts were isolated, and conspiracy thinking was regarded as a form of paranoia and mental illness. These people had little way of reaching each other or getting feedback from like-minded individuals, let alone of finding lots of new conspiracies to read about and believe in. But now that conspiracy theories are only a few keystrokes away, thanks to the internet, they are proliferating at a rate that has never been seen before—because now they feed on and reinforce each other. For example, a 2007 poll[24] found that more than 30% of Americans thought that "certain elements in the US government knew the [9/11] attacks were coming but consciously let them proceed for various political, military, and economic motives" or thought that these government elements "actively planned or assisted some aspects of the attacks." Thanks to relentless conspiracy mongering by the media, polls[25] report that 51% of Americans think that a conspiracy was behind Kennedy's assassination; only 25% agree with the demonstrated reality that Lee Harvey Oswald acted alone.

One of the worst things about conspiracy theories is that they are nearly always airtight, acting like a religion or ideology that refuses to submit to testing and falsification. Every debunking or piece of contrary evidence is viewed as an attempt to "misinform the public," and the lack of evidence is viewed as the result of a government cover-up. Thus conspiracy theorists are very antiscientific, because their closed view of the world will not accept outside information that doesn't fit their core beliefs, much as in a religion or cult. There is much about conspiracy groups that resembles religious cults: suspicion of the outside world, self-reinforcement among like-minded individuals, refusal to look at anything that does not fit their worldview, and an almost messianic devotion to the idea that they have the only truth—that everyone else is foolish, deceived, or part of the conspiracy.

One common theme in conspiracy theories is that they operate on the "slippery slope fallacy": if one conspiracy theory is real, then all the others must be as well. If 9/11 is an inside job, then the Illuminati must also be real. If Michael Jackson/Tupac/ Elvis (insert celebrity here) is alive, then NASA is concealing evidence of intelligent extraterrestrials. But the real world doesn't work that way. Scholars and philosophers know this as the "slippery slope fallacy": one event does not necessarily lead to another. If later evidence *does* show 9/11 to have been an inside job (*very* unlikely, though possible), it doesn't follow that Sandy Hook was a false flag operation. Some conspiracy theorists group *all* conspiracy theories into one big one: Every tragedy was caused to distract from the real problems. War was caused to further the plans of the Illuminati (or because two Illuminati bloodlines wanted to duke it out). A world event was staged to distract us all. A celebrity's death was designed to hide his or her whistleblowing. And a smorgasbord of secret societies all operate to further these plans.

People have a much easier time believing that a huge operation of sinister forces is at work to do something they don't like than they do accepting that "stuff happens." To a conspiracy theorist, the idea that evil forces are ruling the world is much more plausible than the reality—that bad things happen and that we don't really have much control over them. Conspiracy thinking is particularly prevalent among people who have a deep hatred for or distrust of government, so it tends to be concentrated on the conservative fringe (as evidenced by Donald Trump and his embrace of a wide range of conspiracies and crazy ideas). But leftists also exhibit a strain of conspiracy thinking in viewing "Big Pharma," "Big Tobacco," "Big Oil," and the like as more powerful than they really are. Of course, we now know that there were conspiracies by Big Tobacco to suppress antismoking research and by ExxonMobil and some other oil companies to fund climate change deniers and suppress research—but the truth came out in time. Conspiracy thinking also declines slightly with education level: people who know more about how the world actually works tend not to believe in them as much.

The key flaw in conspiracy thinking is that it assumes a level of competence and secret-keeping that has never happened in the history of the world. People often get the idea from television and the movies (particularly from shows such as the *X-Files* and from hundreds of conspiracy-plotted movies, chief among them spy flicks) that secret government organizations are very powerful and are very, very good at keeping secrets. But the opposite has been demonstrated over and over again. Watergate was a grand conspiracy, but eventually it was exposed. For 50 years, tobacco companies conspired to keep research about the death toll of tobacco under wraps, but whistle-blowers in the companies leaked their top-secret memos, and eventually industry executives were indicted and brought to court and in front of Congress. Leaked documents have shown that ExxonMobil covered up its own climate change research and funded a wide range of "front" groups and climate denier groups that used innocent-sounding names to hide their connection to energy companies. Lance Armstrong and just a handful of his closest cyclist friends knew about his doping activities, but even this tiny circle of silence was eventually broken. And despite the fear of *omerta*, or death as a penalty for breaking the Mafia's code of silence, sooner or later someone is the weak link—and the crime bosses go down.

Large secret government operations, such as the Bay of Pigs invasion of Cuba, never work as well as they are planned; eventually, someone screws up, and the whole operation is exposed. The Iran–Contra affair was top secret, but eventually a bunch of people made mistakes and the whole thing was revealed and investigated. As Michael Shermer quips whenever a 9/11 Truther speaks, "You know how I know it's not a big government conspiracy that's been successfully kept secret for many years? Because it happened during the Bush administration." Conspiracy nuts have been claiming that FEMA is preparing concentration camps in which to intern opponents of the Obama administration—a laughable claim considering FEMA does not have that capability (as shown by its botched response to Hurricane Katrina in 2005) and FEMA employees are not sworn to secrecy. Nearly everything FEMA does is completely open.

Conspiracy theorists claim that these top-secret organizations are capable of hiding everything, but as the examples of Wikileaks and Edward Snowden show, sooner or later, a weak link talks or blows the whistle, and government secrets are secret no longer. The Freedom of Information Act has given reporters the power to delve into almost any secret organization, especially governmental organizations, so no secret stays hidden for long. Accordingly, the more people and more organizations that would be required to keep the entire thing hush-hush, the less likely it could actually happen.

A big part of UFO and alien lore is all about government conspiracies and cover-ups. As time goes on, however, many military secrets have been declassified, and the truth is now out. The men who worked at Area 51, who were sworn to secrecy, have all talked about what really happened there—including my father, Clifford Prothero, who worked for the Lockheed Corporation. As we demonstrate in chapter 3, Area 51 was about top-secret spy planes, not UFOs and aliens. The people who were actually there at Roswell have spilled the full story now that the details have been declassified (chapter 4). Many UFO sightings from the 1940s, 1950s, and 1960s turned out to be sightings of known aircraft (especially secret spy planes), their history now entirely open to the public (chapter 3). In short, we must apply Ockham's Razor here as well. Which is more probable: these more down-to-earth claims, or that the entire governments of all the world's countries are colluding together to hide all these secrets? They can't even agree on the simplest of diplomatic issues, let alone maintain an international cover-up for decades! No secret stays hidden long, especially in our internet age. Or is it more likely that there is *no one* behind the curtain pulling the levers—that much simpler explanations will probably suffice? That is the simplest hypothesis, and it uses the proper scientific way of evaluating these claims; accordingly, we will not entertain untestable conspiracy theories in this book for long.

John Oliver, on his HBO show *Last Week Tonight*, did a hilarious sendup of the entire conspiracy theory mind-set, especially the crazy conspiracy YouTube videos. In three minutes, he parodied all the excesses of this way of thinking by "proving" the absurd claim that Cadbury Crème Eggs are a conspiracy by the Illuminati. The video is online and definitely worth watching.[26]

MASS HYSTERIA!

Men, it has been well said, think in herds; it will be seen that they go mad in herds, while they only recover their senses slowly, and one by one.

—*Charles Mackay, 1841,* Extraordinary Popular Delusions and the Madness of Crowds

It is clear that humans as individuals are easily fooled into believing weird things because of faulty observation or altered memories, hallucinations, *pareidolia*, false correlation, agenticity, and many other tricks that our brains play on us. Even more revealing (and particularly relevant to the topic of UFOs) is how large numbers of people can easily be swayed to believe false things in what is known as *mass delusion*. As highly social animals, we are strongly influenced by whatever beliefs our group holds, as well as by whatever we hear or see that our neighbors also believe. This was known to social psychologists even before Charles Mackay's pioneering 1841 book *Extraordinary Popular Delusions and the Madness of Crowds*, and many more examples have occurred since then. Mackay pointed to many examples, from the mania of burning women as witches, to the Crusades, to the fads for Mesmerism and other quack medicines, to the way in which economic bubbles and speculation are driven by groupthink and crowd psychology. Indeed, there are many recent examples of whole populations who can be induced to believe something that doesn't exist simply by social pressure and word of mouth.

Robert Bartholomew and Benjamin Radford point to a wide range of examples in their 2003 book *Hoaxes, Myths, and Manias: Why We Need Critical Thinking*. There are also many examples in Bartholomew's 2001 book *Little Green Men, Meowing Nuns, and Head-Hunting Panics: A Study of Mass Psychogenic Illness and Social Delusion*. We are all familiar with the many instances of "witch trials" and burnings, when innocent women were massacred because of a mass delusion centering on witches—and of how the famous 1938 Orson Welles broadcast of a radio dramatization of H. G. Wells's *War of the Worlds* caused a panic among people who thought they were hearing about a real Martian invasion in New Jersey.

Another famous case of making something out of nothing was the "Mad Gasser of Mattoon." First reported in Boutetort County, Virginia, in the 1930s, and then again in Mattoon, Illinois, in the 1940s, both cases involved a rumor going around a community that someone (the descriptions varied widely) was sneaking into people's homes and knocking them unconscious by means of some anesthetic gas. The first accounts in each case were of people who developed choking or coughing and sometimes even passed out. They started explaining this event as something done to them by the "gasser"—and as soon as the newspapers picked up the story and reported it, everyone began making similar claims if they saw anything suspicious in the shadows or thought they smelled gas. Soon there were reports all over the city at exactly the same time, ruling out the possibility that an individual, or even a group of individuals, could pull it off. Eventually, the hysteria died down as the police stopped taking the accounts seriously, the newspapers stopped reporting new cases, and people no longer copied their neighbors. In every case, it is clear that the "mad gasser" never existed; rather, its extraordinary influence was due to mass panic fed by the media, not by any reality.

More than anything, however, the Phoenix Lights demonstrate the fallibility of both eyewitness testimony and human memory as well as the story-telling, pattern-seeking nature of not only our centers of cognition but also our perceptions. The Phoenix Lights appeared in 1997, again in 2007, and a third time in 2008. It should be stressed that, in all three of these occasions, the lights were witnessed over wide areas by many people, all of whom were reputable witnesses. The witnesses were not subject to delusional thinking, nor were they imposing a preconceived agenda on what they saw. These witnesses were not, in the main, UFO enthusiasts—and they really *did* see something unusual in the skies of the southern tip of Nevada and southwestern Arizona, particularly over Phoenix.

The initial sighting of what came to be known as the Phoenix Lights was in Henderson, Nevada, a bit southeast of Las Vegas. This was at 7:55 p.m. Mountain Standard Time on March 13, 1997. The witness in Henderson reported seeing a V-shaped object, about the size of a 747. It had six lights on its leading edge and made a sound like rushing wind. It was traveling southeast.

The next sighting was by an unidentified former police officer in Paulden, Arizona, at about 8:15. He reported seeing a cluster of reddish or orange lights—four lights together and a fifth trailing light. Through binoculars, he could see that each individual light was made up of two light point sources. The lights eventually disappeared over the southern horizon.

At 8:17, callers in the Prescott Valley began reporting an object that they said was solid: it blocked out stars as it passed. One of these witnesses was John Kaiser, who was outside with his wife and sons. He said that the lights formed a triangular pattern. The leading light was white, but the others were red. He said they passed directly overhead, banked to the right, then disappeared to the southeast. Although observers in the Prescott Valley couldn't determine the lights' altitude, they were low. The supposed craft made no sound. Another observer in the Prescott area also saw a V-shaped line of lights, the first three of which were clustered at the point of V, with two lights farther back.

One of those who saw the lights above the Prescott Valley was Tim Ley. He, along with his wife, Bobbi; his son, Hal; and his grandson, Damien Turnidge, initially saw them as five lights in an arc. They soon realized that the lights were moving toward them. As the lights did so, over the next 10 minutes, they resolved into a V shape similar to a carpenter's square, or resembling two sides of an equilateral triangle. The Leys, like other witnesses, reported a huge object, discernable not only by five lights on its leading edge but also because it blotted out stars in the night sky as it passed directly over them. Soon the object appeared to be coming right down the street where they lived, about 100 to 150 feet above them, traveling so slowly that it appeared to hover and was silent.

When the triangular formation entered the Phoenix area, Bill Greiner, a cement driver hauling a load down a mountain north of Phoenix, saw them. He stated that the lights hovered over the area for more than two hours.

Quite a different story from most was told by amateur astronomer Mitch Stanley, who lived in Scottsdale, a suburb of Phoenix. He observed high-altitude lights flying in formation using a Dobsonian telescope giving 43× magnification. After observing the lights, he told his mother, who was present at the time, that the lights were aircraft. According to Stanley, the lights were quite clearly individual airplanes; a companion who was with him recalled asking Stanley at the time what the lights were, and he said, "planes." When Stanley first gave an account of his observation at the Discovery Channel Town Hall Meeting in the presence of all the other witnesses, he was shouted down when he claimed that what he saw must have been what other witnesses saw. Some have claimed that Stanley was seeing the Maryland National Guard jets flying in formation during a routine training mission at the Barry M. Goldwater bombing range south of Phoenix. It is possible that the Phoenix

Lights V was actually a group of planes, as noted by reporter Tony Ortega when describing a similar sighting in southern California:

> Mitch Stanley's sighting jibes well with witness reports that the configuration of the lights changed over time. In Prescott, for example, witnesses claim that one of the lights trailed the rest. Such evidence supports the claim that the lights were separate objects rather than one large craft.[27]

This also fits Ley's observation that the lights' arrangement changed from a crescent to a V. Fortunately, in addition to the testimony of many witnesses, we have videos taken of the 1997 incident.[28] They show a series of lights appearing in the sky one by one, then winking out one by one. In one of the videos, the man shooting it exclaims, "Another one just showed up!" In that video, the first three lights form a line, and then a fourth appears in such a position as to make an angle. In another video, this one without sound, one light appears, then another, then more—up to five lights, then six. These first take a shallow V shape, then become a more or less straight line. Finally the lights wink out one by one. None of the videos shows a solid V-shaped object blotting out the stars as it moves overhead. In fact, in most of the videos, the lights simply hover rather than moving in a discernable direction. It seems that much of what witnesses saw resulted from the perceptual centers of their brains' automatically filling in the spaces between the lights to create a whole object.

The lights appeared again on February 6, 2007, an event that military officials and the Federal Aviation Administration attributed to flares dropped by F-16 Fighting Falcon jets hailing from Luke Air Force Base, out on a training mission. Military flares dropped by aircraft to light a battlefield at night hang from parachutes and hover for quite some time before hitting the ground. Such flares are a possible explanation for the initial 1997 incident.

The last sighting of the Phoenix Lights was on April 21, 2008. Witnesses said the lights formed a vertical line, then spread apart and made a diamond shape. The lights also once formed a U shape. Tony Toporek videotaped the lights. He was talking to neighbors at 8:00 p.m. when the lights appeared. A valley resident reported that shortly after the lights appeared, three jets were seen heading west in the direction of the lights. On this occasion, the Air Force denied engaging in any activity in the area. On

April 22, 2008, a resident of Phoenix told a newspaper that the lights were nothing more than his neighbor releasing helium balloons with flares attached, and the following day a Phoenix resident who declined to be identified confessed that he had indeed perpetrated a hoax, attaching flares to helium balloons and releasing them from his back yard. However because such activity would have been done over a residential area during a record drought, the alleged hoaxer chose to remain anonymous.

Journalist Benjamin Radford noted that although people often confess to crimes they haven't committed, several clues support the hoax allegation.[29] First, the lights moved independently—rising together, then drifting apart as they gained altitude. They also drifted to the east, in the same direction as the wind. Second, had the lights been part of a single flying metallic structure, they would have shown up on radar—but they did not. However, were they flares attached to balloons, they would be invisible to radar. Witnesses reported that the lights went out one by one after being visible for between 15 and 30 minutes. This, again, is consistent with what we would expect to see if the lights were flares. Finally, one of the confessed hoaxer's neighbors, a Mr. Mailo, actually saw the lighting of the flares at about 8:00 p.m., just before the lights were reported. Flares, whether dropped by parachute—as the Air Force claimed was the case in both 1997 and 2007—or released from the ground as a prank, also fit the pattern of lights' appearing and winking out individually as seen in the 1997 videos.

One of us (Callahan) was tangentially involved in a similar prank while he was attending the Chouinard Art School of the California Institute of the Arts, then located in the MacArthur Park neighborhood of Los Angeles, between 1964 and 1968. He had gone out with friend for coffee in the evening, when they came upon some fellow students sending up much cruder craft than the helium balloons with road flares of the Phoenix hoaxer. These balloons were made of large laundry bags, the open ends of which were attached to crosspieces of balsa wood. Attached to these were several birthday cake candles. Once lit, the candles filled the laundry bags with heated air, and the inflated bags slowly rose into the night sky. Once they had gained sufficient altitude, the laundry bags, diffusing the light, appeared as indistinct glowing spheres that hovered or slowly moved in erratic paths. Callahan and a friend then continued down Eighth Street pointing at the lights and telling people that they were UFOs.

One final problem for those who believe that the Phoenix Lights were the leading edge of a V-shaped spacecraft (even aside from its silence and slow speed) is the failure, particularly in light of the many witnesses who saw the lights and reported them to the authorities, of jets from nearby airbases to scramble and intercept the strange intruder.

SUMMARY

For scientists to take the existence of UFOs and aliens seriously, we must weed out all the crummy evidence that cannot withstand scientific scrutiny. The largest component of such evidence comes from "eyewitness" accounts. The last few decades of scientific research into the "believing brain" shows that our perception and memory are not perfect video cameras; rather, we perceive things selectively, miss detail, or enhance and even add to our memories the longer we think about them. Thus eyewitness testimony is vanishing as reliable evidence in courts of law and is next to useless as proof of the existence of UFOs or aliens.

The human brain has many blind spots that cause errors in our interpretation of data. We tend to force random noise into patterns that are meaningful to us but that don't really exist (pareidolia or "patternicity"), we tend to falsely assume that two events that happen to coincide in time are causally related ("false correlation"), and we tend to see invisible powerful forces or agents behind the events that happen in life ("agenticity"). All these errors in perception must be critically analyzed and teased out of any evidence that someone advances to support the existence of aliens or UFOs.

In particular, we need to be very careful about the strong effects that mass hysteria, social pressure, and media reports have on making masses of people "see" and "hear" things that weren't real. As the New England airship hoax of 1909 shows (see preface), large numbers of "eyewitness accounts" are no more reliable than individual accounts. Accordingly, "UFO sightings" by multiple witnesses (who are often prompted by the media and fueled by the power of suggestion) are no more believable than those originating with a single individual.

3

Area 51: What Is *Really* Going on in There?

We are driving west in a black GMC Yukon Denali SUV across the "Extraterrestrial Highway" (Nevada State Highway 375), about three hours north from Las Vegas. The road itself is unremarkable—miles and miles of a ribbon of asphalt cutting across barren desert of mesquite and Joshua tree yuccas, with no signs of life anywhere. Occasionally the road rises up from the low flats to cross a small mountain range, with jagged rocks exposed on all sides, completely devoid of vegetation. During the summer, the temperatures here stay above 100 °F for weeks on end, and almost no one comes through here. In the winter, the daytime temperatures are more comfortable, but at night it gets bitterly cold, especially if the desert winds are howling through the area. It's also above 4,400 feet elevation here, so some winters are cold enough that snow will accumulate on the high desert surface, and snow may persist on the peaks well into the spring.

After you pass through the tiny towns of Alamo and Ash Springs (last gas station for 150 miles or farther) on U.S. Highway 93 and turn west on to Highway 375 (the "Extraterrestrial Highway"), you drive about 15 miles until you reach Hancock Summit, a mountain pass over barren rock that is the highest place in the region. You can get out of your car and look to the southwest, but all you will see is the Groom Range to your west. Area 51 is down in the valley beyond, and there is no other spot

FIGURE 3.1. Satellite view of Area 51, showing Groom Lake and the airstrips and buildings. (By NASA [Public domain], from World Wind, developed at NASA Ames Research Center by Chris Maxwell, Randy Kim, Tom Gaskins, Bruce Lam, and project manager Patrick Hogan.)

in any direction where you could see the base from the paved road. You can make the strenuous hike to Tikaboo Peak to the south and see parts of the base on the other side of the range without incurring the wrath of base security, but this isn't much more informative.

The isolation of the base was deliberate. Its location was purposely chosen to be as remote as possible. It is impossible to see from the paved road no matter how high up you drive. Until recently, the only way to see the base was by airplane, and the entire airspace above the base is restricted—so unless you want to be chased away by fighter jets, you don't try to fly over it with a civilian craft. Then, when spy satellites and especially Google Earth became available to anyone with a computer, it was possible to get satellite images of the base. But all these show (Fig. 3.1) are a series of airstrips and a bunch of buildings down on the valley floor next to the dry Groom Lake bed. Sure, you can see that there is military activity there, but this reveals nothing that is worth getting excited about. It looks just like any other desert military airstrip.

Driving southwest and down from Hancock Summit, somewhere between mile markers 34 and 35, you come to a right curve, where the paved highway veers off to the northwest. Going straight ahead to the southwest is Groom Lake Road, the primary entrance to the base, deliberately left unpaved and unmarked and hard to find so that only authorized people will drive this way. If you veer off down this dusty road, there are sensors buried beneath it that will let them know you're coming. After about 13 miles, you come to the first signs that warn you that this is a closed military installation and that photography is not permitted. Meanwhile, on the hills above the road are several lookout spots where the "camo dudes" (as some people call the security personnel) are watching you from their white Jeep Cherokees or Chevy pickup trucks with high-powered binoculars. If you stop and look around, you'll see not only their vehicles but also security cameras on posts with a closed-circuit TV feed so that the entire base security can watch you. Finally, at 13.8 miles from the unmarked turnoff from Highway 375, the last sign warns you that you are on the base perimeter (Fig. 3.2). The signs warn you not to go further, noting that photography is not permitted and that the use of deadly force is authorized. These folks aren't kidding! The perimeter is marked not by a big fence but rather by a series of large orange posts spaced about 50 yards apart on the base borderline. As long as you stay in your vehicle and don't cross the line, you'll be okay. Lately, the guards have even begun to tolerate photography of the signs and the base perimeter. But if you drive past the signs or get out and walk too close to the line of orange posts, they will swoop down, arrest you, and turn you over to the Lincoln County Sheriff, where you will have to pay a steep fine for trespassing on a restricted military base.

If you were allowed to pass this perimeter fence, you would drive another 0.8 miles through a canyon through the mountains before you even got to the actual guard post. Tourists and casual UFO fans never get this far, because the checkpoint is well inside the restricted area. And if you're driving the road, you need to pull completely onto the shoulder to let the big white bus speeding out of the base pass you as it brings

FIGURE 3.2. The warning signs at the Groom Lake Road entrance to the base. (Photo by D. R. Prothero.)

authorized personnel out. It typically leaves the base about 4:40 each weekday afternoon.

Hikers try to sneak inside the base perimeter and get photographs from the mountains that overlook Groom Lake, but most are caught quickly before they ever get there. If you are lucky and stick around long enough, you'll see their Pave Hawk security helicopters swoop overhead, looking out for hikers trying to sneak past the perimeter. At one time, these choppers would swoop down low and "sandblast" trespassers with the downwash from their rotors to drive them back and discourage them, but that practice is no longer allowed.

Driving past the unmarked entrance to Groom Lake Road, about five miles farther to the northwest on Highway 375 (between mile markers 29 and 30), there's a small mailbox by the side of the wide graveled parking area south of the main highway. This is the legendary "black mailbox" so popular in the lore of UFO fans. Oddly, it's not black at all (Fig. 3.3) but rather a dull dirty white, and it has so many stickers from all over the world and so much graffiti scrawled on it that the color is almost obscured. The "black" refers to the original mailbox that was here before 1996, a regular black-painted county mailbox that you could buy in a local

FIGURE 3.3. The "black mailbox," showing its current condition and graffiti. (Photo by D. R. Prothero.)

hardware store. Despite all the legends about it, the truth is much less glamorous. It's not a mailbox for the base at all! Naturally, the military has its own system for delivering mail and would never use a mailbox accessible to the public. Instead, it's the mailbox for the only rancher in this area of the Tikaboo Valley. He has to cope with a constant problem of UFO fans going through his mail looking for "top-secret military posts" and breaking his mailbox open—even shooting at it! The original mailbox was eventually removed and auctioned off to a UFO fan for $1,000, and the current mailbox has been there since 1996. It is tightly padlocked and bulletproof, so the only thing tourists can do is decorate its outside. But the wide gravel pullout around the mailbox is a favorite place for UFO fans to hang out at nighttime and even camp overnight, trying to get a glimpse of the lights of anything flying in the skies.

Finally, if you travel about 40 miles from the eastern end of the "Extraterrestrial Highway," you reach Rachel, Nevada, the only "town" in the entire area. Calling it a town is generous: the 2010 census counted only 54 residents, all of whom live in rows of trailers just south of the highway, baking in the desert sun. Most of the residents are retirees living cheaply on fixed incomes. There is no post office, and the few children in town must ride buses all the way to Alamo, Nevada, for school.

The only thing worth seeing in Rachel is the legendary "Little A'Le'Inn," a quirky restaurant, motel, and bar that is the meeting point for all the UFO fans who pass through the area. It has a big tow truck out front next to the sign with its crane and tow hook holding a "flying saucer" made out of old satellite TV dishes (Fig. 3.4). Inside, the walls are lined with UFO articles and paraphernalia, and the ceiling over the fully stocked bar is thickly plastered with dollar bills that have been stapled up there for an unspecified reason. The menu is mostly typical coffee-shop fare, although the "Alien Burger" is quite tasty. (Its meat is not alien, of course but rather probably from some of the cattle that wander across the entire area of open range). Here's how one interview described the place:

> A'Le'inn owner Connie West says that Rachel plays host to people of every shape, size and nationality. With seven rooms for lodging, the inn, restaurant and gift shop has no typical customer. "I've got people from all walks of life coming out here. Some pretty normal, some out there," says West. "People bawl their eyes out—they've been waiting their whole life to be here. We are also a watering hole for the military, or anyone who comes down the highway." She once had to kick out two men who showed up wearing nothing but silver and green spray paint. "I live eight miles from the most publicized top secret military installation in the world," explains West. "Of course I see strange things."[1]

FIGURE 3.4. The "flying saucer" and tow truck outside the Little A'Le'Inn. (Photo by D. R. Prothero.)

The "Little A'Le'Inn" has been featured in many movies and television shows made about Area 51, including the hilarious film *Paul*. It features British actors Nick Frost and Simon Pegg as UFO tourists going on a pilgrimage from the Comicon Convention in San Diego to Area 51 in a big RV. There they meet the escaped alien "Paul" (a CG character voiced by Seth Rogen), who hitches a ride with them, and the rest of the movie is a wild chase with the "men in black" from the base trying to recapture Paul. The "Little A'Le'Inn" is also featured in *The X-Files* season 6 episode "Dreamland II"; in *Louis Theroux's Weird Weekends*, season 1, episode 2; and in the video game *Grand Theft Auto: San Andreas*.

The movie *Independence Day* pretends to have been filmed in Area 51, but it was actually filmed in the California desert, and there are no shots of actual features around Area 51, such as the Little A'Le'Inn. Ironically, once the Nevada Commission on Tourism decided to give Highway 375 the moniker "Extraterrestrial Highway," the producers and some of the cast of *Independence Day* took part in the dedication ceremony of the renaming of the highway as a promotion for their upcoming 1996 movie—even though the film had been made in another state.

If the lack of any interesting features at the main Groom Lake Road entrance fails to impress the UFO tourists, there is always the other entrance to Area 51: the North Gate, also known as the Rachel Back Gate. If you follow the dirt road going south out of Rachel for about 10 miles, you will reach it. There you encounter the warning signs, a fence, and then a real guard post at the gate. Tall posts tower above the gate, equipped with security cameras that can pan all around and see the landscape for miles. For many years, there was a simple guard post building with one guard inside (usually watching television westerns when no one was near) and security vehicles parked nearby. Because the ground is flat for miles, no vehicles need to be parked on hilltops, because there is nowhere to hide—you are easily visible by means of security cameras and sensors. There are three black buildings numbered 997, 998, and 999 and a barbeque behind the guardhouse for the security staff to use while relaxing. In October 2000, the old guard shack (now abandoned) was upgraded to a larger facility, which had a pair of double gates so that anyone passing the first gate would be trapped between the first and second gate while his or her security credentials were being checked. Then in January 2011, the guard post was upgraded again to a building about three times the size of the old guard post. It is much more complete, with facilities for many different personnel as well as room to accommodate all their monitors for security cameras and store their equipment.

If you pass a certain point without authorization, the security forces will swoop down on you. As in the case with the Groom Lake Road entrance, these guys are not messing around! They will arrest you and call the Lincoln County Sheriff to haul you away. The signs also say that they are authorized to use deadly force to stop you, although so far no one is known to have been shot for going too close. But in 2012, a British reality series called "Conspiracy Road Trip: UFOs" decided to test its limits. The show follows five young Britons who are true believers in some unconventional idea (such as creationists or 9/11 "Truthers") and then follows them cross-country as they visit areas that provide the best evidence to challenge their beliefs. After the usual stops at the Black Mailbox and the Little A'Le'Inn, they tried driving right into the

North Gate with their minibus, filming all the way (about minute 38:00 in the episode[2]). The guards, who were not amused, arrested the entire crew and cast in the bus and held them for several hours. They finally released them to drive away late at night but confiscated all their cameras and their footage. The finished episode, which aired in 2012, ends with a tiny video clip, apparently shot in the dark using a cell phone camera that the guards missed, sheepishly explaining what happened. It seems amazing, but apparently people don't realize that these security guards are dead serious!

So be forewarned: curious UFO tourists can wander around the perimeter and get their thrills approaching the various entrances to Area 51, but if you cross the line, you'll be sorry!

DREAMLAND

In the past few decades, this perfectly ordinary military base in the middle of the desert in southern Nevada has taken on mythic status. Most military bases have tight security, and only authorized military personnel and their contractors are allowed on base. This particular base is top secret, with much tighter security than most military land. Not only is it surrounded by a secured perimeter and motion detectors in the ground, but also the guards travel the perimeter regularly and have video security cameras monitoring everything that comes near the fence. It is also located in one of the most remote areas of sparsely populated Nevada, more than three hours' drive north of Las Vegas. Because there is no way to see the base from the paved road—or even from the highest peaks outside the base, except Tikaboo Peak (a long hard desert hike)—it can be viewed only from the air or from space. Naturally, so high a level of secrecy has led to all sorts of speculation about what happens there, and an entire industry of books and movies and television shows need only mention the phrase "Area 51" to conjure up ideas of aliens or some kinds of weird government experiments.

First of all, let's clarify one misconception: the proper name of the base. Some common names for the base are "Groom Lake" or "Homey Airport," as seen on civilian aeronautics maps, or the longer "Nevada Test and Training Range" used in CIA documents,[3] but the older CIA documents do indeed use the term *Area 51*. The name "Area 51" came from the old grid system the Atomic Energy Commission used in the 1950s and 1960s to map the Nevada Test Site and the associated air bases. The original grid system numbering did not go as high as 51, but the Groom Lake area was purchased later and added to the secured perimeter of the base near Area 15 of the original grid; it is speculated that reversing the numerals in 15 provided the 51 for the newly annexed area.[4] The area has acquired additional security nicknames used to hide the true nature of the place, such as "Paradise Ranch," the deceptive name that Lockheed Aircraft designer Kelly Johnson used to attract workers to projects, according to Alexander Aciman in *Time* magazine:

> Area 51 was cheekily nicknamed Paradise Ranch, so that intelligence officers and government employees wouldn't have to tell their wives that they were moving the family to a rather large fenced-off area in the desert.[5]

Other code names include the CIA name "Watertown" (a reference to Watertown, New York, birthplace of CIA Director Allen Dulles), "Dreamland Resort," "Red Square," "The Box," and just "The Ranch."[6] After the U.S. Air Force took over the base from the CIA in the late 1970s, the name "Area 51" was discontinued, and it was called Detachment 3, Air Force Flight Test Center (or simply Det. 3, AFFTC). As such, it was a remote operating location of the Air Force Flight Test Center at Edwards Air Force Base, California. Today the official name of the Groom Lake base appears to be the National Classified Test Facility. The Test Wing (Det. 3, AFFTC) is still the primary occupant of the site.[7]

There is also an Area 52. It is another name for the secret airfield and testing facility near Tonopah, Nevada, about 70 miles (110 km) northwest of Groom Lake.[8] Many of the same aircraft that were developed and tested at Area 51 have their official base of operations at Area 52, especially the stealth aircraft.

For many decades, the activities in the base were top secret, so most of what was written about it was sheer guesswork or based on the reports of people who had worked there and spilled some of their information. The secrecy only helped spur on a lot of baseless speculation, especially by those who assumed that UFOs must be involved if the government was hiding its activities so strenuously. But in July 2013, in response to a Freedom of Information Act request (first filed in 2005), the CIA released nearly all the documentation about Area 51 and its activities; what's more, nearly all of the past work on the "Dreamland Resort" has been revealed. This documentation, along with the interviews of veterans of those projects compiled in recent books, dispels a lot of the mythology that grew up around the base.

Before World War II, Groom Lake was a silver and lead mining area, with several large mines that were worked from the 1870s to the 1890s. By World War II, however, the entire region was taken over by the federal government as a convenient, remote, inaccessible military base in which the U.S. government could do its most secretive and dangerous things: nuclear testing, bombing, and testing top-secret aircraft. Area 51 was originally a small airstrip within the Nellis Bombing Range. To the southwest is the much larger area reserved for the Nevada Test Site. Of the nearly 1,000 nuclear tests conducted by the Department of Energy, 739 were conducted at the Nevada Test Site. At the far southwest edge of the Nevada Test Site (not far from Pahrump, Nevada), is Yucca Mountain, an old mining area that for decades has been planned as a nuclear waste disposal site for the U.S. nuclear waste now stored in unsafe locations. However, years of political fighting have prevented this site from opening, because many environmental groups regard it as unsafe and likely to contaminate the groundwater with nuclear waste. Also enclosed within this large area of federal grounds is the Nellis Bombing Range Test Site (attached to Nellis Air Force Base, on the northeast outskirts of Las Vegas).

In 1942, just after the start of World War II, the main airstrips at Groom Lake became known as Indian Springs Air Force Auxiliary Field and were used to practice bombing runs. In 1955, the CIA began its long-term contract with Lockheed Aircraft to build high-performance spy planes to keep tabs on the Soviet Union. Up to that time, much of testing of high-performance aircraft (such as the X-1, one of the first to break the sound barrier) was done at Edwards Air Force Base, north of Lancaster, California, in the western Mojave Desert, as featured in the movie *The Right Stuff*. (Edwards Air Force Base has become more familiar recently as the alternative landing site for Space Shuttle missions.) CIA Director Richard Bissell, Jr., was worried that the testing of the top-secret aircraft could not be conducted at Edwards because of its exposure to the prying eyes of the public. He was especially concerned that aircraft flying away from the base would be spotted easily. Thus Lockheed's legendary aircraft designer Kelly Johnson and others sought a more inaccessible base. They searched Nevada, looking at sites, as Johnson tells it:

> We flew over it and within thirty seconds, you knew that was the place . . . it was right by a dry lake. Man alive, we looked at that lake, and we all looked at each other. It was another Edwards, so we wheeled around, landed on that lake, taxied up to one end of it. It was a perfect natural landing field . . . as smooth as a billiard table without anything being done to it." Johnson used a compass to lay out the direction of the first runway. The place was called "Groom Lake."[9]

The CIA then asked the AEC to add the land to its existing grid from the Atomic Test Site, and the designation "Area 51" became official. Shortly thereafter, the original airstrip, trailer homes, and few shacks that were used in World War II were transformed into a much bigger base. In three months during 1955, the base expanded to include a long paved runway with three hangars and a control tower, plus accommodations for the Lockheed and CIA personnel working on the base, including a movie theater, mess hall, trailer homes, a water tower, and fuel tanks. By July 1955, there were regular flights bringing Lockheed personnel and supplies up from Burbank, California, where Lockheed's main plant was located. They had already developed the legendary U-2 "Dragon Lady" spy plane at their secret facility near Valencia, California, known as the "Skunk Works." Now the U-2 was being carried on top of a C-124 Globemaster II cargo plane to its test site at Groom Lake. Shortly thereafter, there was a regular shuttle of Lockheed personnel, flying out of Burbank Monday morning on a Douglas DC-3 and returning Friday afternoons after a full week working at the base. These shuttle flights are now known as the "Janet" flights, and they take off from McCarran Airport in Las Vegas in a regular daily pattern to bring contractors out there and back. The "Janet" flights have special priority with air traffic control, and their planes are unmarked and the passengers dressed to be inconspicuous so that no one realizes where they are going or what their purpose is.

Clifford Ross Prothero (one of our fathers) worked for Lockheed Aircraft for his entire career. Originally trained as an artist, he started by building the legendary P-38 Lightning fighter plane during World War II, and then he became one of the founding members of the Technical Illustration Department. All were skilled artists and draftsmen. Their

FIGURE 3.5. The U-2 spy plane: an Air Force U-2 Dragon Lady flies a training mission. (U.S. Air Force photo by Master Sgt. Rose Reynolds.)

job was to take the blueprints of the aircraft engineers and turn them into anything that was needed: color illustrations, manuals for repair and service, assembly line guidebooks, and eventually big multivolume proposals to the federal government to build aircraft such as the Supersonic Transport. Cliff Prothero worked on every aircraft that Lockheed built during his career, which began in 1941 and lasted until he retired as manager and head of the Department of Technical Illustration in 1976. During the 1950s and 1960s, he carried a top-security clearance badge and told us that he could say nothing about where he was working or what he was doing. Most of the time, he came back each night very late, but I remember several times when he was gone all week and my mother was not allowed to tell us where he had gone. Only after he retired did he tell me about those long flights out to Area 51.

Lockheed's U-2 (which my dad worked on) was developed by Kelly Johnson and other Lockheed engineers at the Skunk Works in the early 1950s and was first flown in 1955 at Area 51. By 1957 it was in service for the CIA and the U.S. Air Force. The U-2 (Fig. 3.5) looked essentially like a jet-powered glider with long, straight, narrow wings that allowed it to soar with minimum energy input and benefited from all sorts of technological innovations that allowed it to fly at 70,000 feet, well beyond the range of Soviet radar, missiles, or aircraft of the time. In the nose and in the belly of the plane were a variety of sensors, including a high-resolution camera that could spot even tiny objects on the ground. Its gliderlike shape meant that even if the engines quit, the pilot could glide for 300 miles or farther to make a safe landing (as happened on more than one occasion). But it also was very light and easily buffeted by strong crosswinds, which made it very tricky to fly as well. Nevertheless, it was secretly taking spy photographs of the Soviet Union, China, Cuba, and many other places after its maiden flight in 1957. It took aerial photos for every Cold War task after that, spying on various Communist countries; verifying that Castro had Soviet missiles during the Cuban missile crisis of 1962; and photographing Vietnam, Cambodia, Laos, and many other regions during times of conflict. It is still in service, being used in some specialized purposes in the wars in Afghanistan and elsewhere—although most spy reconnaissance is done using satellite networks and drone aircraft now.

However, the belief that the U-2 could not be tracked by Soviet radar or hit by enemy missiles turned out to be false. During the 24th U-2 mission over the Soviet Union on May 1, 1960, the Soviets tracked the flight of the plane piloted by Francis Gary Powers. One of their three SA-2 surface-to-air missiles (SAMs) exploded near the plane, damaging it and causing it to crash; Powers ejected and was captured. (Another SAM hit a Soviet plane trying to intercept the U-2.) The "U-2 incident" became a worldwide crisis as the Soviets angrily demanded apologies from the United States. For a while the CIA used a cover story about the flight's having been a NASA test. They did not realize that Powers had survived, something everyone had thought impossible. Soviet Premier Nikita Khrushchev let U.S. lies pile up until he revealed that Powers was alive and had confessed, upon which the Eisenhower administration had to do a lot of backtracking. The Paris Summit between the United States and Soviet Union was canceled after Khrushchev demanded a U.S. apology and Eisenhower refused to make one. But U-2 flights over the Soviet Union were discontinued, and its use was restricted to areas where there was no chance of its being hit by missiles. (Powers was eventually released in exchange for a Soviet spy, an event depicted in the 2015 Tom Hanks–Steven Spielberg movie *Bridge of Spies*.)

FIGURE 3.6. (A) The Lockheed A-12, which eventually turned into (B) the SR-71 Blackbird spy plane. (U.S. Air Force photos.)

Even before the U-2 incident, the CIA was seeking a spy plane that could fly even higher above the Soviet missile ceiling and that was sturdy enough and fast enough to outrun interceptor planes as well. This effort was further accelerated after Gary Powers was shot down and the vulnerabilities of the U-2 thus revealed. This became the CIA's "Project OXCART," the top-secret program to develop the next-generation spy plane, Lockheed's A-12 (Fig. 3.6A). This evolved into the YF-12, the eventual predecessor of the legendary SR-71 Blackbird spy plane, itself the most successful aircraft of its kind ever built (Fig. 3.6B). Project OXCART greatly expanded the facilities at Area 51 as more and more personnel were stationed there to conduct test flights. For the A-12, Kelly Johnson abandoned the delicate "jet glider" model of the U-2 by building a sturdier, more powerful aircraft that could fly at Mach 3.35 at elevations exceeding 95,000 feet. Even if it could be detected, it was too fast to be caught by either missiles or other aircraft and too sturdy to be brought down by a nearby missile explosion. By 1962, the A-12s built in Burbank had been shipped out to Groom Lake, where test flights began over the airspace of Nellis Air Force Base (so that only other military pilots could see it—and they were bound by secrecy agreements). The loss of Gary Powers's manned aircraft emphasized the risk to pilots, which resulted in the D-21 program, code-named "Tagboard." Attempts to turn the A-12 into a drone aircraft never worked as planned and were quietly scrapped.

In addition to being the testing ground for several generations of top-secret spy aircraft, Area 51 was often used to test the capabilities of captured Soviet aircraft and evaluate their performance without letting Soviet spies see them. In 1968, Project "HAVE DOUGHNUT" tested a captured MiG-21 that had come into Israeli hands from an Iraqi defector pilot who refused to use napalm on Kurdish towns. From these test flights, U.S. Air Force pilots were able to gauge the relative strengths of their F-4 Phantom IIs against the MiG-21, which proved advantageous in Vietnam. Project "HAVE DRILL" was a code name for the study of two captured Syrian MiG-17s that accidentally landed in northern Israel and that were shipped to the CIA and U.S. Air Force for study. Project "HAVE FERRY" was another captured MiG-17 loaned from Israel to the United States in 1969. "Top Gun" pilots from Miramar Air Base near San Diego (portrayed in the movie Top Gun) flew out to do mock dogfights

FIGURE 3.7. The A-177 Nighthawk stealth fighter. (U.S. Air Force photo/Staff Sgt. Aaron D. Allmon II.)

with these MiG planes to hone their skills and learn the strengths and weak spots of their enemy's aircraft. The sky over Area 51 became an even more restricted air space on the aeronautical maps, marked in red ink—hence the nickname "Red Square."

Finally, the Groom Lake "Area 51" base played a crucial role in the development and testing of the A-117 Nighthawk, the "Stealth Fighter" (Fig. 3.7). Code-named "HAVE BLUE," Lockheed first began developing the plane in 1978, and it was first test-flown out of Groom Lake in 1982. The F-117 became operational in 1983, and its squadron was based in "Area 52," the Tonopah Test Range airfield. The entire plane was designed to minimize the effects of radar waves bouncing off it, complete with radar-absorbing panels on the "skin" and a shape designed to minimize the aircraft's radar footprint. The plane first saw combat in the Balkan wars of 1999, during which one was shot down by a surface-to-air missile and the secrecy surrounding it broken. The first aircraft into Iraq during the 1991 and 2003 Persian Gulf wars were stealth craft, which were able to evade Iraqi radar as they delivered laser-guided bombs to their targets. Although these aircraft had been spotted, the Air Force officially refused acknowledge the existence of the F-117 until 1988, and in 2008 it retired the plane because the F-22A Raptor was due to be delivered.

SO WHAT ABOUT THE UFOS AND ALIENS?

The long history of secrecy surrounding Area 51 has certainly contributed to its mystery, and naturally this has led some people to think that the government was hiding more than just top-secret spy and stealth aircraft. Simply that the CIA and the military went to such great lengths to conceal what they were doing feeds such suspicions and makes the conspiracy-minded more convinced of their beliefs. The popular media (especially such television shows as the *X-Files* and the fake "documentaries" about Area 51 on cable television, as well as many movies, especially *Independence Day*) have created fully fleshed-out versions of what the screenwriters think is going on behind that curtain of secrecy. Screenwriters are interested only in a gripping story that will sell their scripts and hold an audience, so they have no interest in making a movie about the truth; they are only telling a good tale. Their influence on the culture is overwhelming, so enough people have seen these shows enough times that no amount of revelation of the more mundane explanation of secret spy planes will convince them that they are believing a Hollywood fantasy.

In addition to the speculation and fantasies from the outside world, more than one member of the high-security staff at Groom Lake was more than happy to let the myths propagate.[10] After all, the alien/UFO story works as an excellent misdirection if people are thinking about these fantasies and not thinking about the reality of secret stealth aircraft and spy planes. Consider the following:

- During the 1950s–1970s, civilian aircraft flew no higher than 40,000 feet (and they still don't), and military aircraft only a little bit higher. This was a time of an unusually large number of UFO reports in the area around southern Nevada. Right after the U-2 began flying, sightings by civilian pilots became increasingly common, and with more than 2,850 OXCART A-12 flights alone, there were a lot of high-speed, high-altitude craft flying over the civilian airspace. Not only were these craft top secret, but they were also traveling more than 2,000 miles per hour at higher than 80,000 to 90,000 feet, much higher and quicker than any civilian airliner could fly. All the civilian pilots would have seen is an unusually shaped craft, or possibly only a string of lights on its wings, moving much faster and higher than any plane they knew about. Some reports about the "fiery" craft were probably the silvery wings of the U-2 or A-12 reflected in the setting sun. The Air Force's "Project Blue Book" checked a long list of UFO sightings and found that most reports in southern Nevada could be explained by spy planes[11] (although investigators could not divulge the information at the time). Veterans of Area 51 who were interviewed later agreed with this assessment. According to sources in Annie Jacobsen's book about Area 51,

> the shape of OXCART was unprecedented, with its wide, disklike fuselage designed to carry vast quantities of fuel. Commercial pilots cruising over Nevada at dusk would look up and see the bottom of OXCART whiz by at 2,000-plus mph. The aircraft's titanium body, moving as fast as a bullet, would reflect the sun's rays in a way that could make anyone think, UFO.[12]

· The light show visible above the base, especially in the 1960s–1990s, during testing of the A-12, the SR-71 Blackbird, and the A-117 Nighthawk, is most parsimoniously explained by these oddly shaped spy aircraft flying at night. All were designed to fly in nighttime, and all would look strange with the spacing and shape of their lights and their unusual capabilities of speed and maneuverability. In addition, there may have been additional craft not mentioned in the conventional accounts, such as still-classified stealth helicopters (used in the raid to kill Osama bin Laden[13]) and other VTOL (vertical takeoff and landing) craft that easily explain why so many people see these strange craft at night over the area. No UFOs are required. Indeed, this was vividly demonstrated when the reality show episode "Conspiracy Road Trip: UFOs" filmed at the Black Mailbox in 2012.[14] Participants got hugely excited when a string of lights appeared in the sky over Area 51 and they heard a sonic boom—only to realize that the lights were aerial flares and the sonic boom must have been from a night-flying aircraft.

· In 1989, a man named Robert Lazar shocked the world with the claim on television that he had worked at Area 51 and a secret

facility called "S-4." He claimed that the government had at least
nine alien spacecraft in its possession; that the base contractor,
EG&G, had hired him to "reverse-engineer" the alien craft to
discover their secrets; and that the aliens had powered their craft
with a heavy substance he called "element 115." This report started
the entire industry of people claiming that there are alien craft
and bodies hidden on Area 51, where the base scientists are trying
to create new aircraft and make new discoveries of alien secrets.
But as soon as the story was publicized, it began to unravel,
because Lazar was a big liar.[15] He claimed to have master's degrees
from Caltech and MIT, but records show that he attended neither
institution. The Air Force and Los Alamos National Laboratories
also disavowed him despite his claims that he worked for them.
Naturally, paranoid conspiracy theorists see this as further
evidence of cover-up, but in later interviews Lazar began to recant
his story, and in interviews on the 25th anniversary of his initial
claims, Lazar said he no longer involves himself in the topics
about UFOs.[16] More important, things that he claimed (that can
be independently checked) show that he doesn't know what he's
talking about. For example, element 115 on the periodic table[17]
was first discovered by Russian scientists in 2004, then confirmed
by Swedish scientists at Lund University in September 2013. It's
called ununpentium ("one one five"), and it bears no resemblance
to the properties of the substance that Lazar claimed to have
studied in alien craft.

· In 2012, journalist Annie Jacobsen published a best-selling book,
*Area 51: An Uncensored History of America's Top-Secret Military
Base*. Most of the book consists of the now publicly documented
history of the U-2, A-12, SR-71, F-117, and other projects that
we know occurred there. Her sources were the Roadrunners,
a once secret club of retired Area 51 employees who meet on a
regular basis, typically in a secret location. They were now able
to speak to her since the secrecy has been lifted and the original
CIA documents have been released. Unfortunately, Jacobsen's
training in science and technology was pretty limited, and the

book has been lambasted[18] by scientists and engineers (many who worked in Area 51) for its many blunders and simple mistakes that any scientist or engineer would never make (for example, sound moves at 700 mph, not 100 mph; shotguns don't shoot bullets; Burbank Airport, then called "Lockheed Air Terminal," is not 9 miles west of Northridge but rather 19.9 miles east—and these are only some of her many mistakes and misstatements). Her most startling claim in an otherwise sober and well-researched book, however, came in the last chapter. There she quotes an unnamed source who said that the 1947 Roswell UFO crash (see chapter 4) was actually a Soviet "flying saucer" built by captured Nazi rocket scientists and that the "aliens" inside were actually children who were deformed due to Nazi breeding experiments! These, according to her source, were spirited off to Area 51, where her source claims he saw them in 1951. Needless to say, such a claim is preposterous, and even the UFO fanatics have disavowed it. There is no evidence of such Soviet aircraft expertise, nor of any Nazi breeding experiments—let alone the rest of her claim (see the problems with the Roswell myth in chapter 4). When Jacobsen spoke before the Skeptics Society in 2012, both of us (Prothero and Callahan) were among a number of scientists and skeptics who questioned her strenuously about this point. Many of the others were ex-military pilots or JPL engineers and scientists, and under the questioning, her story began to fall apart. More important, the people she originally interviewed among the Roadrunners who were all there, and who saw everything, disavowed her claim and said that no such "Soviet–Nazi deformed youth saucer" was ever at Area 51.[19] Considering our rules of parsimony and the strong evidence against this extraordinary claim, two other possibilities are much more likely: that her single "source" was pulling her leg or turning senile (he would have been in his 80s or older) or that she made it up to spice up sales for the book. Either way, she got to laugh all the way to the bank: the book was on the best-seller lists for months, and so another ridiculous story is out there fooling people who don't know any better.

SUMMARY

After all this huge legend built up by television and movies and UFO believers concerning Area 51, the truth turns out to be much more mundane. It was certainly a top-secret installation with lots of projects that the CIA and Air Force would not declassify until 2014; they practiced all sorts of misdirection and subterfuge to keep people guessing. But now that the CIA has released all the formerly top-secret documents, there is no smoking gun in there to suggest that anything other than spy planes and stealth aircraft were being tested and developed. A huge number of early UFO reports are now explained by high-altitude supersonic planes that civilian pilots did not know about. So, too, are the weird patterns of lights that fly out of Groom Lake at night. The Lazar stories turns out to be from an unreliable source caught in numerous lies. Jacobsen's "Soviet–Nazi deformed youth flying saucer" story has never been corroborated by a second source, so there is no reason to take it seriously.

What we *do* have is huge number of pages of just-declassified CIA documents that lay out what really did happen in Area 51. What we *do* have is the Roadrunners and other former Area 51 personnel, now released from their oaths of secrecy, saying that there were *no* alien craft or bodies being stored there. These men had the highest possible security clearance, and not one saw any aliens. I asked my dad before he passed away in 2004 what he thought of the stories, he having been there in the 1960s. He laughed and said that it was all hokum—myths fostered by people who couldn't accept the idea that the secrecy was all about hiding spy aircraft, not aliens.

4

The Roswell Incident: What *Really* Happened?

Roswell, New Mexico, is a quiet little city in the scorching southeastern plains of New Mexico. With a population of more than 48,000, it is actually the fifth-largest city in the state, and it serves as the primary hub for businesses in southeastern New Mexico. It was named an "All-American City" in 1978–79 and in 2002. The limestone bedrock creates lots of local sinkholes, such as Bottomless Lakes State Park about 15 miles (24 km) southeast of Roswell. The area also has lots of limestone caverns; the Carlsbad Caverns are just 154 km (96 miles) to the south. Just south of Carlsbad Caverns are the ancient Permian reef rocks of Guadalupe Mountains National Park, on the border with the Texas Panhandle. Despite the desert conditions in the summer and mild winters, there is a lot of irrigation for farming in the region, as well as dairy farms and cattle ranches, some manufacturing, and some oil production from fields on the fringe of the productive Permian Basin in adjacent Texas. The New Mexico Military Institute is located there, and at one time large Cold War military bases were located nearby as well.

If you drive down the main street, it looks much like any other small city in the plains of west Texas or eastern New Mexico—with one major exception. All throughout the town are signs and statues and other objects (Fig. 4.1) that remind you that Roswell is "Alien City." One of its largest sources of income is thousands of tourists who come each year to

FIGURE 4.1. Roswell, New Mexico, has many signs, sculptures, and other objects designed to appeal to its UFO tourism. (A) A series of cut-out characters on the edge of the town, with the UFO and aliens giving the American family a jump for their old truck's battery. (Courtesy Dennis Begin and *RVwest Magazine*.) (B) Even the lampposts have eyes (© iStock.com/zrfphoto). (C) Alien-related merchants are everywhere (© iStock.com/mixmotive). (D) Flying saucers are a popular theme. (Courtesy International UFO Museum and Research Center.)

FIGURE 4.1. *Continued*

gawk at UFO paraphernalia, visit the "International UFO Museum and
Research Center" built in an abandoned movie theater downtown, or
follow signs on Highway 246 northwest of town to the "UFO crash site"
(closer to Corona, New Mexico, than it was to Roswell). Every summer

in late June or early July, Roswell hosts its "UFO Festival," which brings huge crowds eager to see the sights, hear guest speakers, hang out with fellow UFO believers, take part in the "Alien Chase" 5K, 10K, and other distance runs—and spend *lots* of money in town. Ask people whether they've heard of Roswell, New Mexico—if they have, it's always in connection with the alleged UFO crash site.

Just like the towns near Loch Ness, which make many pounds sterling from Nessie tourism; Lake Champlain, which brings gawkers looking to see "Champie," the alleged lake monster; or the areas in Oregon and northern California (especially Willow Creek, California), which have lots of attractions catering to the Bigfoot believers, there is profit to be made by publicizing legends and paranormal stories connected with your town. Even if they really don't believe in UFOs, the people of Roswell have nothing to gain from openly doubting the paranormal nature of the Roswell incident in 1947. But residents with tourist dollars at stake are not the most reliable source for the truth about Roswell. Instead, we need to look critically at the evidence, bearing in mind that most eyewitness evidence is worthless, remembering that contradictory evidence weakens the likelihood of something's really having happened, and using Ockham's Razor ruthlessly. Did a UFO really crash there—or is there a simpler explanation?

PROJECT MOGUL

Let's go back to 1947 and remember what was happening at the time. World War II had ended two years earlier, but the United States was in the midst of a "Cold War" with the Soviet Union. We were justifiably worried about spies and about the Soviets' getting nuclear weapon capabilities. We now know that there were indeed spies in the U.S. nuclear labs and that the Soviets were well along in developing their own nuclear device, which they successfully tested in 1949. One of the biggest research efforts was focused on methods to detect when the Soviets were testing nuclear weapons. Multiple efforts were tried, and eventually it turned out that an array of seismographs could pick up the distinctive vibrations of a nuclear blast and distinguish them from earthquakes (and an air sampling program was designed to pick up traces of nuclear fallout in the atmosphere).

Before this was established, however, the military and its scientists tried another approach: sensing devices held up by array of balloons that would float at a certain level in the atmosphere where the sound waves from the blast could be detected. The idea was originally proposed by the brilliant scientist Maurice Ewing, who had founded the Lamont–Doherty Geological Observatory at Columbia University in New York (where Prothero was trained). It had a series of microphones and a disc-shaped antenna that beamed the signals down to receiving stations on the ground. The entire effort, which was top secret, was dubbed "Project Mogul." Gildenberg,[1] who worked on Project Mogul, provides the complete details of how this top-secret initiative was launched and how it led to all the legends.

He writes that his mentor, Professor Charles Moore of NYU, was in Alamogordo in June and July 1947, and they launched a number of Mogul balloon arrays to test their behavior and capability.[2] Mogul Flight 4 was launched from Alamogordo on June 4, flew east, and was hit by lightning in a thunderstorm and crashed on the Foster Ranch northeast of Roswell (closer to Corona, New Mexico). On June 14, Foster Ranch foreman Mack Brazel found debris made of paper, rubber, and foil, but ignored it.

The story would have ended there but for an extraordinary coincidence. Ten days later, "businessman pilot" Kenneth Arnold was flying his small plane up in Washington and Oregon, and spotted several aircraft in formation "like the tail of a kite," with a motion he described as being like that of stones or saucers skipping across a pond. The press garbled his original story, and it came out that there were "flying saucers" over Oregon and Washington, creating a national hysteria about these strange UFOs. There were more than 1,000 reports all over the country, especially over Independence Day weekend, as people primed by this one story saw "flying saucers" everywhere.

On July 5, Mack Brazel drove into Roswell and learned of all this news. He returned to the ranch and looked over the debris, then collected some of it. He had seen weather balloons land on his property before, but nothing with shiny foil attached. Returning to town on July 7, he told the Roswell sheriff "kinda confidential like" that he might have wreckage of one of these reported flying saucers. The sheriff contacted the Roswell Army Air Force Base, which was the nearest military facility but which had no connection or knowledge of what was going on in the top-secret experiments out in Alamogordo. Intelligence Office Major Jesse Marcel and Counterintelligence Officer Captain Sheridan Cavitt were sent to investigate, and they gathered up a lot of the debris and brought it back to be flown out to the Fort Worth Army Air Force Base for examination by higher-level authorities.

On July 8, 1947, Marcel's office arranged for Walter Haut, the public information officer for Roswell Army Air Force Base, to issue a press release stating that soldiers from the 509th Operations Group had recovered a "flying disc" (which apparently referred to the disc-shaped antenna). The press release read as follows:

> The many rumors regarding the flying disc became a reality yesterday when the intelligence office of the 509th Bomb group of the Eighth Air Force, Roswell Army Air Field, was fortunate enough to gain possession of a disc through the cooperation of one of the local ranchers and the sheriff's office of Chaves County. The flying object landed on a ranch near Roswell sometime last week. Not having phone facilities, the rancher stored the disc until such time as he was able to contact the sheriff's office, who in turn notified Maj. Jesse A. Marcel of the 509th Bomb Group Intelligence Office. Action was immediately taken and the disc was picked up at the rancher's home. It was inspected at the Roswell Army Air Field and subsequently loaned by Major Marcel to higher headquarters.

The following day, this story from the *Roswell Daily Record* is typical of what the press reported:

> The balloon which held it up, if that was how it worked, must have been 12 feet long, [Brazel] felt, measuring the distance by the size of the room in which he sat. The rubber was smoky gray in color and scattered over an area about 200 yards in diameter. When the debris was gathered up, the tinfoil, paper, tape, and sticks made a bundle about three feet long and 7 or 8 inches thick, while the rubber made a bundle about 18 or 20 inches long and about 8 inches thick. In all, he estimated, the entire lot would have weighed maybe five pounds. There was no sign of any metal in the area which might have been used for an engine, and no sign of any propellers of any kind, although at least one paper fin had been glued onto some of the tinfoil. There were no words to be found anywhere on the instrument, although there were letters on some of the parts. Considerable Scotch tape and some tape with flowers printed upon it had been used in the construction. No strings or wires were to be found but there were some eyelets in the paper to indicate that some sort of attachment may have been used.[3]

As soon as the press release went out on the Associated Press wire, it made national news and was a sensation. Inquiries came from around the world, and J. Edgar Hoover of the FBI sent his own agents to investigate. A telex sent to the FBI from a major in the Eighth Air Force said:

> The disc is hexagonal in shape and was suspended from a ballon [sic] by cable, which ballon [sic] was approximately twenty feet in diameter. Major Curtan further advices advises [sic] that the object found resembles a high altitude weather balloon with a radar reflector, but that telephonic conversation between their office and Wright field had not [UNINTELLIGIBLE] borne out this belief.

Most of the military officers who originally saw the debris had no knowledge of Project Mogul. They did know that weather balloon arrays often had devices hanging down from them to facilitate their tracking by radar, although the actual Mogul reflectors were unfamiliar to them and led them to doubt the "weather balloon" story. But as the papers reported on July 8 in Texas, General Roger M. Ramey of the Eighth Air Force in Forth Worth examined the debris, as did Warrant Office Irving Newton; they confirmed that it looked like a "kite" or reflector used to track balloons from the ground. Another news release from Forth Worth came out, this time with the famous photos (Fig. 4.2) of General Ramey holding the flimsy debris of foil, balsa wood, and tape.

To be very clear, we have the testimony of many of the military personnel who saw the debris, all of whom confirmed that it was nothing more than foil, balsa wood, tape, rubber, and some aluminum connecting rings. These include not only General Ramey, Warrant Officer Newton, Major Marcel, and Captain Cavitt but also Colonel Albert Trakowski, who was head of Project Mogul at the time. He presented sworn testimony for the 1995 Pentagon report.[4] He also testified that he spoke to Colonel Marcellus Duffy, who was head of Project Mogul before him, who said that the debris resembled the Project Mogul flight apparatus. Project Mogul wasn't declassified until 1972, so many military personnel were unaware of what the Mogul array looked like, but Mogul insiders knew. Other witnesses notice the distinctive tape used on the Mogul radar reflectors and the toy company that made them, with its distinctive floral print. (UFOlogists call these flowers "alien hieroglyphics.") Brazel's daughter Bessie noted the aluminum rings in the wreckage, which were the kind of connectors used in Project Mogul. Cavitt described the "black box," which matches the kind of battery box used on Project

FIGURE 4.2. General Ramey at Fort Worth, Texas, holding up the debris from the Roswell crash. (© Photo Researchers, Inc./Alamy stock photo.)

Mogul. After all this investigation, the Official Air Force Report concluded that "the material recovered near Roswell was consistent with a balloon device and most likely from one of the MOGUL balloons that had not been previously recovered."[5]

As Gildenberg points out,[6] at the time of the original incident, none of the eyewitnesses reported anything other than foil, balsa wood, rubber, and aluminum rings: no alien bodies or alien autopsies, no flying discs or saucers or any type of spacecraft, no multiple crash sites. In fact, in 1947, the word *UFO* hadn't even been coined yet, and the first garbled account that led to the "flying saucer" myth was only weeks old. All these elements of the story, along with the additional implications of huge conspiracies by the military and the government, "Men in Black," and all the rest, were added much later. Considering that the debris was part of a top-secret project and that the military officers who recovered it had no knowledge of Project Mogul, they were remarkably honest and open about what they found—not secretive in the least.

Meanwhile, Project Mogul grew into Project Skyhook and eventually Project Blue Book, which launched more and more balloon arrays (more than 2,400) into the upper troposphere and stratosphere, where they gathered data—and occasionally were reported as a UFO.[7] For Operation High Dive, some of these balloons had gondolas, which then launched test dummies to drop through the sky and monitor the conditions as they fell. These "bodies falling from the sky" were quickly recovered by the military, but some were reported as UFO incidents. Finally, in 1960, Captain Joseph Kittinger jumped from a balloon at 102,800 feet (19 vertical miles), almost on the edge of space, and Guildenberg was among those monitoring his dive. Kittinger accelerated in freefall for 4 minutes and 38 seconds, reaching a velocity of 614 mph (92% of the speed of sound). About 14 minutes after he jumped, he deployed his parachute and landed safely.

THE MYTH IS BORN

In fact, months after the incident got into the newspapers, the story quickly died. It was largely forgotten for the next 31 years. As late as 1967, the Roswell story was so dead and obscure that when UFOlogist Ted Bloecher reported on the 1947 UFO Sightings Wave (spurred by the Arnold report that was garbled as "flying saucers"), there is no mention of Roswell in either June or July 1947.

Project Mogul was eventually declassified in 1972, so the secrets behind the crash were public knowledge. But the key event was the 1977 release of Close Encounters of the Third Kind, with its humanoid "gray aliens," which raised UFOs and aliens to the top of the cultural zeitgeist. It was not until 1978, some 31 years after the Roswell Incident, that the entire mythology of the "Roswell UFO crash" began to emerge. Major Jesse Marcel was now out of the Army, working as an obscure television repairman in Houman, Louisiana, when UFO researcher Stanton T. Friedman contacted him and interviewed him. Marcel could no longer remember the details of what had happened after more than 30 years had passed, but he saw an opportunity, especially as he became famous with stories in the *National Enquirer* and in the documentary film *UFOs Are Real*. As the story grew and grew, he was interviewed and filmed over and over again. Naturally, each time he told the story, it got better and more colorful. According to Gildenberg, it was originally "a modest adventure recalled with difficulty to a great whopper of a tale remembered in vibrant detail. Slowly, the debris he remembered gained magical properties, such as imperviousness to sledgehammers and fire. A movie based on his claims was produced with Marcel himself as the lead character, played by the perpetually mysterious Kyle MacLachlan (of *Twin Peaks* fame)."[8]

In 1980, the story exploded with the publication of *The Roswell Incident* by Bermuda Triangle promoters Charles Berlitz and William Moore. Meanwhile, as Friedman and later UFO researchers kept building up the publicity for the story, more and more "witnesses" claimed that they had seen things not corroborated by anyone who actually saw the wreckage in 1947. A whole set of new stories by "witnesses" (most of who cannot be confirmed to have had any direct connection to Roswell in 1947) were introduced into the mythology, and the story grew and grew. In 1992, Friedman and Don Berliner ran an ad in search of more witnesses "in order to strengthen their case for government knowledge of what they call 'the truth behind almost 50 years of UFO sightings.'" This kind of ad was guaranteed to bring cranks and attention-seekers out of the woodwork, who would have no evidence of having any close connection to the Roswell incident—but it filled the UFO books with lots of pages of worthless "testimony" nonetheless.

The Berlitz–Moore version of what happened at Roswell was so successful that soon other paranormal writers saw the money potential and began writing other accounts to cash in. In 1991, Kevin Randle and Donald Schmitt wrote *The Truth about the UFO Crash at Roswell*, which had 100 new "witnesses" (somehow able to precisely remember events from almost 50 years ago—and even though few people actually lived near Roswell) and changed the storyline to introduce sinister new elements. The new material must have helped, because they sold 160,000 copies and the authors laughed all the way to the bank.

Then Friedman got back in the act with his own book, *Crash at Corona* (1992), coauthored with Don Berliner. Friedman somehow found additional new "witnesses" and doubled the number of flying saucers to two and the number of aliens to eight—two of whom allegedly survived and were hidden by the U.S. government. Finally, in 1997, Lieutenant Colonel Philip J. Corso wrote *The Day after Roswell*, claiming that while he was stationed at Fort Riley, Kansas, in July 1947, five trucks with 25 tons of cargo entered the base from Fort Bliss. According to him, he was shown alien bodies from an "air crash." This book was dissected line by line by the late researcher Philip Klass, who showed that it was full of inconsistencies and factual errors. Even Roswell believers such as Friedman attacked Corso's book, calling it pure fiction.

The mythology of the Roswell incident just kept growing and growing, with more and more inconsistencies and conflicting accounts from "witnesses" whose actual connection to the incident was tenuous at best. Most of the "witnesses" claim to be repeating what others told them—which is hearsay evidence, inadmissible in a court of law. The Berlitz–Moore book has testimony of only 25 people, of whom only 7 can claim to have seen the debris, and only 5 of whom allegedly touched it. The Randle–Schmitt version has some 300 people who were interviewed, only 41 of whom can be "considered genuine first- or second-hand witnesses to the events in and around Roswell or at the Fort Worth Army Air Field" and only 23 of whom can be "reasonably thought to have seen physical evidence, debris recovered from the Foster Ranch." Of these, only seven have asserted anything suggestive of otherworldly origins for the debris, according to Pflock in *Roswell: Inconvenient Facts and the Will to Believe*. There are all sorts of problems with these supposed "witnesses": most are only reporting hearsay and second-hand accounts, many have made demonstrably false claims or have given multiple contradictory accounts, and some are dubious deathbed confessions made by elderly witnesses who were easily confused. Even for those who were possibly direct witnesses there are problems: most are trying to remember events from 31 or more years earlier, and as we have already discussed, human memory is very fallible and very likely to add information later on based on things that the "witness" has been exposed to, especially with the Roswell mania in the media contaminating witnesses' actual memories.

For example, Jesse Marcel has serious issues with credibility, his account having changed over time and he having gone from being

a military officer with high levels of responsibility to a television repairman who needed money and publicity. In the Berlitz–Moore book, Marcel is reported to have said, "Actually, this material may have *looked* like
tinfoil and balsa wood, but the resemblance ended there . . . . They took
one picture of me on the floor holding up some of the less-interesting
metallic debris . . . . The stuff in that one photo was pieces of the actual
stuff we found. It was not a staged photo." But when skeptics point out
that this is the same material positively identified as part of the Project
Mogul balloon experiment (and even UFO advocates agree that it is),
Marcel then changed his story, claiming that the material shown in the
photo was not what he had recovered. A closer look at Marcel shows that
he has a history of exaggeration and embellishment. He claimed to be
a pilot and to have received five Air Medals for shooting down enemy
aircraft—but this is a lie. That his Roswell account kept changing, especially when there was big financial motive for making it more colorful,
suggests that he lies in this regard, too.

Even more damning to Marcel's accounts are the interviews of Captain Sheridan W. Cavitt, the only other person who was with Marcel
when the crash site was first investigated. Marcel says that the rancher,
Brazel, led them to the crash site; Cavitt says that he never met Brazel.
Marcel says that they spent the night at Brazel's ranch house; Cavitt
denies it. Marcel claims that they tested the debris with a Geiger counter; Cavitt says that they had no Geiger counter. Marcel claims that they
traveled in an International Carryall SUV; Cavitt says that they rode in a
Jeep. And Marcel claims that they returned to the site the next day with
some MPs, which Cavitt denies. Given that Cavitt was one of the only
people to go to the original site and witness the original material and
carry it back to base, his recollections are far more relevant than those of
the dozens of "witnesses" who were nowhere near the site and who never
saw anything firsthand.

The problems and discrepancies with the various Roswell accounts
have created problems even in the UFO community. The Center for
UFO Studies (CUFOS) and the Mutual UFO Network (MUFON),
the two biggest UFO fan clubs, have held several conferences trying to
resolve the differences on the assumption that most of what the "witnesses" are telling is true and that there were really aliens and spaceships

at Roswell that were covered up. Rarely do they raise the possibility that most of the "witnesses" are unreliable or do not have any credible connection to Roswell and that the whole thing is mostly a network of lies and exaggerations.

This is reinforced by how the believers reject critics and skeptics as well as the only reliable evidence: the original reports and photos from 1947, which don't fade or change over time as human memories do. The original scientists behind Project Mogul (Charles Moore, Gildenberg, and others) met with UFOlogists such as Friedman and Berlitz around 1992, offering to set them straight, but the UFOlogists didn't want to hear from them.[9]

Finally, in response to all the pressure put on Congress by UFOlogists, both the General Accounting Office and the Air Force conducted their own investigations. The first report found no evidence of anything other than debris from Project Mogul.[10] The second report, 50 years after the crash, concluded that the reports of alien bodies were probably poorly remembered accounts of military accidents with injured or killed personnel, conflated with the stories of the dummies from Operation High Dive and blended with the cultural memes and hoaxes that have overwhelmed any vague old memories.[11] Naturally, the UFO advocates dismissed all this careful work as part of the government cover-up.

ALIEN AUTOPSIES?

One of the minor players at Roswell was a local mortician named Glenn Dennis, who founded the "International UFO Museum and Research Center" in downtown Roswell in 1991. He vaulted to fame that year (and promoted his museum) by giving an interview in the Randle–Schmitt book in which he claimed that he had witnessed alien autopsies at the Ballard Funeral Home. The long shaggy dog story he gave to Randle and Schmitt seems convincing until you look into the details—which Gildenberg, Klass, and others have done. A short version of the many disconfirming details includes the following:

· Dennis claims he received a call from the base Mortuary Affairs office, but that office did not exist in 1947.

· The Mortuary Affairs office asked whether small caskets were available, but no such caskets were available anywhere near Roswell, and the medical supply officer (Lorenzo Kimball) in the story would have known this.

· Dr. Johnson officially pronounced the aliens dead, but Dr. Johnson was a radiologist, not a pathologist, so he would not have been assigned this job.

· There was a big redheaded colonel running the hospital—yet that colonel (Colonel Ferrel) did not take command until 1954.

· The surviving alien walked up a ramp in the rear of the hospital, but there was no such ramp in 1947.

· The autopsy took place in an operating room suite, yet no such suite existed in 1947.

· The nurse who allegedly pushed Dennis out and told him that they were operating on aliens could never be tracked down, and lame excuses were made for her absence; Dennis kept changing her name each time he was told the previous name hadn't panned out.

· The tall, lanky Captain "Slatts" Wilson in Dennis's account never existed. A nurse Lucille "Slatts" Slattery was short, had a different surname, and arrived at the base a month after the incident. Another officer, Captain Idabelle Miller, who was tall and thin, married someone named Wilson, but did not work at the base until 1956.

· The pediatrician, Captain Frank Nordstrom, did exist but didn't come to Roswell until 1951.

Finally, we have Professor Emeritus Lorenzo Kimball, who was the medical supply officer at the Roswell Army Air Force Base in 1947. He would have had access to any "alien autopsy" (or would have at least heard about it), and he knew all the personnel and logistics of the base. He has refuted nearly every detail of the Dennis account.

Nevertheless, the "alien autopsy" account has flourished as part of the Roswell lore and is featured as a prominent exhibit in Dennis's "International UFO Museum and Research Center" (Figs. 4.3, 4.4). After appearing in the Randle–Schmitt book (despite all its problems and

FIGURE 4.3. The Roswell International UFO Museum and Research Center. (By PunkToad [CC BY 2.0 license (https://creativecommons.org/licenses/by/2.0/)], via Flickr.)

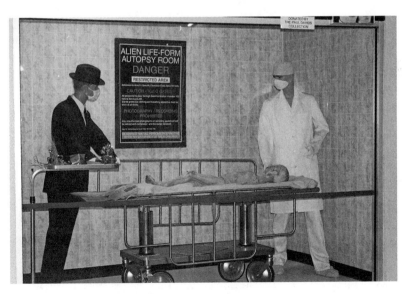

FIGURE 4.4. Display of the "alien autopsy" scene in the International UFO Museum. (Courtesy International UFO Museum and Research Center.)

discrepancies), it has been repeated and embellished with many more accounts by people who have no demonstrable firsthand access to what happened in Roswell in 1947.

The temptation to embellish and fake stories was so great that in 1995, a London-based video entrepreneur by the name of Ray Santilli promoted "secret" film footage that claimed to show an alien autopsy being conducted by U.S. military doctors. The sensational film sequence was soon featured in many UFO documentaries and gained great notoriety. But when critics looked closer, they found evidence that it was all a hoax. By 2006, Santilli had admitted that it was mostly a staged dramatization but claimed that it was based on original foot-age, now lost. He claimed that some of the frames in the hoax were real shots, but nobody can get him to reveal which ones. The claim was so preposterous that 2006 comedy film *Alien Autopsy* mocked the entire charade. The entire affair exposes just how easy it is to create fraudulent material and hoaxes that the media will quickly snap up, pay for, and promote—no questions asked. They will perpetuate the mythology without thinking twice about ethics or journalistic responsibility to fact check and report the truth.

THE VERDICT

B. D. Gildenberg, whose firsthand knowledge of the events of 1947 goes a long way toward debunking most of the tales, thoroughly got to the bottom of the Roswell myth. As he analyzed the stories, the original verified accounts have now been amplified so that there are now as many as 11 reported alien recovery sites, most of which bear almost no resemblance to what is known for a fact in 1947. Some of this prolifera-tion of sites may be from other military plane crashes that occurred in the area through 1950 or from the sightings of the falling dummies. Pflock wrote that

> the case for Roswell is a classic example of the triumph of quantity over quality. The advocates of the crashed-saucer tale ... simply shovel everything that seems to support their view into the box marked "Evidence" and say, "See? Look at all this stuff. We must be right." Never mind the contradictions. Never mind the lack of independent supporting fact. Never mind the blatant absurdities.[12]

Korff pointed out that there are lots of incentives to creates false sto-ries about Roswell, especially the money that people will pay to hear it:

> [The] UFO field is comprised of people who are willing to take advantage of the gullibility of others, especially the paying public. Let's not pull any punches here: The Roswell UFO myth has been very good business for UFO groups, publishers, for Hollywood, the town of Roswell, the media, and UFOlogy . . . . [The] number of researchers who employ science and its disciplined methodology is appallingly small.[13]

ROSWELLIAN SYNDROME

The entire Roswell incident is now considered the perfect case of a myth that was created in modern times. Charles Ziegler argues that the Roswell accounts are a classic example of a traditional folk narrative. Ziegler pointed to six distinct narratives. Each is transmitted by way of different storytellers, using a core story modified from the original wit-ness accounts. These tales were then shaped and molded by the UFO community. Eventually, the UFO researchers found additional "wit-nesses" to add to the core narrative. Any accounts that conflicted with the core beliefs were repudiated or simply omitted by the "gatekeepers." Still other authors then retold the narrative in its new form. And then the whole thing repeated, cycling over and over again—and still does.

Joe Nickell and James McGaha even gave a name, "Roswellian Syn-drome," to any such examples of myths that are propagated by the media. They identify five stages of the process, which is a pattern seen in many other myths and legends which grow in the media—Incident, Debunk-ing, Submergence, Mythologizing, and Reemergence and Media Band-wagon Effect:

> Incident. *On July 8, 1947, an eager but relatively inexperienced public information officer at Roswell Army Airfield issued a press release claiming a "flying disc" had been recovered from its crash site on an area ranch).*[14] *The next day's Roswell Daily Record told how rancher "Mac" Brazel described (in a reporter's words) "a large area of bright wreckage" consisting of tinfoil, rubber strips, sticks, and other lightweight materials.*
>
> Debunking. *Soon after these initial reports, the mysterious object was identified as a weather balloon. Although there appears to have been no attempt to deceive, the best evidence now indicates that the device was really a balloon array (the sticks and foiled paper being components of dangling box-kite–like radar reflectors) that had*

gone missing in flight from Project Mogul. Mogul represented an attempt to use the airborne devices' instruments to monitor sonic emissions from Soviet nuclear tests. Joe Nickell has spoken about this with former Mogul Project scientist Charles B. Moore, who identified the wreckage from photographs as consistent with a lost Flight 4 Mogul array.

Submergence. With the report that the "flying disc" was only a balloon-borne device, the Roswell news story ended almost as abruptly as it had begun. However, the event would linger on in the fading and recreative memories of some of those involved, while in Roswell rumor and speculation continued to simmer just below the surface with UFO reports a part of the culture at large. In time, conspiracy-minded UFOlogists would arrive, asking leading questions and helping to spin a tale of crashed flying saucers and a government cover-up.

Mythologizing. This is the most complex part of the syndrome, beginning when the story goes underground and continuing after it reemerges, developing into an elaborate myth. It involves many factors, including exaggeration, faulty memory, folklore, and deliberate hoaxing.

Re-emergence and Media Bandwagon Effect. In 1980 the story resurfaced in the media with publication of the book The Roswell Incident. Its authors were Charles Berlitz (who had previously written the mystery-mongering best seller The Bermuda Triangle, containing "invented details," exaggerations, and distortions) and William L. Moore (who was a suspect in the previously mentioned "MJ-12" hoax, as well as author of The Philadelphia Experiment, an expanded version of another's tale that itself proved to be a hoax). The Roswell Incident's book jacket gushed: "Reports indicate, before government censorship, that occupants and material from the wrecked ship were shuttled to a CIA high security area—and that there may have been a survivor!" It adds that ". . . Berlitz and Moore uncover astonishing information that indicates alien visitations may actually have happened—only to be hushed up in the interest of 'national security.'"

The book is replete with distortions. Consider rancher Mac Brazel's original description of the scattered debris he found on his ranch—strips of rubber, sticks, tinfoil, tough paper, and tape with floral designs—the same as shown in photos and consistent with a Mogul balloon array with radar reflectors. However, Berlitz and Moore impose a conspiratorial interpretation, saying that in a subsequent interview Brazel "had obviously gone to great pains to tell the newspaper people exactly what the Air Force had instructed him to say regarding how he had come to discover the wreckage and what it looked like." In fact, Brazel quite outspokenly insisted, "I am sure what I found was not any weather observation balloon," and he was right: the debris was from a Project Mogul array, much of it foiled paper from the radar targets.

Berlitz's and Moore's The Roswell Incident launched the modern wave of UFO crash/retrieval conspiracy beliefs, promoted by additional books (e.g., Friedman and Berliner 1992), television shows, and myriad other venues. Roswell conspiracy theories were off and running, typically linked to strongly anti–U.S. government attitudes. The Roswellian Syndrome would play out again and again.[15]

SUMMARY

Gildenberg wraps it up beautifully:

Pulling back to look at the big picture, one must appreciate the slapstick aspect
of the story's basic premise. According to UFO theorists, hyper-intelligent
aliens, having crossed light years through the yawning interstellar void with the
super-advanced technology, reach the earth, only to promptly blunder into the
desert floor—either once, or several times, or perhaps like a flock of lemmings
or beached whales! These inconceivably sophisticated craft, perhaps capable
of transcending basic physics, failed to avoid flattening themselves on the state
of New Mexico as a result of: very scattered thunderstorms (verified by actual
weather data); confusion caused by our marginal 1947 radar; an inability to evade
a primitive 1947 WSMR missile; or a collision with a balloon. After 55 years, there
isn't a shred of evidence that a saucer crash occurred, but there is evidence far
beyond a reasonable doubt that a series of regular terrestrial events did. The truth
about Roswell is that it was always a marginal case, and for most of its history
all the interested parties agreed. Although it has been recently and profitably
been built into a pop phenomenon, real evidence has only diminished with every
investigation—including those by the most respectable UFOlogists. Small towns,
enterprising individuals, and (most of all) book publishers and TV producers have
kept the Roswell Industry alive, but the real-life foundation for the Roswell legend
is now known in detail. Roswell is the world's most famous, most exhaustively
investigated, and most thoroughly debunked UFO claim. It's far past time for
UFOlogists to admit it and move on. Those who hope to discover alien life are
going to have to look where the aliens are—which is (if anywhere), somewhere
else. Perhaps outer space would be a good place to start.[16]

5

Close Encounters of the Second Kind: Physical Evidence of Alien Contact

TOUCHED BY AN ALIEN?

A California surgeon claims to have removed several alien implants from the bodies of abductees. After being abducted by gray aliens, Betty Hill draws, from memory, an accurate star map the aliens showed her, pinpointing the Zeta Reticuli star system as their home. A UFO seemingly ejects a strange metallic object made of an alloy unknown on Earth. Another UFO, sighted at night in rural Kansas, lands briefly, leaving behind a mysterious glowing ring. Another, landing in an English forest, leaves in its wake radiation levels 10 times the normal background level. A woman is hospitalized with symptoms of radiation sickness after a close encounter with a UFO. Weird skulls, seemingly of aliens or of alien/human hybrids, turn up in various parts of the world. A mathematician and an astrophysicist discover a mathematical code embedded in our seemingly random "junk" DNA. Surely all this constitutes a body of palpable physical evidence of extraterrestrial visitation—or does it?

The late J. Allen Hynek (1910–86), astronomer and UFOlogist, created what is called the Hynek Scale, which lists various levels of close encounters with UFOs—the first level, or CE-1, being sightings of UFOs from within 150 meters (which we've discussed in other chapters). Close encounters of the third kind (CE-3), made famous by the Steven Spielberg movie of the same name, involve actual interactions with extraterrestrials. Abduction by extraterrestrials was eventually listed as CE-4, close encounters of the fourth kind, an elaboration of close encounters of the third kind. In this chapter we will deal with close encounters of the second kind—environmental effects and other physical evidence allegedly left behind by UFOs. Certainly, if a significant number of the tales of alien abductions are true, the extraterrestrials visiting Earth should have left behind considerable trace evidence of their presence, such as those claimed in the preceding paragraph. Let us consider these in some detail.

DIRECT PHYSICAL EVIDENCE OF ALIEN ABDUCTION

Implants

Considering the malleable nature of memory and the fact that most memories of alien encounters emerge only well after the event, and then only through the highly suspect procedure of hypnotic regression, actual physical evidence of some sort would be far more convincing than any narrative. Those who claim to have been abducted, along with their supporters, argue that such evidence exists. A common motif of alien abductions is that of implants. The abductees often view these as tracking devices, not unlike those used to tag and track wild animals. The late Dr. Roger Leir, who specialized in removing alien implants, offered this explanation of their purpose in a 2014 video from a program called *Third Phase of Moon*:

> It looks like to me that the entire human race is being genetically re-manipulated again; because, if you look at children born within the last 60–75 years, they are not the same humans as a person of my age. In fact, a gentleman by the name of John White, who used to put on UFO conferences in the state of Connecticut years ago, recognized this, and he came up with the term *Homo noeticus*, instead of *Homo sapiens*, and the term means "new humans." If you were to listen to what a child has to say today and ask them or get the feeling for where they get their knowledge from, it's not from books. It's not from TV. It's not a computer. They already know. It's a different kind of human being.
>
> So you're doing genetic manipulation on a population. You want to know exactly what's going on genetically without having to abduct the individual again. So you would do it remotely, like—as I stated previously—like what you do with lesser animals.[1]

One of us (Callahan) has to say that based on his stint as a substitute teacher in the Pasadena and Burbank Unified School Districts, he has yet to see any examples of these *wunderkinds*. Most of those he dealt with, at all grade levels, seemed to get what knowledge they had—which often wasn't much—from pop culture via television, movies, and sensationalist internet sites. These acerbic observations aside, Dr. Leir's *Homo noeticus* children sound much like the extraterrestrial-engendered mutants of Midwich in the 1959 movie *Village of the Damned*, based on John Wyndham's 1957 novel *The Midwich Cuckoos*. The suspicion that Dr. Leir might well have developed his notions of alien genetic manipulation from dubious sources is bolstered by his reference to theories of previous manipulation of human genes by extraterrestrials. In fact, he cites the late Zecharia Sitchin (see chapter 8) as a source. It is also evident from Dr. Leir's references to "the missing link" that his knowledge of evolution is superficial and outdated. Still, one could argue, he is a surgeon and certainly has expertise in the removal of foreign objects from the body—or does he? Here is what skeptic author Joe Nickell had to say of Dr. Leir:

> Many of the removals have been performed by "California surgeon" Roger Leir. Actually, Dr. Leir is not a physician but a podiatrist (licensed to do minor surgeries on feet). His office includes UFO magazines for patients to read and displays of "bug-eye alien dolls." Leir was accompanied by an unidentified general surgeon (who did not want to be associated with UFO abduction claims). The latter performed all the above-the-ankle surgeries.[2]

Nickell quotes Dr. Virgil Priscu, associated with the Department of Anesthesia at Kaplan Hospital, Hebrew University, and Hadassah, Jerusalem, who notes that foreign objects can easily accidentally be superficially embedded in the human body without the individual in question realizing it. Nickell goes on to note that Priscu tried to gain access to some of the alleged implants:

> He challenged Dr. Leir's associate, a hypnotherapist named Derrel Sims, to produce specimens, or at least color slides of them for analysis at a forensic medicine institute but reported he received no cooperation. Dr. Priscu also noted the lack of scientific peer-review process in the case of implant claims. Although he is himself an admitted UFO believer, he states, "I also firmly believe that meticulous research by competent persons is the way to the truth."[3]

Regarding hypnotherapist Derrel Sims, his own website states the following:

> Inspired by the haunting memory of his own "implantation" at the age of 12, the Alien Hunter quietly began collecting alien artifacts that allegedly came from the human body. Some of these were surgically removed and then given to him by patients, while others were expelled naturally from the body. Sims sees implants as a relatively rare phenomenon, perhaps affecting less than 1% of the abductee population. Yet in 1995, he finally felt the evidence was compelling enough to formally document the removal of alleged objects from two of his most promising cases.[4]

This admission that he has a "haunting memory of his own implantation" hardly inspires confidence in Sims's objectivity. In another portion of his website,[5] Sims says that his first abduction incident took place when he was between three and a half and four years old. In his account he relates the common experience of waking from sleep only to find himself paralyzed, then realizing that there was an alien presence in his room. Both the likelihood of a small child's memory being prone to imaginative alteration and the likelihood that he was experiencing sleep paralysis make this basis for his belief in alien abductions and implantations highly dubious.

A common story about implants is that government labs test them, report that they are of extraterrestrial origin, then retract those findings and insist that they never came up with such results. Often the implants are then inexplicably lost. When Susan Clancy tried to convince abductees that their memories of abduction could easily be false reconstructions, she got this response from one of the abductees:

> After I'd finished there was a long pause. Then one of the abductees turned to Budd and said, "Tell her about the metallic tube that came out of Jim's nose, the one that went down the sink before you both could catch it."[6]

Of course, sinks have U-shaped traps, and the trap in this case should have caught the metallic tube in question. For it to be lost before the late UFOlogist Bud Hopkins (1931–2011) could catch it, someone, Johnny-on-the-spot, would have to have turned on the faucet, full blast, to make sure the implant was blown out of the trap and irretrievably lost.

Although Dr. Leir states in the video that most implants are inserted superficially, to be able to send radio signals and act as tracking devices, an anonymous abductee referred to as "Judy" (see chapter 7) tells a different story about her alien implants:

> And he was putting in what appeared to be laser objects in my brain. He put one in each temple, one in here (points to a spot between her eyes) and one in the back. And I was told I'd be able to heal people who were ill.[7]

What "laser objects" might be is difficult, if not impossible, to say, and we strongly doubt that "Judy" knows. Judy's healing powers, the documentary implies, involve the laying on of hands, a claimed miraculous form of healing going all the way back to the earliest days of Christianity if not earlier.

As to the origin of alien implants, the motif of aliens inserting devices into people's heads by which to control them can be traced to the 1953 science fiction classic *Invaders from Mars*. A Martian flying saucer lands and buries itself in a sand pit behind the house of a scientist, then sucks people down into the spaceship like an ant lion sucks down ants. The Martians then implant control devices at the base of their captives' skulls, recognizable by the clot of dried blood at the entry point in the back of the neck. The sand pit capture scenes and those showing the needle descending into the neck of the victim, with the transformation of those under control from nice people into destructive automatons, were all creepy and memorable. The Martian mind control also fed on fears of brainwashing that came out of the Korean War (June 25, 1950–July 27, 1953), which was just ending when the movie was released on April 22, 1953—fears that were kept alive by the paranoia of the ongoing Cold War.

Betty Hill's Star Map

A more indirect form of physical evidence is Betty Hill's drawn star map, based on her recovered memory of the original she allegedly saw aboard the alien craft (Fig. 5.1). It supposedly shows, quite accurately, the star Zeta Reticuli as the home of the "grays." There may well be problems with Zeta Retuculi as the home system of grays, because it is a binary

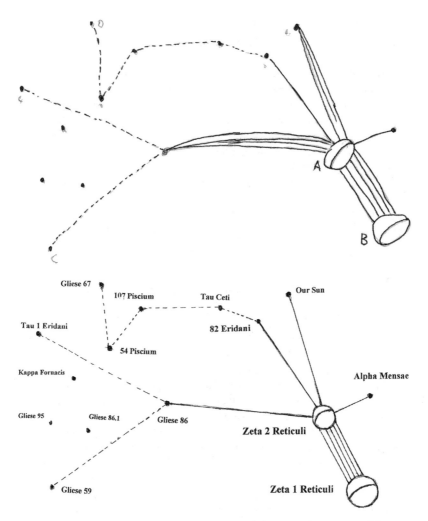

FIGURE 5.1. Star maps of the Zeta Reticulan Grays: (A) Betty Hill's star map. (B) Marjorie Fish's version of Betty Hill's star map. (Both redrawn by T. D. Callahan.)

star—actually two stars gravitationally bound together. Another problem might well be the age of that particular star system. Planets orbiting binary stars can have characteristics that would be disruptive to the development of life. For example, they might be forced to orbit both stars and thus have unstable seasonal periods, or they might be forced into a figure eight orbit and be heavily radiated when passing between the two stars. However, Colin Johnston of the Armagh Planetarium notes the following about Zeta 1 and Zeta 2 Reticuli:

> They orbit a common centre of mass approximately midway between them and are separated by at least 3750 AU (about 0.06 light-years). By the standards set by objects in our own Solar System, this is a huge distance; about a hundred times as far as Pluto is on average from the Sun. A beam of light from one of the stars would take three weeks to reach its companion.
>
> This is great enough a separation for each star in the duo to have its own planetary system. Unlike some other binary systems where the stars are closer, the skies of any Reticulan planet would not feature two suns. From the planets of one star, the other star would be a brilliantly bright star about 30 times as bright as Venus looks in Earth's sky and would be visible even in the daytime sky. Each star could have its own habitable zone (or Goldilocks zone, a region where there is enough warmth from a star to allow water to persist as a liquid on the surface of a planet). Another myth about the Zeta Reticuli system I have encountered states that emissions from the two stars contribute to a dangerous level of background radiation in their vicinity. The two stars are too far apart for this . . .[8]

The age of the two stars, Johnston says, is another story:

> What are the prospects for life in this star system? Older sources claim these stars to be relatively old (6–8 billion years old is often quoted) but more recent evidence acquired by analysing the stars' light with spectroscopes suggests these are young stars, possibly only two billion years old (compared to our own Sun's 4.5 billion year age). Their youth may count against the possibly of higher forms of life existing on Zeta Reticulan worlds. After all[,] there has been life on Earth for most of the planet's existence, but for most of this time terrestrial life has been single-celled micro-organisms, whereas multicellular life, plants and animals and so on, appeared only in the last 600 million years. If the story of life on Earth is typical (and who knows for sure), I would not bet on there being any little grey men, or indeed ladies, on any Zeta Reticulan planets.[9]

To be fair, Johnston says that there is still some controversy about the age of the two stars. However, he also notes that based on spectrographic analysis, the two stars are deficient in metals. This could make the formation of rocky Earth-type planets unlikely.

Even if we accept that the Zeta Reticuli system could be the home of intelligent life forms, there is a glaring problem with the star map: It isn't really the one Betty initially drew. In 1968, a woman named Marjorie Fish of Oak Harbor, Ohio, read *The Interrupted Journey*, the 1966 account of the Hills' abduction written by John G. Fuller. Fish, an amateur astronomer, was intrigued by Betty's map. She assumed that one of the 15 stars on the map had to represent the Earth's sun, so she constructed a three-dimensional model of nearby sunlike stars using thread and beads, basing stellar distances on those published in the 1969 *Gliese Star Catalogue*. She found, by studying thousands of vantage points over several years, that the only one that seemed to match the Hill map was from the viewpoint of the double star system of Zeta Reticuli. A comparison of Fish's star map with Betty's original crude drawing shows that the original is essentially useless in determining anything about the grays' supposed home star. Fish's comparative sophistication as an amateur astronomer is evident in her map, which is essentially her own invention.

However, we shouldn't just dismiss Marjorie Fish's map. As Colin Johnson points out, she really was quite scrupulous and went to great pains to make a correct star map. First, she made a number of very sensible assumptions. Among these are that because the map was supposed to show trade routes, the stars shown on the map were likely to have planetary systems that included planets that could harbor carbon-based life. Accordingly, she eliminated stars that were not on the main sequence of stellar evolution, because habitable planets orbiting them wouldn't likely survive the stars' transition to red giants. She also eliminated variable stars, because their huge temperature variations would make it difficult for life to arise on planets orbiting them. Third, she eliminated stars in the class F4 and higher, because these would have shorter lifetimes than our sun, thus allowing less time for life to arise and develop. She also eliminated multiple star systems in which the stars were too close together to allow stable planetary orbits. Finally, she eliminated M-class red dwarfs, because any potential planets orbiting these stars would be tidally locked—that is, one hemisphere of the planet would be permanently facing the star and the other permanently facing away, making one half of the planet excessively hot and the other excessively cold. After eliminating these types of stars, Fish was left with 46 stars:

> Using data from the 1969 edition of the *Gliese Catalog of Nearby Stars*, for nearly five years Fish painstakingly constructed several three-dimensional models of the Sun's stellar neighbourhood from wire and beads. She viewed these from every possible angle, hoping to find a pattern matching the Hill map, a long and

very difficult process. It is impossible to criticise the effort Fish made. Eventually she found almost a perfect match! It seemed that the map drawn by Betty Hill accurately depicted the stars near our own. All the stars lay roughly on the same plane and the aliens apparently came from the Zeta Reticulum system. The view point was from slightly above the star Zeta 2 Reticuli[10].

Basing her map on the best data available in 1969, Marjorie Fish created a very believable map for her day. Unfortunately for the validity of that map, that knowledge was eventually proven to be erroneous:

> Based on obsolete data, Marjorie Fish's interpretation of Betty Hill's map has been shown to be wrong. In the early 1990s the European Hipparcos ("High precision parallax collecting satellite") mission measured the distances to more than a hundred thousand stars around the Sun more accurately than ever before. Some turned out to be much further away than previously thought. Other research has looked at stars included in Fish's research. Two, 54 and 107 Piscium, have been revealed to be variable stars, while Gliese 67 and Tau 1 Eridani are in fact close binaries. Then some stars discounted by Fish have turned out to be potential abodes for life after all, for example Epsilon Eridani is not after all a binary star. Using Fish's own assumptions and more up to date data, six of the 15 stars chosen by her must be excluded.[11]

Eventually, Fish herself agreed that her original map was wrong based on the new data:

> The Fish interpretation falls to pieces at this point, and at some point Fish herself confirmed this . . . . As far as I know, this was first pointed out in an article "Goodbye Zeta Reticuli" by Brett Holman, published in the November 2008 *Fortean Times* and the silence which has followed from supporters of the Hill's account has been telling. Using more up-to-date details of nearby stars, one could probably come up with a reasonable match with completely different stars (no UFOlogist appears to want to do this), but that is beside the point, the whole idea that Betty Hill was accurately reproducing something she really saw is flawed.[12]

Had Marjorie Fish's map been a bit off, according to data available in 1969, and then been validated by later data, her interpretation of the very vague map originally drawn by Betty Hill would have been a stunning support for the objective reality of the Hills' abduction. By the same token, that it was invalidated by newer data reinforces the view that the Hills' abduction was manufactured after the fact by false memories generated under hypnotic regression.

FIGURE 5.2. The Bob White artifact, actually a stalactite formed from residue accretions on the inside of a foundry grinder. (*UFO Hunters*, season 3, episode 7, "UFO Relics.")

The Bob White Artifact

There is, however, one mysterious object—much more than just an imbedded sliver of metal—that could be physical evidence of an alien spacecraft. In 1987 the late Bob White and a female companion were driving from Denver, Colorado, to Las Vegas, Nevada. Somewhere in the desert near the Nevada state line, White and his friend saw a huge object hovering near the ground about 100 yards away. When White's companion flashed the car's headlight high beams on the object, it shot into the sky but ejected an object, which White later found and kept. It is metallic, roughly conical in shape, symmetrical, and rather organic or crystalline, as opposed to looking like a manufactured object (Fig. 5.2). In a video titled "Bob White Case UFO Object," White tells how he took the object to the government lab at Los Alamos, Nevada and there contacted two experts—whose names are, unfortunately, bleeped from the video:

> I had a contact to go to Los Alamos and there I met _____, who was the top metallurgist, and Dr. _____, who was called in on special occasions. And he was really excited about this, because he hadn't seen anything like it before. He said to me, "This is something I've been looking for all my life." He said, "This is definitely extraterrestrial."[13]

Later, however, Dr. _____ refused to talk further with White, who then suspected a cover-up. The video later features Chris Ellis, a solid-state physicist with expertise on semiconductors. In the video Ellis states the following:

> We found that the object is an aluminum alloy of unknown origin. This is not like anything we've seen before. There are some very unusual metals in the object that you typically do not find in other alloys.[14]

So the video presents a scenario in which a man finds a strange object, seemingly ejected from an alien spacecraft, and takes it to Los Alamos. There government experts (conveniently unnamed), at first enthusiastic about it, later become secretive. Finally, a physicist and expert on semiconductors reports that the object is made of an alloy of unknown origin—a sort he's never seen before. This would seem to be stunning evidence of both an alien UFO visitation and a shady government cover-up.

The operative word here, however, is "seem." As it turns out, the explanation for the object's origin is prosaic and not in the least bit otherworldly. In an October 12, 2011, article on *eSkeptic*, the online magazine of the Skeptics Society, Pat Linse, with retired steel foundry quality control supervisor, Ean Harrison, showed the origin of the object and why it is of a previously unknown alloy:

> The object in question is made of accreted grinding residue. It forms in a manner similar to a common stalagmite when metal castings are "cleaned" on large, stationary grinders.[15]

Harrison goes on to say that these stalagmites have to be broken off from time to time and that he owns several of them, which he uses as landscape ornaments. He also states that the stalagmites can be an exotic mix of stainless steel, manganese, aluminum, and other metals, depending on what has been fed into the grinder; hence the exotic alloy found by Chris Ellis. As an expert in semiconductors, but one ignorant of the vagaries of the industrial technology of foundry grinders, Ellis was baffled by what he took to be a deliberately formed exotic alloy but which was in fact a random, unintentional conglomerate of various metals. As the *eSkeptic* article's subtitle states, "sometimes all it takes is finding the right expert."

The Delphos Ring

One piece of physical evidence often touted by those who believe UFOs are extraterrestrial spaceships is the Delphos ring. One website puts it like this:

> This case has never been debunked, and is still considered as one of the very best physical trace cases involving a UFO landing.[16]

Here's the story of the Delphos ring. On the night of November 2, 1971, Ron Johnson, then age 16, was out with his dog, Snowball, tending his family's sheep near the town of Delphos, in northern Kansas, when he saw a UFO shaped like a mushroom (that is, like the cap of a mushroom and with a truncated portion of its stem) hovering just above the ground about 75 feet away. It appeared to be about six to eight feet in diameter and had bright, multicolored lights. These were so bright that Ron was nearly blinded by them. He went inside his house and called his parents. When they came out, they saw the craft ascending into the heavens. The Johnsons then noticed a glowing ring on the ground about eight feet in diameter, seemingly at the point where the craft had landed. Ron's mother touched the glowing substance, and her fingers went numb. The next day the Johnsons found that the ring was still there, though it was no longer glowing. Instead, the soil of the ring had a dry crust that shed water. The Johnsons photographed the ring both during the night, when it was glowing, and on the following day. Plants would not grow in the ring, though analysis of the soil taken from it showed that it was not radioactive. Analysis also revealed the presence of *Nocardia*, a funguslike, filamentous form of bacteria, in the ring.[17] Further analysis of the soil, as noted in an article in a UFO journal, showed that the soil of the ring contained calcium oxalate—which made up the white, fibrous substance of the ring—calcium carbonate, and a humic substance (that is, some component of humus, the dark brown organic material in soil derived from decomposed plant and animal remains and animal excrement). In fact, the humic substance in the Delphos ring turned out to be fulvic acid. This analysis also showed that the ring soil had not been exposed to heat.[18]

At first, it would seem that the Delphos ring was something anomalous and unearthly. However, there are clues in the evidence just revealed that point to a fascinating, but quite terrestrial, explanation of the nature of the initially glowing ring. Consider first that the glowing substance wasn't radioactive and hadn't been exposed to heat. Had the ring been created by the landing gear of an extraterrestrial spacecraft or the exhaust from its engines, we would expect the glow to perhaps be from radiation given off by decaying fission products. We would also expect that the numbness felt by Ron Johnson's mother was a symptom of radiation effects and that she would have subsequently developed radiation burns. Even were there nothing radioactive in the ring, we would expect the ring to have been created by the heat of the spacecraft's engines. However, someone arguing that the ring was left by an alien spaceship could contend that its lack of radioactivity and heat effects points to a technology unknown to us.

Three clues point to a very earthly explanation for the Delphos ring. One is the presence of *Nocardia* in the ring soil. Another clue is the identification of the white substance in the ring as calcium oxalate. The third is the fulvic acid in the ring's soil. *Nocardia*, as UFOcasebook.com notes, is frequently found in conjunction with certain fungi, particularly those of the genus *Armillaria*, and the species *Armillaria gallica* is indigenous to the central and eastern United States. Hence, we would expect to find it in Kansas. *Armillaria* fungi have a number of interesting properties. One of these is that they produce calcium oxalate as a byproduct of their metabolism:

> Calcium oxalate is the most important manifestation of this in the environment and, in a variety of crystalline structures, is ubiquitously associated with free-living, plant symbiotic and pathogenic fungi.[19]

So the crystalline white fibers in the ring, which were made of calcium oxalate, were likely produced as a metabolic byproduct by a fungus.

This is significant both because calcium oxalate is a toxic irritant that causes, among other things, numbness and because calcium oxalate is of low solubility and can absorb water to form hydrates. Thus it will likely stay dry when rained on, because the moisture that dampens the soil around it will be absorbed into the hydrated crystalline material. This will, of course, inhibit plant growth. Thus the calcium oxalate explains the numbness of Mrs. Johnson's fingers upon touching the glowing substance, the ring's resistance to wetting, and the inhibition of plant growth in the ring. Even apart from the production of calcium oxalate, fungus rings can generate layers of soil that resist wetting:

> Under certain conditions, and with certain fairy ring fungi, a ring of dead grass develops. Some of the responsible fungi have been shown to penetrate and kill root cells resulting in dead rings of grass. In addition, the mycelia of some fairy ring fungi are reported to be hydrophobic, creating a water-impervious layer resulting in drought-stress problems for the grass. Once the soil under this mycelial layer becomes dry, it is very difficult to wet, and the roots of the grass plant die.[20]

Thus the mysterious property of the ring to shed water is quite in keeping with the nature of fungus rings. Other substances produced by fungi and other microbes are humic acids, such as fulvic acid, one of the principle components of humus[21]—hence the fulvic acid in the ring's soil.

Fungi often grow outward from a central point, forming an ever-expanding ring as the older filaments die away and newer filaments grow outward. This is why their fruiting bodies—their mushrooms and toadstools—often form what are called "fairy rings." So nothing unearthly or even mechanical was required to form the Delphos ring.

Another property of the *Armillaria* fungi is that their mycelia are bioluminescent: they give off light at night. Fungi form fruiting bodies, such as mushrooms, for the dispersal of spores, but they spend most of their lives in a threadlike form called a mycelium. The expanding threads of the mycelium lie under the surface of the soil. So it's not readily apparent that there is fungal life in the rings. Thus, at night, when the mycelia give off light, the ground itself seems to be glowing, as one student of bioluminescent fungi notes:

> Among CE-2 cases I have had the opportunity to see and/or read about, include tales about glowing marks on the ground, "phosphorescent patches," or "ghostly

lights" in the forests. Fungi are capable of lighting up the woods. Bioluminescent
mycelium, spores, and fruitbodies of some mushroom species (*i.e., Armillaria,
Mycena, Omphalotus, Panellus*) which usually grow in wood, soil, and leaf litter
[*sic*]. The mushrooms produce a non-pulsing light, which attracts insects, which
spread fungal spores.[22]

Bioluminescence both in animals and fungi results from the oxidation
of a compound called luciferin by the enzyme luciferase. This results in
a release of energy in the form of light. In fungi, this can have an anti-
oxidant effect while also attracting insects. Bioluminescence in fungi has
been a well-known phenomenon for centuries, variously referred to as
"fairy lights" and "foxfire."

Thus the Delphos ring doesn't require debunking so much as understanding. Every aspect of the physical evidence found in the wake of the
sighting—the water-resistant, finger-numbing, growth-inhibiting white
crust made largely of calcium oxaltae; the fulvic acid in the ring's soil;
the ring's nocturnal glow; even the ring itself—has an earthly, but quite
intriguing explanation.

Yet what was the UFO, shaped like a truncated mushroom, that gave
off light so bright as to be virtually blinding? And could it have had any
connection with the glowing fungal ring? One possibility is that it was an
oblate sphere of ball lightning, a rare and little understood atmospheric
phenomenon. Whereas most lightning is in the form of linear discharges
that flash in the sky and are gone in a fraction of a second, ball lightning
takes a roughly spherical form and often lasts for several seconds, often
moving horizontally and sometimes passing through glass windows
before finally exploding or simply dissipating. These balls of plasma can
be as small as a pea or several meters across. Although ball lightning is
most often generated during thunderstorms, it can also appear in the
absence of storm conditions. What is particularly intriguing is that there
is a special relationship between lightning and fungal growth, as noted in
a paper by a group of Japanese electrical engineers.[23] Noting that lightning
has traditionally been associated with increased mushroom growth, the
paper goes on to describe the development of a system of artificial electrical stimulation to grow cultured mushrooms. Accordingly, ball lightning
hovering over the ground where the ring occurred could well have stimulated the growth of *Armillaria* fungi to produce the glowing Delphos ring.

FIGURE 5.3. Crop circles. (A) Savernake Forrest, Wiltshire, UK, in 2006. (By Ian Burt [CC BY-SA 2.0 (http://creativecommons.org/licenses/by-sa/2.0)], via Flickr.). (B) Near Diessenhofen, Germany. (By Hansueli Krapf (Own work) [CC BY-SA 3.0 (http://creativecommons.org/licenses/by-sa/3.0)], via Wikimedia Commons.)

Crop Circles

Crop circles or formations, usually circular patterns made by flattening crops—generally cereals—have been attributed both to UFOs and to paranormal forces (Fig. 5.3). Those who believe that these are from UFOs usually see them as messages from extraterrestrial beings. Because these "messages" are in the form of circles and other geometric designs rather

than mathematical formulae or writing, the nature of the communication seems to be rather cryptic. The medium of these supposed messages,
bent and flattened wheat, also seems an odd choice. Reviewing the book
Confessions of an Alien Hunter by Seth Shostak of SETI, skeptic Michael
Shermer has this to say:

> Shostak notes that crop circles are a very poor means of communication because
> they represent only a few hundred bits of information, 1,679 bits in the most
> complex crop circle to date, which is less than a paragraph of text! If ETIs
> are advanced enough for interstellar space travel, why resort to using wheat
> fields, which are only ripe a couple of months a year, and then the crop-circle
> communication is quickly mowed down by angry farmers![24]

Typically, crop circles are formed in a single night, appearing suddenly at dawn. They began appearing, mainly in the county of Wiltshire,
southwest England, in the 1970s, then spread to other countries, though
always in modern secularized states. There are, for example, no reports
of crop circles appearing in predominantly Muslim countries. This concentration of crop circles in more secularized societies coincides with the
fact that New Age movements are also usually found in such cultures,
often forming as a reaction to secularization and its inherent materialism. Although crop circles eventually spread across the developed world,
they initially appeared, and the majority of them still occur, in a very
small area: Wiltshire County, England, a region famous for its many
ancient religious sites. These include Stonehenge, the Avebury stone
circle, Silbury Hill, and West Kemet Long Barrow. If crop circles are
either landing marks of UFOs or cryptic messages from extraterrestrial
beings, why are they so fixated on Wiltshire County, England? But New
Agers and neo-pagans would both be likely to focus on a region rich in
ancient megalithic monuments.

If anything, crop circles are an excellent demonstration of a belief
system maintained in the face of strongly disconfirming evidence. This
evidence came first in the 1991 announcement, by Douglas Bower and
David Chorely, that they were responsible for the Wiltshire crop circles.
They said they had been inspired by a 1966 report from a farmer in Tully,
Queensland, Australia, that he had seen a UFO rising from a swampy
area, where, in its wake, the plants had been flattened. Bower and Chorely
said they had made the patterns using simple tools consisting of a plank

of wood, rope, and a baseball cap fitted with a loop of wire to help them walk in a straight line:

> To prove their case they made a circle in front of journalists; a "cereologist" (advocate of paranormal explanations of crop circles), Pat Delgado, examined the circle and declared it authentic before it was revealed that it was a hoax.[25]

In the face of this rather devastating revelation, crop circle advocates said that although some circles might well have been made by hoaxers, others were the result of either UFO activity or paranormal forces. One argument they made was that many of the designs were far too complicated and sophisticated to have been made in a single night by a group of people using simple tools. In response to this, a group of crop circle artists, the Circle Makers, headed by John Lundberg, did execute a very geometrically complex and demanding pattern in a single night, in the middle of summer, when there were very few hours of darkness in England. The creation of this pattern was documented in a 2010 video made for National Geographic Television.[26]

Crop circle advocates, however, point to a famous crop circle that mysteriously formed in broad daylight in July 1996. This formation, called the "Julia Set" because it resembled a fractal by that name, was spotted at 6:15 P.M. by a pilot of a private airplane and his passenger as they were flying over Stonehenge. The plane had flown over the same spot 45 minutes earlier without either the pilot or his passenger seeing the formation. According to a 2009 article by Colin Andrews, a crop circle researcher named Lucy Pringle reported that after she gave a talk in 2009, a friend of hers telephoned her:

> [A] friend of hers had been in a taxi and had mentioned to the taxi driver that she had just been to a fascinating talk on Crop Circles. "M.," the taxi driver, said: "I saw one appear opposite Stonehenge." Lucy soon realized the formation being discussed was the Julia Set of 1996.[27]

According to the story, the formation of the Julia Set took place under a swirling cloud of mist in 20 minutes. At first glance, it would seem that a number of witnesses saw the paranormal formation of a complex crop circle. However, consider that the recollection of the taxi driver was in 2009, concerning an incident that purportedly took place in 1996, 13 years before. Also, consider that Lucy Pringle was reporting on a phone call she received from a friend, referring to what another friend said a taxi driver told her. In other words, the tale of the swirling cloud of mist creating the Julia Set is a fourth-hand story told by people who are never identified—13 years after the fact.

However, what of the pilot and his passenger (both, again, unidentified) who saw the Julia Set, fully formed, 45 minutes after they had first flown over the same area opposite Stonehenge and had seen nothing? Can people be blind to something happening before their eyes? As a matter of fact, they can. This effect is referred to as inattentional blindness. When we are focused on a given task, our minds edit out perceptions not related to the task. A classic demonstration of this is the "invisible gorilla test" developed by Daniel Simons and Christopher Chabris in 1999. In this test, subjects are instructed to watch a short video in which two teams are passing basketballs back and forth between members of their same team. One team is wearing white T-shirts. The other is wearing black T-shirts. The subjects watching the video are told to count the number of passes made by members of the team wearing white T-shirts. During the video, a person wearing a black-furred gorilla suit walks into the scene, pauses, faces the camera, thumps his chest, then walks the rest of the way through the scene and off camera. Most who watch the video do not see this, being too focused on watching the team with the white shirts and having screened out anyone dressed in black. Then the subjects are shown the video a second time. This time, however, the words *Gorilla Incident: Second Look* appear on the screen before the action takes place. Now, alerted by those words, everyone sees the person in the gorilla suit. The pilot flying over Stonehenge, focused on his flight, and his passenger, focusing on the aerial view of Stonehenge, could easily have missed seeing the Julia Set as it was being made only to notice it later.

What's more, the Julia Set crop circle near Stonehenge might not have even been made during the day. In an online article, Michael Lindermann writes about how he interviewed crop circle artist Rod Dickinson, who claimed that the Julia Set had actually been made the night before:

> I pointed out that the Stonehenge Julia Set had evidently appeared in broad
> daylight, in a span of perhaps 45 minutes.

RD: "That isn't true," Rod insisted. "It was made the previous night, by three people, in about two and three-quarters hours, starting around 2:45 am (on Sunday morning, July 7). It was there all that day. When that doctor flew over, he just didn't see it the first time. That happens a lot. His report was wrong. He just didn't see it."

ML: "You mean, it sat there next to that highway all day, and no one saw it? Are you kidding?"

RD: "If you went there, you'd see how the field slopes down and away from the road. The formation was in a kind of bowl, below the level of the road. Going by in a car, you couldn't see it. You would have to get out and walk toward it and look down into that bowl-shaped area to see it."[28]

Thus, if what Dickinson says is true, the inattention on the part of the pilot was not that he failed to see the crop circle being made but rather that on his first pass over Stonehenge, he didn't see the crop circle that was already there. The whole story, told second- or third-hand, years after the fact of the Julia Set forming under the whirling mist, remains so much hokum.

Another daylight creation of a crop circle appeared in a video made by John Wave in 1996. In it, two white lights hover over a field, and a crop circle formation takes place below the moving lights. We hear John Wave whispering excitedly, "This is amazing!" This would seem to be a stunning record of a paranormal or UFO creation of a crop circle in broad daylight, and crop circle believers highly touted it. Then Wave revealed that the video was a hoax. He began with an already completed crop circle and, using a computer, faked the formation of it under the moving lights—which he, of course, created in the computer. Wave's filmed confession included footage of the creation of the images in his original video. The reaction of crop circle believers to this dramatic debunking is telling. Crop circle researcher Charles Mallet said of the John Wave video:

There's no actual evidence to say official [sic] that that piece of footage is faked. That represents a genuine piece of footage showing a genuine crop circle forming in a very few seconds.[29]

Other crop circle believers charged that Wave's confession and his demonstration of how he faked the video were all part of an elaborate government cover-up. In this case, at least, true belief trumped strongly disconfirming evidence. Belief notwithstanding, however, crop circles fail as physical evidence of UFOs.

Radiation and More at Rendlesham Forest

Late in December 1980, an incident occurred that has been referred to as "the United Kingdom's Roswell." It took place in Rendlesham Forest, Suffolk, England, near the Bentwaters RAF base, which at the time was being used by the U.S. Air Force. On the night of December 26, 1980, a strange light appeared to descend into the forest. The USAF sent a party out to investigate the lights, concerned that the lights belonged to a downed aircraft. Instead, when the men entered the forest, they saw lights that seemed to move away from them through the trees. Eventually, one of those in the party, Staff Sergeant Jim Penniston, who initially saw the lights about 70 meters away, came upon their source, an alien craft with a triangular base—and even touched its surface:

> The UFO was pyramidal in shape, approximately nine feet by nine feet by nine feet at the base and standing about six to seven feet in height. The exterior was black in color and to his touch felt warm. There seemed to be static in the air. Surprisingly, the covering was like smooth fabric, and not metallic as is often portrayed. He noted the edges of the craft had colors of red, blue and yellow and that these lights would move about. On one side was glyphlike [sic] writing that he could not decipher. Penniston noted these symbols seemed to be part of the craft, not something engraved into the material. In his written notes taken at the site he indicated there was no apparent landing gear.[30]

Penniston went on to say that after moving away from him, the craft shot into the air and that its departure was witnessed by about 80 individuals.[31] One of those accompanying Staff Sergeant Penniston, Airman First Class (A1C) John Burroughs, said the animals in nearby farms were making frantic noises and that he heard what sounded like a woman's scream.

The next day, three indentations were found at the spot where Penniston had touched the craft. The following evening, December 28, the deputy base commander, Lieutenant Colonel Charles I. Halt, investigated the spot for radiation with a Geiger counter and got at least one reading as high as 0.1 milliroentgens per hour (mR/H)—which, though not harmful, is about 10 times higher than natural background radiation.

Lieutenant Fred Buran, who led the initial search party, said that Staff Sergeant Penniston, after encountering the strange lights, seemed agitated:

After talking with him face to face concerning the incident I am convinced that he saw something out of the realm of explanation for him at the time.[32]

Lt. Buran went on to say that Penniston was a sound, reliable witness. So a reliable witness not only saw but touched an alien spacecraft; one of his companions heard strange animal noises, including something like a woman's scream; and the deputy commander of a U.S. Air Force base subsequently found radiation readings at the craft's landing site that were 10 times the normal background level. This would seem unshakable evidence of a genuine UFO landing.

However, as with the Bob White artifact, "seem" is the operative word. This is the full-blown myth of the Rendlesham Forest UFO as it developed in the years following the 1980 incident. First of all, the uncanny shriek A1C Burroughs heard might well have been from a muntjac deer in the forest, which are known for their loud, shrill bark when alarmed, and their alarm could as easily be caused by men with flashlights walking through the forest at night as by a UFO. As to the source of the lights, one of those in Penniston's party, A1C Edward N. Cabansac wrote the following in his witness statement in early January 1981:

After we had passed through the forest we thought it to be an aircraft accident. So did CSC as well. But we walked a good 2 miles past our vehicle, until we got to a vantage point where we could determine that what we were chasing was only a beacon light off in the distance.[33]

What, then, are we to make of Staff Sergeant Jim Penniston's actually touching the UFO? In his witness statement—which, though neither signed nor dated, was probably made at the same time as Cabansac's—he tells a far different story from that of touching the alien craft (emphasis added):

When we got with in [sic] a 50 meter distance. [sic]. The object was producing red and blue light. The light was steady and projecting under the object. It was lighting up the area directly under extending a meter or two out. At this point of positive identification I relayed to CSC Sgt. Coffey. Positive sighting of object . . . 1 . . [sic] color of light and that it was definitely mechanical in nature. This is the closes [sic] point that I was near the object at any point.[34]

So initially, Penniston said he got no closer than within 50 meters of the object. It was only later that he claimed to have touched the alien spacecraft. This is not the only extraordinary claim he made regarding the encounter:

> In 2010 it was revealed by Jim Penniston that he initiated a download of information when he touched the pictorial glyphs on the craft of unknown origin during his investigation in the early morning hours of December 26th, 1980, in Rendlesham Forest. When Sgt. Penniston put his hand on the etched symbols, which felt like sandpaper compared to the rest of the smooth molded surface, everything became a brilliant bright white. He could neither see nor hear. He was alone in a brilliant bright white light.[35]

The website telling of this transmission of information goes on to say the following (parenthetical material in the original):

> The following day, back in his room Jim was "seeing" ones and zeros (1's and 0's) in his minds eye. Troubled by the revolving flashing images of ones and zeros he received from touching the glyphs, he felt compelled to write them down in a notebook. . . .
>
> Upon completing the transfer to notebook paper, the image in his mind disappears. Jim puts the notebook away with his belongings and doesn't think too much about it again until the year 2010. In 2010, during a casual conversation with a researcher, he mentions the codes and displays the notebook. The researcher immediately recognizes the ones and zeros as binary code and sets out to help Jim decipher it.[36]

According to the website, Penniston also wrote of seeking treatment from a hypnotist for posttraumatic stress disorder (PTSD), with the following result:

> 1994 Sept. Hypnosis was accomplished. Talked about time travelers and the receipt of the binary code which was given from contact with the craft. Treatment continued for the PTSD. No help on the sleep issues.[37]

So now it appears that the encounter wasn't with a spacecraft after all but rather with a time machine. Considering that this revelation came through hypnotic regression, it hardly seems credible.

For all that, we still have the unusual radiation readings at the site taken by Lt. Col. Halt. Certainly the fact that they were 10 times the normal background radiation is the strong physical evidence of something out of the ordinary that we are seeking in this chapter? Alas for believers, science writer Ian Ridpath has published a handwritten memo, dated April 15, 1994, by Nick Pope, one of the main proponents of the theory that the Rendlesham Forest incident was an encounter with a UFO:

> Spoke to Giles Cowling at the Defence Radiological Protection Service re the radiation readings recorded at the time of the Rendlesham Forest incident. 0.01 would be the general level of background radiation, so the 0.1 reading is about 10 times what would be normal. [Ian's note: this would be true if 0.1 were a steady level, but as we know it refers only to a random peak.]
>
> However, military radiation detectors are geared for high level readings, so low-level readings may be difficult to record accurately, as the scale will be small at the bottom of the meter.* We don't have details of what instrument was used.
>
> It is just possible to hoax such an event. A university lab might well have some radioactive source with a very short half life, and could use it so as to give readings which would not be recorded a few days later.
>
> The level of radiation of 0.1 is completely harmless.
>
> *Especially if the needle was fluctuating.[38]

So the radiation at the supposed landing site turns out not to have been 10 times above the background level after all. Rather, the needle of a detector geared for much higher radiation level readings, hence not really accurate for readings in the range of 0.01 to 0.1 milliroentgens per hour, gave a fluctuating reading—that is, a momentary, localized spike of 0.1 mRH.

To sum up, we have a supposedly reliable witness, Penniston, who has changed his initial testimony from being at best just within 50 meters of the spacecraft/time machine to having touched it and then to having received a message in binary code from it; and we have no physical evidence, in the form of heightened radiation levels, that an alien spacecraft landed in the Rendlesham Forest on December 26, 1980. The Rendlesham Forest incident is a story that grows in the telling, as folklorist Thomas Bullard notes:

> In the elaborated versions the original three guards lost consciousness when the UFO flashed a light on them, while on the second night parties of troops saw the shadowy figures of occupants inside a translucent UFO, or in yet another version, the UFO landed and aliens emerged to communicate with a general.[39]

Considering all these inflated tales, can anyone take this supposed UFO seriously?

Radiation Sickness, Melting Vinyl, and Scorched
Asphalt: The Cash–Landrum Incident

Just days after the initial sighting in the Rendlesham Forest, two women, accompanied by the seven-year-old grandson of one of them, had a close and fiery encounter with a UFO at night on a road in Texas. On the evening of December 29, 1980, Betty Cash and her friend, Vickie Landrum, along with Vickie's grandson Colby Landrum, were driving home to Dayton, Texas, in Cash's fairly new Oldsmobile Cutlass after having dined out. At about 9:00 p.m., as they were driving through Piney Woods, they suddenly found their way blocked by a craft hovering just over the road. According to the descriptions given, as recorded in John Schuessler's book on the incident, Cash and Landrum were southbound on Texas state highway FM 1485 when, at 9:00 p.m., they encountered the UFO. Its initial location, based on the same descriptions, was just south of Inland Road, approximately at 30.0926°N, 95.1109°W.[40]

The craft was large, about the size of a local water tower, and shaped like a vertical diamond. In artists' reconstructions of the event, the craft is portrayed as being shaped like two square-based pyramids joined at their bases, hence the diamond shape. Both the upward and downward points of the craft were truncated and flat. The craft stood on a pillar of flame shooting out of the bottom. This flame was quite hot, and Vicki told Betty to stop the car lest they be burned.

Vickie Landrum, being a born-again Christian, at first interpreted the flame-spewing craft as a sign of the second coming of Jesus and told her grandson, Colby, "That's Jesus. He will not hurt us."[41] Cash, however, was somewhat alarmed by the apparition and considered turning the car around to flee. However, the two-lane road was too narrow for her to turn her large car around and still stay on the pavement. It had recently rained, so she feared that the car would get stuck on the dirt shoulders.

Both Cash and Landrum got out of the car to investigate; but Colby was terrified, so Vickie Landrum got back inside the car to comfort her grandson. Cash remained outside, seemingly mesmerized by the apparition. However, she could not bear the intense heat for long and began to get back inside. The metal of the car had become so hot that she had to use her coat to protect her hand from being burned by the door handle. The interior of the car was also hot. When Landrum touched the dashboard, her hand, pressed into the now nearly melting vinyl, left an imprint.

At this point, the UFO began to ascend, leaving behind a scorched section of road. As it rose above the treetops, a group of helicopters surrounded it. Cash and Landrum claim to have counted 23 helicopters, which they were later able to identify as tandem-rotor CH-47 Chinooks—military helicopters. Now that the road was clear, Cash drove on. Cash and Landrum said that about 20 minutes had passed between their first sighting and the craft's departure.

Cash took the Landrums home and went to bed. That night, all three of them experienced the same symptoms, though Betty Cash experienced them with far greater intensity than did either Vickie or Colby Landrum. These symptoms were nausea, vomiting, diarrhea, general weakness, inflammation and swelling in the eyes, and a feeling as though they were suffering sunburn. Although Vickie and Cole's symptoms were passing—though both suffered lingering weakness, some sores and hair loss—Betty Cash's got worse. Along with the nausea and vomiting, she developed many large, painful blisters. Eventually, she had to be taken to a hospital emergency room on January 3, 1981. At this point, she couldn't walk and had lost large patches of skin and clumps of hair. She was released after 12 days, but her condition hadn't significantly improved, and she had to return to the hospital for another 15 days. A radiologist, it is claimed, examined the medical records of Cash and the Landrums and believed there to be strong evidence that they had suffered damage from ionizing radiation—that is, radiation sickness.

Betty Cash and the Landrums weren't the only ones to witness the diamond-shaped UFO and the military helicopters. A Dayton police officer, Detective Lamar Walker, and his wife, claimed to have seen 12 Chinook-type helicopters near the same area where the Cash–Landrum event allegedly occurred and at about the same time, though they didn't witness the UFO. Moreover, an oilfield laborer, Jerry McDonald, happened to be in his backyard, in Dayton, when he witnessed a large UFO flying over his head. At first he thought it was the Goodyear airship, but he quickly realized that it was something else. "It was kind of diamond shaped and had two twin torches that were shooting brilliant blue flames out the back," he said. As it passed above him, he saw that it had two bright lights on it and a red light in the center.[42]

It would seem that this is unshakable evidence of a close encounter of the second kind with what seems to be an extraterrestrial flying craft. Besides multiple witnesses, there were a scorched area at a very precisely known spot on a specific rural highway, a portion of the Oldsmobile's vinyl dashboard that was heated so much that Vickie Landrum left her handprint in it when she touched it, and—most striking of all—Betty Cash's symptoms of radiation sickness. However, on closer examination, the evidence begins to unravel. The handprint wasn't found, and nobody photographed it initially. We actually have no evidence of this assertion.

The scorch marks on the highway were, likewise, never found. In fact, the location of the incident is hardly as dramatically pinpointed as originally claimed. The main investigator of the alleged incident was John F. Schuessler, a director of MUFON (Mutual UFO Network), who was also affiliated with other UFO organizations. Texas Department of Health investigator Charles Russ Meyer wrote the following concerning the location of the incident:

> I then asked Mr. Schuessler if he had pin-pointed the location of the siting [sic]. Mr. Schuessler stated that due to the late hour and the ladies' emotional state they could only state that they believed they saw the object on the straight portion of FR 1485 between a beer joint and some kind of highway warning sign.[43]

This description, "between a beer joint and some kind of highway warning sign," is glaringly different from the precise position in the final story of "just south of Inland Road, approximately at 30.0926°N, 95.1109°W." Because we have no real idea where the incident actually took place, both searching for the scorch mark and reliably attributing any scorch mark found on the highway to the UFO incident is hopeless and pointless.

Attributing Betty Cash's symptoms of blistering, hair loss, nausea, vomiting, and diarrhea to radiation sickness is also problematic. In an online article, Brad Sparks, UFO investigator and cofounder of Citizens Against UFO Secrecy, points out that in cases of radiation sickness in which there is an immediate onset of severe symptoms, the dose of ionizing radiation is so high as to be lethal (capitalization in the original):

Let me begin with the bottom line first, even though the supporting evidence and argument will follow and I risk being misquoted without the backup: Had Betty Cash really experienced an ionizing radiation exposure with such rapid onset of symptoms she would have been DEAD within several DAYS, uniformly 100% FATAL, yet instead she survived for 18 years and so have the Landrums, who would also have been dead within DAYS. There is no "almost" about it, you can't argue that she "almost" died therefore it must have been ionizing radiation because she AND both Landrums ALL THREE should have been DEAD within DAYS if it was a massive dose of ionizing radiation. You can ask any radiation biologist who specializes in the study of ionizing radiation and they will tell you that a 1,000–2,000+ rem dose has NEVER been survived by ANYONE in history—remember, there were thousands of radiation cases at Hiroshima and Nagasaki.[44]

Sparks rightly notes that in cases of sublethal, total body exposure to ionizing radiation, the gastrointestinal symptoms of nausea, vomiting, and diarrhea don't begin until one to two days after exposure, followed by a latent period of about two weeks, which is then followed by a return of the gastrointestinal symptoms, along with hair loss and hemorrhaging under the skin. Because Betty Cash's gastrointestinal symptoms began about a half hour after the incident and lasted for weeks without any latent period, they don't fit the pattern of sublethal radiation sickness. Finally, Sparks notes a symptom that directly contraindicates radiation sickness:

If I recall correctly the medical data, Betty Cash had an elevated WBC (white blood cell) count which is not consistent with ionizing radiation exposure which depresses the white cell count.[45]

White blood cells are extremely sensitive to radiation and are killed before any symptoms appear. An elevated white blood cell count would indicate an infection, either viral or bacterial, depending on what types of white blood cells are most prevalent.

Thus Betty Cash's symptoms could have been caused by an infection that was coincidental to the UFO encounter. Another possible cause of her symptoms is severe sunburn or sun poisoning. The symptoms of sun poisoning are skin redness and blistering, pain, tingling, swelling, headache, fever and chills, nausea, dizziness, and dehydration. Because Betty would have been unlikely to experience sun poisoning in December, even in a sunbelt state, her symptoms, though not being the result of ionizing radiation, might still have been caused by the heat and light of the UFO's column of flame. If so, they would be physical evidence validating the UFO incident. However, the reason Sparks had to say of Cash's white blood cell count "If I recall correctly . . ." is that there is some confusion about what her exact symptoms were, as Robert Scheaffer notes:

> However, Betty Cash's medical records have never been released, on grounds of "privacy." Fair enough, but you cannot simultaneously cite alleged medical symptoms as proof of a UFO encounter while refusing to release the medical records that might confirm or refute the claim. So long as the medical records remain "private," anecdotal accounts of what they contain are worthless.[46]

Schaeffer and some others have asserted that Betty Cash and Vickie Landrum fabricated the entire story in a crude attempt to get money out of the federal government. The two attempted to sue the U.S. government for $20 million over the health effects of the incident, based on the government's failure to protect them from the UFO:

> They were represented pro bono by UFO lawyer Peter Gersten, who attracted much attention in 2012 by his announced plan to leap from Bell Rock in Sedona at the moment of the winter solstice Mayan apocalypse (but fortunately he had second thoughts, and is still alive). The suit was dismissed by a U.S. District Court judge in 1986 on the grounds that U.S. government involvement had not been demonstrated. It ought to be quite straightforward to trace a fleet of 23 Chinook helicopters flying over the United States. Much effort has been expended to trace such helicopters, to no avail. The U.S. military simply didn't have a fleet of that many Chinook helicopters in one place, nor did any private firm.[47]

Despite the absurdity of the Cash and Landrum lawsuit, some support for their story can be found in the report of a strange diamond-shaped UFO spouting flames on the part of Jerry McDonald and, to a lesser degree in Lamar Walker and his wife's saying they saw 12 Chinook helicopters flying near where the incident is alleged to have taken place. But a large, flaming, diamond-shaped UFO, with its attending fleet of Chinook helicopters—which are quite noisy—should have attracted more attentions than one couple seeing the helicopters and another lone witness seeing the UFO. Regardless of what happened in the skies over Texas on the night of December 29, 1980, as a source of physical evidence, the Cash–Landrum incident is inconsequential.

ALIEN SKULLS?

There can certainly be no more convincing evidence of extraterrestrial visitation that the aliens themselves. A live alien would be spectacular proof, but a dead one will do. Because the supposed film of an alien autopsy produced in 1995 by Ray Santilli is an admitted hoax, we must search elsewhere for the remains of extraterrestrials. There are a number of skulls that various UFOlogists have claimed were either skulls of aliens or of alien/human hybrids, though the latter claim is highly questionable on the basis of genetics. Let's consider the evidence.

The Paracas Skulls

The Paracas skulls from the Paracas peninsula, Peru, are strangely elongated and look as if they are the product of some weird mutation. They were initially discovered in 1928 by Peruvian archaeologist Julio Tello and belong to the Chavin culture (900–200 BCE). Actually, the elongated skulls were produced by artificial cranial deformation. In infancy the bones of the human skull are quite malleable, and the skull can be deliberately deformed, by the use of bands and boards, to constrict the skull in one direction, causing it to elongate. A number of ancient and indigenous cultures practiced deliberate deformation and elongation of the skull. Among these were the Huns (Fig. 5.4C), the Sarmatians of the Russian steppes, the Mangbetu tribe of central Africa, and many peoples from the Americas. The latter include the Chinook from the northwestern United States, the people of the Nodena Phase of the Mississippian culture (from 1450 to 1600), the Maya, and several groups in Peru, including the Inca. Elongated skulls have also been found on the Mediterranean island of Malta. Thus the strangely shaped Paracas skulls shouldn't pose an enigma. However, early in 2014, Brien Foerster, assistant director of the Paracas History Museum, claimed that DNA testing had shown that the Paracas skulls were not human. Articles supporting this claim also stated that the volume of the cranium in the Paracas skulls was as much as 2.5 times that of modern humans.[48] All this would be stunning physical evidence supporting claims of an ancient alien visitation—were any of it true. In fact, the Paracas skulls have a

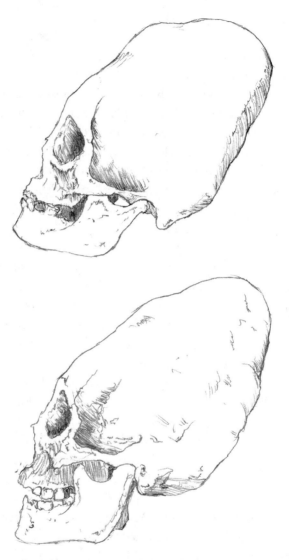

FIGURE 5.4. Examples of deliberate cranial deformation. (A) One of the Paracus skulls, Chavin Culture, Peru, *ca.* CE 100; original held by Paracus History Museum, Paracus, Peru. (B) Elongated skull from the megalithic hypogeum of Hal Saflienti, Malta, *ca.* 3000 BCE; original held by National Museum of Archaeology, Valietta, Malta. (C) Skull of a Hun, southern Kazakhstan, first to third century CE; original held by the Institute of Ethnography, Moscow and St. Petersburg. (A–C drawings by T. D. Callahan.) (D) One of the Mangbetu of Central Africa. (Courtesy Tropenmuseum, part of the National Museum of World Cultures [CC BY-SA 3.0 (http://creativecommons.org /licenses/by-sa/3.0)], via Wikimedia Commons.)

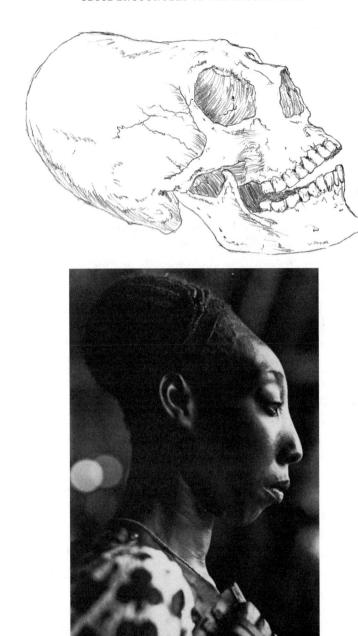

FIGURE 5.4. *Continued*

cranial capacity averaging around 1,600 cubic centimeters (cm³). The average cranial capacity for modern humans is 1,450 cm³, whereas that for Cro-Magnon men was 1,600 cm³. Some skulls of *Homo sapiens* have a cranial capacity as high as 1,800 cm³. Because the Paracas skulls are all from males, and because men have, overall, larger brains than women (more brain tissue being required to handle larger bodies), their cranial volume is well within the normal range for modern humans.

As to the DNA analysis, Foerster never names the geneticist who supposedly made the analysis in any of his reports or in the reports of anyone supporting his claims. Foerster's own credentials are quite shaky. It turns out that the Paracas Historical Museum, of which he is assistant director, is a commercial venture rather than a bona fide museum. It is part of Foerster's business, "Hidden Inca Tours," which specializes in tours supporting the notion of ancient alien astronauts and evidences of paranormal activity. Foerster has also coauthored a number of books with David Hatcher Childress, one of the pseudo-experts on the History Channel's *Ancient Aliens* series—a college dropout posing as an archaeologist. Leaving aside Foerster's dubious credentials and unfortunate connections, let us consider just how he concluded that the DNA from the Paracas skulls was alien. Foerster sent one of the skulls to the late Lloyd Pye (1946–2013), the main proponent of the "Starchild" skull, which we will look at shortly. Pye had a BS degree in psychology, made his living writing for television shows, and was a paranormal investigator. Thus he was not the geneticist whom Foerster claimed found the alien DNA in the Paracas skull. Pye sent the skull to Dr. Melba Ketchum, a Texas veterinarian and genetics researcher. However, her company, DNA Diagnostics, deals only with dog and cat DNA. Ketchum is best known for having claimed to be in possession of samples of Bigfoot DNA. She further claimed to have had an article about the Bigfoot DNA published in a peer-reviewed journal, *De Novo*. But it turns out that *De Novo* was a fake journal that published only Ketchum's article, which was never peer-reviewed. When Eric Berger, a reporter for the *Houston Chronicle*, sent some of the supposed Bigfoot DNA to a geneticist friend of his, the DNA turned out to be that of an opossum.[49] If Ketchum were the "geneticist" who claimed to have found alien DNA in the Paracas skull, then Foerster would seem to be on very shaky ground with his

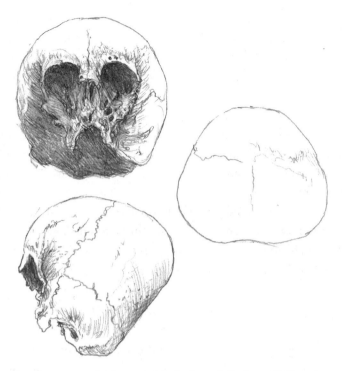

FIGURE 5.5. Front, top, and side views of the hydrocephalic "starchild" cranium. (Drawing by T. D. Callahan.)

claim. However, Foerster hasn't named Ketchum as the source of his alien DNA claim. Accordingly, the geneticist in question remains anonymous, and the claim remains both questionable and, unless he releases more material from the skulls to a reputable geneticist, untestable.

Lloyd Pye, the intermediary between Foerster and Ketchum, also claimed to have found an alien skull of sorts—to be specific, a hybrid between gray aliens and humans (Fig. 5.5). The skull in question is that of a child between four and five years old, with an extremely large cranium. Pye called it the "Starchild." Its history and provenance are quite sketchy. Pye said he obtained the skull from Rex and Melanie Young of El Paso, Texas, in February 1999. They told him that it was found in a mine 100 miles southwest of Chihuahua, Mexico, sometime during the 1930s. The remains consist mainly of the bulging cranium and a small part of the face, including a small portion of the maxilla (upper jaw).

Despite Pye's claims that it was the skull of a human/gray alien hybrid (again, biologically unlikely), DNA testing showed it to be fully human and Native American. The bulging cranium was the result of congenital hydrocephalus.[50] The name of the disorder is from the Greek words for "water" and "head." Hydrocephalus is a medical condition resulting from abnormal accumulation of cerebrospinal fluid in the brain. This causes increased pressure inside the skull. In cases of congenital hydrocephalus, the build-up of fluid occurs before the bones of the skull have fused, and the increased pressure causes the cranium to greatly enlarge. The modern treatment for hydrocephalus is to surgically install a shunt that drains the excess cerebrospinal fluid from the brain. Before shunts were available, the condition caused mental instability, tunnel vision, and convulsions. In technologically primitive cultures, this would likely lead to an early death—as, sadly, it must have done in the case of Pye's "Starchild."

The Rhodope Skull and Adygea Skulls

Both the elongated Paracus skulls and the hydrocephalic "Starchild" are recognizably human, though deformed—either genetically or deliberately. What is purported to be the skull of an extraterrestrial being, discovered in the Rhodope Mountains of southern Bulgaria, is definitely not human. Several websites have touted this find, the Rhodope Skull, as something for which scientists can't account. Most of these sites display the skull, and it certainly looks anomalous—that is, if one assumes that it is a complete skull. However, beginning with how the skull was discovered, along with when it was discovered, one can find holes in this seemingly concrete and solid bit of evidence.

Let's take a closer look the Rhodope skull. All you need to do is type those two words into any search engine and you get a string of paranormal sites that make extraordinary claims about it. The breathless hyperbolic prose of this website are typical, and most of the other sites seem to be verbatim copies of this same text:

> This skull was found in the Rhodope Mountains in Bulgaria by a man from Plovdiv, a city located 90 miles (150 kilometers) southeast of the capital city of Sofia. The man, who chose to remain anonymous discovered the skull buried alongside a small elliptical metal object.

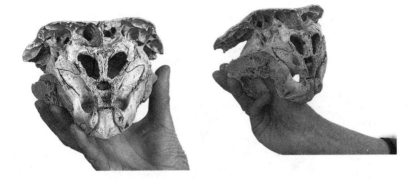

A close up of the frontal view of the specimen, with the various anatomical features labeled (see text).

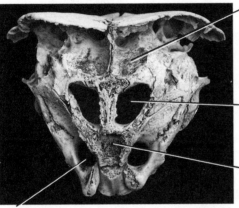

1. The strange "pockets" above the "eyes" are the broken sinuses of frontal bones of a skull. Frontal bones of many animals are flat outside and are penetrated by a set of sinuses. The external structures of a bone distinguish cattle skulls from skulls of other large mammals.

2. The weird "eyes" are actually a zone of a damaged ethmoid bone (olfactory fossa of braincase), a place of passage of olfactory nerves from a brain to a nasal cavity.

3. The "nose" is just the partially damaged body of a sphenoid bone.

4. The odd paired holes that appear to be a "mouth" are actually the orbitorotundum foramen (an aperture for some other cranial nerves and vessels).

FIGURE 5.6. The Rhodope skull and Adygea skulls. (A–B) Two views of the specimen. (A–B courtesy Bulgarian News Agency, BTA.) (C) Labeled view of the "face" of the specimen (see text for details). (D) Photo of a goat skull broken in the same way that the Rhodope skull is broken, showing that the holes in the skull are the same. (C–D courtesy Alexei Bondarev, P. Linse, and *Skeptic Magazine*.) (E) If you break the snout off of a cow or goat skull, you get the "Rhodope alien skull." (Courtesy P. Linse and *Skeptic Magazine*.) (F–G) Two views of one of the Adygea skulls, held at the Belovodye Museum in Kamennomostsky, Russia. (F–G courtesy Andrey Zherlitsyn and the Belovodye Museum, http://www.belovodye-adygea.com.) (H) Labeled view of the "face" of the specimen. (Courtesy Alexei Bondarev, P. Linse, and *Skeptic Magazine*.) (I) West Caucasian tur, *Capra caucasica* (GFDL [CC BY-SA 3.0 (http://creativecommons .org/licenses/by-sa/3.0)], via Wikimedia Commons.) (J) Diagram showing how the Adygea skull fragment would fit within the head of the West Caucasian tur. (Courtesy P. Linse and *Skeptic Magazine*.)

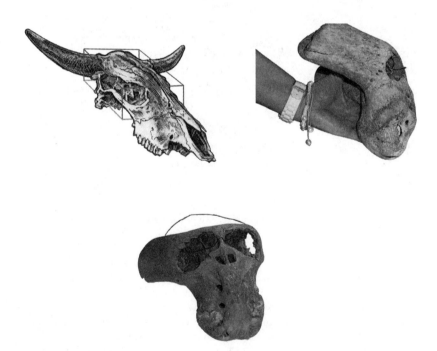

FIGURE 5.6. *Continued*

One of the Adygea skulls, with the anatomical features labeled.

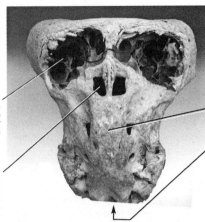

1. The "eye sockets" are the open sinuses of frontal bones of a skull roof. Further back, these sinuses go to horn cores.

2. The "nose holes" are openings in the ethmoid bone that pass into the braincase.

3. The bony "nose" is the surface of the base of an occipital bone. Behind it there was a brain stem. The holes around the bone are intended for cranial nerves and have nothing to do with the mouth.

4. The hole at the base, which was also called a "mouth," is actually the foramen magnum, through which the spinal cord passed.

FIGURE 5.6. *Continued*

His interesting story began after having a dream in which five humanoids wearing yellow metallic clothes directed him to a spot in the Rhodope Mountains, near the border between Bulgaria and Greece. On the 21st of May, 2001 he headed to the spot the strangers had indicated and surely enough, the skull was there.

His finding generated a lot of excitement and controversy and many ufologists consider it evidence for the existence of aliens and their presence on Earth. He initially presented his find to a narrow circle of people interested in the paranormal but the skull did not remain secret for long.

The skull was examined by archaeologist Katya Malamet of the Bugarian Academy of Sciences and by Professor Dimiter Kovachev, the Director of the Paleontology Museum in Asenovgrad. Both of them have said they have never seen anything like it before.

The skull is as small as a human baby's but its bone structure is lighter and thinner and it weighs only 250 grams (8.8 ounces). It has six cavities that researcher think belonged to the alien's sensory organs. When alive, this being possessed six eyes or perhaps other completely unknown organs. Interestingly, it has no mouth hole where we would expect to find one.

Professor Kovachev is positive it is not a fossil and does not resemble any hominid skull known to science. "We couldn't find any analogy or correlation with anything from the past 30 million years," Kovachev said.

Skeptics believe the skull belongs to a still unidentified animal species although its features do not resemble those of any other animal. Skeptics, on the other hand, believe that the Rhodope skull is just (another) elaborated hoax, or that it might have belonged to a species that remains unidentified.[51]

No, skeptics do not think it's a hoax. And of course it doesn't look human, and anthropologists don't recognize it—because it *isn't* human! But that doesn't mean we can jump to the conclusion that it's some alien skull instead. As soon as anyone trained in animal anatomy or paleontology (which teaches how to recognize animals from broken fragments) looks at the specimen, the answer is obvious. It's a partial skull of an animal with the facial region broken off. In most mammals, the facial skeleton and braincase are connected by several narrow apertures for the nerves, and some skull bones have internal cavities familiar to us as sinuses. The facial skeleton is much less durable than the braincase and typically is quickly destroyed by forces of the nature. The joint between the braincase and facial skeleton is rather weak. The braincase slowly collapses, and finally the skull roof collapses. In the course of destruction, skulls lose all facial bones and a part of the bones of the braincase, and the details of the internal structure of the skull is exposed. The "eyes" and other strange openings on the front of the skull are not eyes at all but rather some of the exposed sinuses and cavities in the facial region, snout, and braincase.

The first clue for the trained anatomist is the top of the skull, which comprises frontal and parietal bones that normally form the top and back of an animal skull, just around the eye sockets. Their rough broken sutures and the spongy bone tissue beneath them can be clearly seen, along with their irregular suture along the midline, as seen in dorsal (top) view. The rest of the features are as follows (Fig. 5.6C):

- The strange "pockets" above the "eyes" are the broken sinuses of frontal bones of a skull. Frontal bones of many animals are flat outside and are penetrated by set of sinuses. The external structures of a bone distinguish cattle skulls from skulls of other large mammals.
- The weird "eyes" are actually a zone of a damaged ethmoid bone (olfactory fossa of braincase), a place of passage of olfactory nerves from a brain to a nasal cavity.
- The "nose" is just the partially damaged body of a sphenoid bone.
- The odd paired holes that appear to be a "mouth" are actually the orbitorotundum foramen (an aperture for some other cranial nerves and vessels).

A glance at the top of the skull shows small bony horn cores, so this is some sort of bovid, and the size and immaturity (as shown by the unfused sutures) and other characteristics of the top of the skull show that it is just the broken skull of a calf—not some sort of alien being. Regrettably, no one will be able to do further analysis on the skull, it having been lost.

Yet another instance of bumbling amateurs seeing aliens in ordinary broken animal skulls is the case of the Adygea skulls (Fig. 5.6F, G). They are housed in a small museum in the town of Kamennomostsky (Каменномостский), in the Republic of Adygea, which is a federal subject of Russia, located near the Black Sea. One website says this about them:

Vladimir Malikov said that two years ago, cavers had found two unusual skulls in a cave on the mountain Bolshoi Tjach (Большой Тхач), which is about

50 miles southeast of Kamennomostsky. One of the two skulls is very unusual. Malikov says that the presence of the hole at the bottom of the skull where the spine attaches, proves that this creature was walking upright on two legs. It is also very unusual that the skull does not have a cranial vault as with humans. It also has no jaws. The whole head is one fixed bony enclosure. The large eye sockets arches back, and then we have horn-like extensions. He has sent photos to paleontologists, but they could not explain it.[52]

Well, if they really sent it to paleontologists, those scientists must not have had any familiarity with mammalian anatomy, because it was apparent to both Alexey Bondarev and Prothero that the specimens were the badly water-worn and broken cranial regions of a hoofed mammal, probably a goat. The not-so-mysterious anatomical features are identified in Fig. 5.6H.

- The "eye sockets" are the open sinuses of frontal bones of a skull roof. Further back, these sinuses go to horn cores. Judging from the photos, the shape of the horns and occiput is characteristic for a wild Caucasian goat that lives in the region (West Caucasian tur, *Capra caucasica*—Fig. 5.6I).
- The "nose holes" are openings in the ethmoid bone that pass to the braincase.
- The bony "nose" is the surface of the base of an occipital bone. Behind it there was a brainstem. The holes around the bone intended for cranial nerves and have nothing to do with the mouth.
- The hole at the base, which was also called a "mouth," is actually the foramen magnum, through which the spinal cord passed. The impression of a vertical "alien face" arises because of the incorrect orientation of a skull (if the internal structure and base of the skull take as face). Also, there is a curvature of the axis of the skull of goats.

So once again, the internet is full of bad information. "Alien skulls" turn out to be broken and misinterpreted skulls of ordinary calf and Caucasian goats. Another mystery solved—but don't count on the followers of aliens and paranormal phenomena to pay any attention to their fantasy's debunking. On the internet, and in the world of believers in the paranormal, false information and bad ideas live forever:

> Unfortunately, the skull disappeared before scientists had the chance to perform DNA analysis and carbon dating. Whoever the owner is today, he obviously doesn't want to share his possession with the rest of humanity, even though it could be evidence for the existence of aliens.[53]

So here we again have the evasion we so commonly find when looking for specifics in extraordinary claims involving direct physical evidence of UFOs.

The Sealand Skull

One last allegedly alien skeletal find is the Sealand skull (Fig. 5.7). It looks generally human and is (probably) close to human in size. However, its eye sockets are enormous in ratio to the size of the skull itself. They resemble those of either a tarsier or an owl monkey, both of which are nocturnal primates. The skull's upper canine teeth are elongated to the point of being fangs.

The skull, complete with its lower jaw, was allegedly discovered by workers repairing sewer pipes in a house in Olstykke, on the island of Sealand, Denmark. It is also claimed that the skull was sent to the Niels Bohr Institute in Copenhagen, where it was dated, using carbon-14 dating, to between CE 1200 and 1280.[54] Perhaps the author of the online article making this claim is unclear about where the skull was carbon-dated, because the Niels Bohr Institute had ceased carbon-dating by the middle of the 1960s, and the Sealand skull, according to the same article, was discovered in 2007. A more likely possibility is that the carbon-dating claim is bogus. Although photographs of the skull are displayed on many websites, none of them indicate where the skull resides. Not surprisingly, there is also no information regarding DNA tests on the skull.

Compounding the Sealand skull's dubious provenance is the possibility that it might be a bit of ceramic artistry as opposed to actually being made of bone. One of those posting on a forum noted, among other things, that in photographs of the skull, the interiors of the eye sockets appear to have been artificially darkened. The poster also pointed out

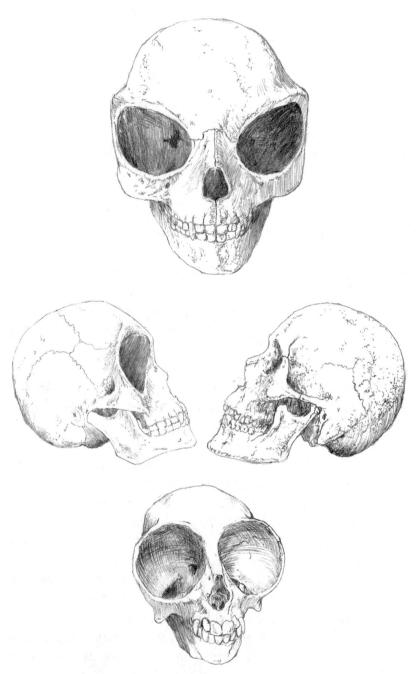

FIGURE 5.7. (A–B) The Sealand skull. (C) Brachycephalic human skull. (D) Skull of an owl monkey. (Drawings by T. D. Callahan.)

that the condyle of the lower jaw, which should be a rounded knob for articulation with the skull, is sharp and flat. It's not likely to have worked to open and close the jaw.[55] In fact, in the available photos of the skull, it's impossible to tell whether the zygomatic arch, which stretches, in human skulls, from the ear to a junction with the cheekbone, is fully separated from the temporal bone of the cranium above the ascending ramus of the lower jaw. That's the part of the lower jaw that articulates with the skull. This separation of the zygomatic arch (or process) from the temporal bone of the cranium is necessary to allow two of the four muscles that move the jaw, the temporalis and the external pterigoid, to attach to both the skull and the jaw. These two muscles stretch through the opening between the temporal bone and the zygomatic arch. So if the zygomatic arch isn't separated from the cranium of the Sealand skull, it's unlikely that its jaw would work well, and it's likely that the skull is a fake.

Seen in profile, the skull is extremely short. Human skulls in which the cranium is so shaped are called brachycephalic. In itself, this doesn't constitute a problem and is in no way abnormal. However, in human brachycephalic skulls, the eye sockets are much smaller in relation to the cranium than they are in the Sealand skull. Considering the seeming depth of the Sealand skull's eye sockets, it's doubtful there is enough room in this skull for the occipital lobe of the brain. The occipital lobe is the visual perception center of the human brain. To accommodate such huge eyes and the visual acuity of any being possessing them, the occipital lobe of this skull should be large. Of course, those contending that the Sealand skull belongs to an alien could argue that its alien brain was structured differently from our own. However, such an argument would take away any chance at a pass/fail test to judge the skull's validity. The huge eye sockets of the Sealand skull have been compared to those of two nocturnal primates, owl monkeys and tarsiers, but although the outer rims of the eye sockets of these two primates are noticeably rounded, those of the Sealand skull are flat and squared-off in a way that looks artificial rather than natural. Of course, because there is no indication of its location on any of the websites dealing with the Sealand skull, there is no way to test whether this is so. Accordingly, the Sealand skull cannot be considered to be evidence of much of anything.

Common to the "Starchild," Rhodope, and Sealand skulls is their lack of provenance. Lloyd Pye purchased the so-called Starchild skull from a couple in El Paso, Texas, who got it in Mexico from someone who possibly found it a mine, possibly in the 1930s. In all the sites touting the Rhodope skull, there's no mention of who now has it, nor is there any reference to any peer reviewed papers on the skull in any refereed scientific journals. We don't even know where the Sealand skull resides or even whether it's really a skull rather than an artifact. The Paracas skulls have a better provenance, but the claim that their DNA is not human is utterly unsubstantiated, and the geneticist who supposedly tested this DNA remains unnamed. Such vagueness smacks of evasion, and evasion smacks of fraud.

ALIEN DNA AND OTHER BIOLOGICAL EFFECTS

Had aliens frequently visited Earth, or were they presently doing so, it would seem likely that they might have left some of their genetic material behind, either in the form of microbes they might have accidentally shed or, if David Jacobs is right, in the form of alien DNA embedded in our own genetic code, particularly among those who are human—alien hybrids. Various UFO websites have claimed evidence of alien microbes in the form of the *Pandoravirus*, as voiced in a 2013 headline on a UFO blog:

> Scientists have found a new virus thought to be the biggest ever seen on Earth. The virus, dubbed Pandoravirus, is one micrometre big—up to ten times the size of other viruses—and only six per cent of its genes resemble anything seen on Earth before.
> This has led French researchers to believe the virus may have come from an ancient time or even another planet, such as Mars.[56]

So is the *Pandoravirus* from Mars or even from another star system? Well—no. *Pandoravirus*, like other giant viruses, such as *Mimvirus, Pithovirus* and *Megavirus*, is parasitic on amoebas, a rather odd choice of food for something alien to this planet. As science writer Christine Dell'Amore notes in an article for *National Geographic News*, the genetic oddities of these giant viruses—particularly *Pandoravirus*, the largest virus found so far (1,000 times larger than typical viruses)—are significant in that they may constitute a new biological domain. To understand what this means, one must remember that traditionally, all life had been divided into two "kingdoms": the animal kingdom and the plant kingdom. These were the largest classification groups in taxonomy. However, it became obvious that both plants and animals are more closely related to each other than either is to bacteria. Green plants (including most algae), animals (including one-celled animals), fungi, and slime molds all have comparatively large cells with well-defined nuclei. Bacteria and blue-green algae have much smaller cells, with nuclear material (such as DNA) that is not separated from the rest of the cell by a nuclear membrane. So life forms called eukaryotes, which have nucleated cells, are now classed in the domain called Eukarya, which contains both the plant and animal kingdoms, as well as fungi and other eukaryotes. Prokaryotes, life forms without nucleated cells, have been further divided into two domains, Bacteria and Archaea, the latter of which are more primitive even than bacteria. Blue-green algae, formerly classified as being in the plant kingdom, are now classified as Cyanobacteria, a phylum within the kingdom Eubacteria, which is in the domain called Bacteria.

Viruses don't fit into any of these three domains, because they do not reproduce by themselves and because they consist of little more than genetic material (either DNA or RNA) surrounded by a protein container. Viruses hijack the DNA and RNA replicating systems and the protein-making apparatus of the cells they invade to reproduce their DNA or RNA, as well as viral proteins—in short, to make new copies of themselves. Giant viruses, such as the *Pandoravirus*, which have many more genes than regular viruses do, might constitute a new domain. Unfortunately, when scientists or science writers use certain words, they might well mean something much more specific than, or at least different from, what nonscientists think they mean. Consider Dell'Amore's use of the word *alien* in her article:

> Perhaps most striking, 93 percent of pandoraviruses' [*sic*] 2,500 genes cannot be traced back to any known lineage in nature. In other words, they are completely alien to us.[57]

In saying these genes are, "completely alien to us," she was not using the term *alien* to mean "extraterrestrial." Rather, she meant that they are radically different from our genes—to the degree that the giant viruses may constitute a new, fourth, domain of biological classification.

With regard to alien biological effects on the human race, those who believe extraterrestrials are either manipulating our DNA or mating with us to produce hybrids often cite the upswing in IQ scores in the 20th century. Those who favor the theory that this is evidence of extraterrestrial influences see children born with these higher IQ scores as more than merely intelligent, as Bullard notes:

> Called the New Children, Millennium Children, Indigo Children, or Star Children, these offspring, born since around 1980, are alleged to be exceptionally intelligent, psychic, and spiritual. They remember past lives or manifest matter-of-fact familiarity with paradise-like places and seemingly divine beings.[58]

These children call to mind Dr. Leir's *Homo noeticus* children. But is there any real indication that children born in the 20th century are smarter than those born before 1901? In a 1996 article in *Newsweek*, Sharon Begley writes the following:

> In a phenomenon overlooked even by scholars who study the nature and roots of intelligence, IQ scores throughout the developed world have soared dramatically since the tests were introduced in the early years of this century—27 points in Britain since 1942 and 24 points in the United States since 1918, for instance, with comparable gains throughout Western Europe, Canada, Japan, Israel, urban Brazil, China, Australia and New Zealand (graph). The rise is so sharp that it implies that the average schoolchild today is as bright as the near-geniuses of yesteryear.... James R. Flynn, a political philosopher at the University of Otago in New Zealand who stumbled upon the phenomenon, told an IQ conference earlier this month. "And there is not a single exception to massive IQ gains."[59]

However, before one assumes that the human race is suddenly producing geniuses, there are some words of caution when it comes to interpreting these findings:

> If IQ measures innate intelligence, then there is a serious IQ generation gap: either a large fraction of today's children are gifted, with IQs of 130 or more, or "almost half of white Americans during World War I had IQs so low [below 76] that they lacked the intellectual capacity to understand the basic rules of baseball," says Flynn.... And there is no evidence that real-world intelligence—an ability to learn faster or make creative leaps or do any of the other things that "intelligence" connotes—is rising at anything like the rate of IQ scores.[60]

Because almost all Americans living at the time of World War I managed to understand the rules of baseball, and because there is no evidence of a sudden upsurge in creativity in the human race, the higher IQ scores, notably occurring in developed nations, would seem to be an artifact of environment (nurture rather than nature) rather than the worldwide eruption of alien/human hybrid "star-children."

Of potentially greater substance is the possibility of a message imprinted in our very DNA. Here is how UFO websites see this news. One such sight posted the headline "Humans' Extraterrestrial DNA: We All Carry a Coded Alien Message Left by the Star Gods."[61] Another website says the following:

> After extensive analysis with the help of other researchers in diverse fields such as mathematics, chemistry and programming, Maxim A. Makukov of the Fesenkov Astrophysical Institute have [*sic*] ventured out and asked if there is a possibility that, what we call "junk DNA" is actually some sort of extraterrestrial code, created by an "Alien engineer."[62]

What these two websites refer to is an article in the May 2013 issue of *Icarus,* a bona fide refereed science journal, by Vladimir I. shCherbaka, of the Department of Mathematics at the al Farabi Kazakh National University, Almaty, Republic of Kazakhstan, and Maxim A. Makukov of the Fesenkov Astrophysical Institute, Almaty, Republic of Kazakhstan, titled "The 'Wow! Signal' of the Terrestrial Genetic Code." The title refers to the name given to a strong narrowband radio signal that seemed to have been projected by an intelligent source outside the solar system. It was detected by Jerry R. Ehman on August 15, 1977, while he was working on a SETI project at the Big Ear radio telescope of Ohio State University. It appears to have come from the northwest of the globular cluster of M55 in the constellation Sagittarius, near the Chi Sagittarii star group. It lasted a full 72 seconds, the window of time the Big Ear was able to detect it. Ehman circled the signal on the computer printout and wrote the comment, "Wow!" on its side, which became the name of the signal itself. The signal has never been detected since. The paper by shCherbaka and Makukov asserts that the equivalent of the "Wow signal"—that is, the sure print of an extraterrestrial intelligence—is embedded as a code in our "junk" DNA: the large amount of DNA we carry that doesn't code for the amino acid sequence of any of our proteins.

Although the article is impenetrable to most laypersons, it appears that shCherbaka, a mathematician, and Makukov, an astrophysicist, have read the mathematical complexity of an intelligent code *into* our junk DNA rather than reading the code *out* of the genetic material. The situation here is reminiscent of what is called Equidistant Letter Spacing, or ELS for short, which was used to find hidden messages in the Hebrew text of the Jewish scriptures (what we who were raised in a Christian context call the "Old Testament"). In the process of supposedly reading messages—often prophecies—out of the biblical text, the mathematicians who formulated ELS actually read messages into it. To get the messages supposedly embedded in the text, they first reconfigured its original lines of irregular length into a rectangular grid, each line of which had an equal number of letters from the Hebrew alphabet— regardless of whether arriving at this grid split words in two. Of course, because the grid's dimensions—the numbers of characters vertical and horizontal—were arrived at arbitrarily, anything read out of the grid was actually read into it by those who formulated it. That much the same process was at work in creating the seeming mathematical patterns in our genes can be seen in the importance the authors placed in the total number of nucleons (protons plus neutrons) in the side chains of the amino acids coded for by DNA codons. The number of nucleons in the side chain has absolutely nothing to do with either the nature of the amino acids or the properties of three ring-shaped compounds that code for each amino acid. Another problem with the alleged code is that of random mutations. Over the at least 3.5 billion years during which the genetic code has been functioning, random mutations—which, along with natural selection, have resulted in the evolution of life from Archaean prokaryotes to human beings—would have greatly degraded any message embedded in our junk DNA.

Both shCherbaka, the mathematician, and Makukov, the astrophysicist, seem to have suffered the common misadventure of intelligent men who make pronouncements on a field in which they have little or no expertise. The result is a paper that, despite the impressive credentials of its authors and the soundness of the journal it was printed in, is essentially without substance.

SUMMARY

Let us sum up the alleged physical evidence of UFOs we've examined in this chapter: The "alien implants" are nothing more than nondescript metallic foreign bodies. Betty Hill's star map—actually made by Marjorie Fish—isn't based on knowledge imparted by the gray aliens that she believed abducted her. Rather, it's based on an obsolete star catalogue. Bob White's alien artifact isn't from outer space; it's from the inside of a foundry grinder. The Delphos ring wasn't made by a spaceship. Rather, it was made by bioluminescent fungi. Crop circles are made by ingenious and artistic hoaxers, not by UFOs or paranormal forces. The heightened radiation in Rendlesham Forest, supposedly the footprint of a UFO landing, wasn't heightened after all. Whatever hospitalized Betty Cash, it wasn't radiation sickness. Of the various "alien" skulls, the Paracas skulls from Peru were elongated by deliberate cranial deformation, the "Starchild" was a hydrocephalic, the Rhodope skull is possibly the sheared-off back portion of either a goat of cow skull—and its whereabouts are unknown—and the Sealand skull, of which we only have photographs, may be nothing more than a clever piece of artwork. The Pandora virus is unusual but not extraterrestrial. The rise in IQ scores in the 20th century isn't reflected in a corresponding rise in intellect and creativity and is more likely a social artifact than a measure of innate intelligence or any indication of extraterrestrial manipulation. The supposed extraterrestrial message embedded in our junk DNA seems to be nothing more than an imposed mathematical artifact, read *into* the DNA rather than *out* of it.

So the alleged physical evidence of extraterrestrial visitation we've viewed in this chapter reveals, among other things, some interesting medical facts (the "implants" and the hydrocephalic "Starchild" skull), a little known aspect of industrial technology (the Bob White artifact), some fascinating information from the biological sciences (the Delphos Ring and the Pandoravirus), a dubious measurement of innate human intellect (rising IQ scores), the misreading of a spike detected by a Geiger counter (the Rendlesham Forest radiation), an unknown ailment (Betty Cash), an odd bit of anthropological data (the Paracas elongated skulls), an intellectual misadventure on the part of two professionals operating out of their fields of expertise (the supposed message in our junk DNA), and not a little deception (crop circles and the Sealand skull). Although much of this is intriguing, it is without substance as evidence of extraterrestrial visitation. If there is credible physical evidence of extraterrestrial visitation of planet Earth, we haven't found it in this chapter.

6

Close Encounters of the Third Kind: Direct Encounters with Aliens

ALIEN ENCOUNTERS?

George Adamski (1891–1965) met a man from Venus in the California desert who told him, telepathically, that his name was Orthon. Not only that, but George also visited other planets in Orthon's flying saucer—or so he claimed. Although he didn't take any interplanetary trips, Frank Stranges (1927–2008) did claim to have spoken to and shaken hands with another Venusian, a fellow by the name of Valiant Thor. Both Orthon and Valiant Thor were quite human in appearance. The otherworldly being that supposedly met Claude Verilhon at a volcanic crater—one of the Elohim, whose name was Yahweh—was also quite human in appearance, as were the Nordic aliens from the Pleiades whom Eduard Albert "Billy" Meier supposedly conversed with and photographed. Dwight York claimed that he himself was a space alien from the planet Rizq.

When people claim to have seen and photographed a UFO, or when they claim to have been contacted or abducted by aliens, are they relating actual occurrences, are they delusional, or are they simply lying? In the case of sightings, we can, for the most part, accept that people have seen something in the sky for which they cannot account, whether there is a mundane explanation for it or not. In the case of alien abductions, most abductees recall the details of their alien encounters only via hypnotic regression. Considering what we know about "recovered" memories and the general unreliability of hypnotic regression, as well as the similarity between UFO abductions and the long folk history of supernatural or fairy abductions, we can logically surmise that these narratives are psychological constructs rather than memories of actual events. However, the contactees just mentioned—George Adamski, Frank Stranges, "Billy" Meier, Claude Verilhon (also known as Rael), abductee Travis Walton—claimed to clearly remember their alien contact episodes, and, what's more, Dwight York actually claimed to *be* a space alien. In addition, Bob Lazar claimed to have directly examined alien spacecraft.

Because the problems and fallacies inherent in so-called recovered memories don't apply to their stories of alien contact, those stories are not likely to be figments of the imagination. Hence, there are only two possible explanations for the narratives. They are either lies or are accounts of actual encounters with extraterrestrial beings or their artifacts. Narratives delivered by way of telepathic transmissions or automatic writing are not that easily categorized as either truth or lies; they may as easily be the product of the medium's imagination, deliberate deceptions or—assuming telepathic communication is possible—the truth.

CRITERIA OF JUDGMENT

How are we to judge whether a claimed encounter with extraterrestrials, a claim of having examined their spaceships, or a claim of telepathic communication is honest or fraudulent? There are three broad criteria by which we can determine the likelihood that such claims of contact are true or are outright lies: the scientific plausibility of the narrative, supporting evidence of the encounter, and an examination of the history, character, and motives of the claimant. Let us consider each criterion separately.

Had supposed contactees or mediums, often people with little scientific knowledge, been given facts unknown to the scientific community at the time but that were later verified, this would have been astounding confirmation of the validity of their contacts. For example, suppose that at a time when most scientific opinion favored the view that Venus was a watery twin of the Earth the space aliens communicating with either Unarius cofounder Ernest Norman or "contactee" George Adamski had told them that the clouds of Venus were mainly made of sulfuric acid and that surface temperatures on the planet were hot enough to melt lead. Had these "contactees" known such facts, later verified by the Soviet *Venera* probes, this would have strongly validated their respective claims. Because Norman's visions of Venus were "transmitted" through trance states, he was probably honest but deluded. Adamski, on the other hand, claimed to have met a real, live Venusian. Likewise, Adamski's Venusian and Billy Meier's Pleiadians were both of the Nordic alien type, as human as any man or woman of planet Earth—as was Frank Stranges's Venusian, Valiant Thor. Considering that intelligent beings evolving on a different world would be unlikely to be identical to humans from Earth, we have strong reason to suspect dishonesty in such cases. If the science is off, the narrative doesn't conform to reality: it's either a delusion or a lie.

Supporting evidence of alien encounters comes in three forms. First, did more than one person experience the same contact, and do all witnesses generally agree on the nature of the extraterrestrial being and the sequence of events in the encounter? Also, were these other witnesses reliably objective and honest? Often groups of people sharing the same deluded belief system also share the same biased view of what they saw or experienced. Second, did other observers, separate from those contacted, detect the arrival or departure of the UFO? Third, was there unequivocal evidence at the landing site of a recent extraterrestrial visitation? Often claims are made of scorched grass at supposed landing sites, indicating radiation effects—but without adequate monitoring of radiation levels or taking of samples using radioisotopes. For example, did an atomic-powered engine leave behind significant traces of fission products, such as strontium-90 or cesium-137? Artifacts left behind by the aliens or given to the contactee, material obviously not of terrestrial origin, would also strongly support the contactee's testimony.

The third criterion, the history, character, and motives of the claimant, can be considered in three parts. First, does the personal history of the supposed contactee indicate that he or she is a person whose testimony we would be likely to trust? Has this person had a past history of deceit? Has he or she committed fraud? Second, is he or she now a person of good reputation, or is his or her word to be doubted? Finally, are there motives of notoriety, wealth, and power to be considered? Has the alleged contactee used his or her narrative of contact to gain attention?

Has he or she made considerable money writing books or touring the lecture circuit, popularizing his or her contact narrative? Has the person perhaps even used that narrative to establish a cult that gives him or her power over the cult's members? Each of these—a past history of deceit, an untrustworthy character, or a strong motive for perpetrating a false narrative—is a red flag indicating a deliberately constructed false narrative aimed at taking in the gullible.

When we find that someone has a questionable past history, is of doubtful honesty, relates a narrative that is both at odds with scientific plausibility and unsupported by other reliable witnesses, and has amassed a fortune based on his or her story—or has even founded a cult with himself or herself at the center—the likelihood is strong that this person is dishonest and that his or her narrative is a fraud. Let us now consider the history of some of the more famous contactees to see whether their stories have the clear ring of truth or trigger the strident alarms of falsehood.

GEORGE ADAMSKI

On November 20, 1952, George Adamski, with six others—two couples, Mr. and Mrs. Bailey and Dr. and Mrs. Williamson, and two women, Alice Wells and Lucy McGinnis—drove out into the California desert to search for UFOs. That night they all saw a large, silvery cigar-shaped craft, and Adamski claimed that he made contact with an extraterrestrial being who communicated with him telepathically, telling Adamski that his name was Orthon and that he was from Venus. Orthon needed to tell Adamski that he was from another planet because he appeared to be fully human. A drawing of Orthon from the Fortean Picture Library, attributed to Alice Wells, shows a tall, handsome man with long, flowing blonde hair—in other words, a classic Nordic alien. In the picture, Orthon is wearing a belted, loose, one-piece turtleneck garment, gathered tightly at the wrists and ankles. We have only drawings of Orthon, because according to Adamski, the Venusian refused to let himself be photographed. Orthon related that the Venusians, who shared Adamski's own religious beliefs, a theosophical pastiche, were concerned about Earth people's use of atom bombs and wanted to warn us of the dangers of nuclear war.

Following subsequent contacts with Orthon and his fellow Venusians, Adamski claimed to have visited both the far side of the moon and Venus, riding in Venusian scout ships. On the far side of the moon, he said, he found a world with breathable air, forests, and snow-capped mountains. Adamski eventually declared that all the planets of our solar system had breathable atmospheres. In 1962 he claimed to have attended a conference on Saturn. There, he said, he rode in electromagnetically levitated cars over vast fields of flowers.

Because Venus has a dense atmosphere of carbon dioxide with clouds of sulfuric acid and a surface temperature hot enough to melt lead, any life form that could have evolved on such a planet would have to have a biochemistry so radically different from our own carbon-based chemistry that Orthon would have had to be encased in a spacesuit to deal with Earth's atmosphere, which would be poisonous to him, and to protect him from our temperatures, which to him would be freezing cold. Also, it's highly unlikely that such a radically different being, possibly with a fluorosilicone-based chemistry (fluorine plus alternating silicon and carbon atoms—see chapter 12 on plausible aliens) would be highly unlikely to look anything like a human being. Perhaps Adamski could be excused for believing Venus was Earthlike. However, even with only a high school education of the level available in the early 20th century, when Adamski was a teenager, he should have known that the moon lacked an atmosphere, that the atmosphere on Mars is thin and devoid of free oxygen, and that the atmospheres of the outer planets are mixes of hydrogen, helium, methane, and ammonia—and that those planets are far too cold for our type of life. Even without knowledge of the nature of gas giants, including that they lack discernable surfaces, his representation of them as habitable and teeming with life is, and was in his day, patently absurd. Furthermore, Saturn's gravity would be crushing to anyone from Earth. Thus Adamski's claims completely fail the test of the first criterion—scientific plausibility.

As to supporting evidence, let us consider the six other witnesses who accompanied Adamski when he went to the desert on November 20, 1952. Didn't they see the cigar-shaped mother ship? Didn't they witness Adamski conversing telepathically with Orthon when the Venusian stepped out of his saucer-shaped scout ship? Weren't they reliable witnesses? As to the first question, all six people seem to have seen an elongated silvery aircraft, which could easily have been a passenger plane seen from an angle at which its wings were not readily visible. As for the witnesses seeing either Orthon or his flying saucer, they didn't. Adamski told them to wait by the car while he went over a hill to investigate. Sixty minutes later, he came running back, excitedly waving his arms and telling them he had met a man from Venus.

Were the witnesses themselves credible? As skeptic Daniel Loxton notes, both Alice Wells and Lucy McGinnis were already Adamski's close friends and devoted followers.[1] They were hardly objective witnesses. The two couples, however, were only recent acquaintances of Adamski's. At first, they would seem to have been sober and objective. Writing in a flying saucer magazine, James W. Moseley had this to say about these witnesses:

> When I first read "Flying Saucers Have Landed," I was impressed by the fact that Adamski's story was backed up by four people (the Baileys and the Williamsons) whom Adamski knew only slightly. Although the text does not explicitly say so, I came to the conclusion (as many readers did, no doubt) that these four were impartial, reasonably conservative, well educated people, not prone to indulge in hoaxes or be easily swayed by a hoax perpetrated upon themselves. I learned, however, from my own investigations, that all four were ardent "Believers" before they made the November 20 contact, and none had any particular educational advantages that would qualify them as expert or impartial observers. In particular, Williamson, though a pleasant young man, admits that he has no degree entitling himself to be called a "doctor," even though he allows himself to be called "Dr. Williamson" throughout the book—just as Adamski, among his friends and admirers (though not in the book) is known affectionately as "professor" without the benefit of any degree.[2]

Arrogating degrees and honors to oneself seems to be common among contactees and founders of flying saucer cults. For example, George King, founder of the Aetherius Society, called himself "Sir" George King, without having been knighted by Queen Elizabeth, and also styled himself as "Dr." King. The young Mr. Williamson to whom Moseley refers is none other than George Hunt Williamson, the inventor of the fictional lost planet Maldek. A number of websites note that though Williamson and Baily, who published *The Saucers Speak* in 1954, claimed to have made short-wave radio contact with alien flying saucer pilots, they actually obtained most of their "messages" from Ouija board sessions.[3] Thus the Williamsons and the Baileys not only did not witness the actual contact between Adamski and Orthon but also were already flying saucer enthusiasts who were ready to believe Adamski's tale and who then claimed to have gained further information from friendly aliens using a Ouija board.

The location in the California desert where Adamski's supposed contact with Orthon took place isn't that far from Mount Palomar, yet the powerful Mount Palomar telescope failed to detect either the giant mother sip or Orthon's Venusian scout ship—yet Adamski and his six friends saw the mother ship with their naked eyes, and Adamski was not only able to track Orthon's craft but also take photographs of it through his six-inch telescope. At the head of his critical article on Adamski, James W. Moseley placed two photos side by side. One was allegedly taken of the Venusian scout ship by Adamski on December 12, 1952. The other, which had originally appeared in *Yankee Magazine* in May 1954, was of a model made from a Chrysler hubcap, a coffee can, and three ping pong balls. The images in the two photos are virtually identical.

Had Adamski been merely an amateur astronomer who happened to come across UFOs during the pursuit of his hobby, the quasi-religious nature of Orthon's communications to him would have been less suspect than it was considering his past involvement with what purported to be an Eastern religion. Before he became involved with Orthon, he founded, in the 1930s, a religious group called the Royal Order of Tibet. The members of his small cult wore purple and gold robes and lived communally in a mansion in Laguna Beach, California. As one might guess from the group's name, it was heavily influenced by Madame Blavatsky's theosophical views. Tibet was in those days sufficiently remote and mysterious to be considered the hidden repository of the ancient wisdom of ascended masters and of the spiritual adepts of Atlantis and Mu. The Royal Order of Tibet, one of many cults based on Theosophy, never did well. By 1940 it had collapsed, and Adamski and some of his followers then attempted to run a communal farm for four years. When this didn't prosper, Adamski moved to Palomar Gardens, a tourist stop below the observatory on Mount Palomar. Palomar Gardens featured a small observatory dome, some cabins, and a café, where Adamski worked during the day serving hamburgers.

This was Adamski's situation when, in 1947, the first newspaper story came out about flying saucers. He seems to have sensed that this was the new bandwagon on which to jump. He wrote, or rather had Lucy McGinnis ghostwrite for him, a science fiction novel titled *Pioneers of Space: A Trip to the Moon, Mars and Venus*, which was published in 1949. Eventually, in 1952, he claimed to have photographed a cigar-shaped space ship, which bore an uncanny resemblance to an illustration of an Atlantean airship from a 1905 science fiction novel titled *A Dweller on Two Planets*. This was written by Frederick Spencer Oliver, who had been a teenager at the time he wrote the novel—which, he claimed, was dictated to him by a spirit from Atlantis.[4] Thus Adamski took the Venusian mother ship from the pages of science fiction. Loxton notes also that Orthon's costume was virtually identical to that worn by Michael Rennie in his role as the morally superior extraterrestrial Klaatu from the 1951 movie *The Day the Earth Stood Still*[5] (see Fig. 10.1). Like Klaatu, Orthon came to Earth to warn its people of the dangers of nuclear war. This plagiarism from science fiction tropes is similar to Barney Hill's "remembering" his abductors as gray aliens after having seen the *Outer Limits* episode "The Bellero Shield." However, Barney Hill's supposed memory surfaced under hypnosis. Adamski's conscious use of these motifs demonstrates an inherent dishonesty at the base of his narrative.

The motives of notoriety and money can clearly be attributed to Adamski. His books on the benevolent space brother from Venus brought him considerable fame and wealth. It was a major step up from flipping burgers for a living.

FRANK STRANGES

As with George Adamski's space brother, Orthon, Frank Stranges's friend, Commander Valiant Thor, was a thoroughly human-looking visitor from Venus. Stranges, however, avoided the problems eventually exposed by the Venera probes by having his Venusian come from *inside* the planet. As noted in our chapter on UFO religions, Stranges also believed the Earth to be hollow.

There are a number of reasons why we know that the Earth and other planets aren't hollow. First, the time it takes seismic waves to travel through and around the Earth, and the pattern of their movement, isn't consistent with what would be the case were the Earth hollow. The evidence from seismic waves points to our planet's having a solid crust and semisolid mantle, a liquid nickel–iron outer core and a solid nickel–iron inner core.

Another problem for the hollow Earth model is gravity. A hollow shell with the known, observed thickness of the Earth's crust would not be able to achieve hydrostatic equilibrium with its own mass and would collapse. Based on the size of our planet and the force of gravity on its surface, its average density is 5.515 grams per cubic centimeter (g/cc). Typical densities of surface rocks are only about 2.75 g/cc. Were the Earth hollow, its average density would be less than that of its surface rocks—that is, less than 2.75 g/cc. With an average density of 5.515 g/cc, the Earth cannot be hollow. That Earth's average density is about twice that of its surface rock is best explained by its having a nickel–iron core. The nickel–iron core also explains Earth's strong magnetic field, which could not be generated by a hollow planet. This completely debunks Stranges's assertion that the Earth and Venus are hollow.

Another part of Stranges's narrative that doesn't hold up is how a being from Venus came to be named "Valiant Thor." Because Thor was a Norse god, it almost seems as if Stranges lifted the "Valiant" portion of the name from Hal Foster's Sunday comics hero, Prince Valiant, who ruled over a Viking kingdom. It might be a measure of the gullibility of those who bought Stranges's tale that none of them ever demanded, "This guy from Venus was named *what*?"

Stranges's assertions of having met Valiant Thor at the Pentagon are utterly unsupported by any other witnesses. Of course, were he alive today, he would, no doubt, assert the requirements of government secrecy for the purposes of security as preventing those witnesses from coming forth. However, this begs the question of how Stranges, an ordinary citizen, could have gotten access to a visitor from Venus residing in the Pentagon. As Phoenix-based skeptic Mike Stackpole pointed out,

> ultimately the whole of Stranges' case rests upon his own anecdotal evidence or pictures and incidents that have been proven over and over again to be frauds.[6]

In his article on Frank Stranges, Stackpole mentioned a 1972 incident that cast an interesting light on the character of a man who traded much on his image as an evangelical Christian minister:

> More important is an incident on 16 September 1972 in which, I am certain, Dr. Stranges wishes the incident to involve a UFO—namely the one that he and a companion were trying to fly. The reason he wishes this plane to be unidentified is because it contained 3,500 pounds of marijuana. The plane did not get off the ground, and Stranges got eight months in jail and three years' probation for possession of marijuana with intent to traffic.[7]

Considering such a past history, it's not unreasonable to attribute desire for fame and money to Stranges as the motives for his Valiant Thor fiction. Stackpole noted that during a lecture Stranges gave in Phoenix, he said that the room he had rented for the night cost him $400. Stackpole estimated that although the room's capacity was 300 people, only between 150 and 200 were in attendance. The "suggested donation" for attending Stranges's lecture was $5. At $5 a head, then, Stranges grossed between $750 and $1,000, for a net profit—once the $400 room fee was subtracted—of between $350 and $600. Assuming that the lecture lasted for two hours with a half-hour for questions and another half-hour for book sales and signing, Stranges would have made between $116 and $200 per hour that evening on admissions alone. Stackpole noted that, in addition, Stranges's wife was selling copies of three of his books as part of the program. Thus Valiant Thor, coupled with Stranges's hollow Earth tales and stories of Nazi flying saucers (see chapter 9 on UFO religions), was a source of a lucrative income for him and his wife.

Thus Stranges's narrative of meeting a fully human-seeming Venusian, who came from the interior of a hollow planet and whom Stranges visited at the Pentagon, is scientifically absurd, nonsensical with regard to government and military security during the Cold War period, and unsupported by other witnesses. Furthermore, Stranges's involvement in drug trafficking casts doubt on his reliability as an honest witness, and his lucrative career lecturing and selling his books was a strong motive for promoting a fraudulent account.

TRAVIS WALTON

In contrast to most narratives of alien abduction, that of Travis Walton was not "recovered" under hypnosis. The basic elements of his story are available on Walton's website, www.travis-walton.com. On the night of November 5, 1975, Walton and six other members of a logging crew, having finished work for the day thinning trees in the Apache–Sigraves National Forest, had piled into a pickup truck and were heading for the small town of Snowflake, Arizona. They saw a bright light in the forest, which they thought could have come from a downed aircraft, stopped the truck, and went to investigate. They found a UFO hovering soundlessly close to the ground, shining a shaft of light down from where it hovered. Walton ran up to it, though the others yelled for him to stay away. Once under the craft in the beam of light, he found himself sucked up into it. When he vanished, the others drove off in a panic. When they regained their nerve and came back, both he and the UFO were gone. Meanwhile, aboard the UFO, Walton found himself lying on an examination table surrounded by gray aliens. He was terrified, jumped off the table, seized what looked like a glass container, and brandished it at them. When they backed away, he ran from the room. Eventually, he found himself in another room, where very human-looking aliens, who are often rendered in illustrations of the incident as extraterrestrials of the Nordic type, found him. When he tried to question them about who they were and where he was, they simply smiled. They then gently forced him to lie on a table and one of them, a woman, placed what looked like an oxygen mask over his face. He lost consciousness and later found himself on the edge of a highway. There he located a phone booth and called his sister and brother-in-law, who informed Walton that he had been missing for five days.

What is verifiable in Travis Walton's story is that the six other loggers spotted a bright light, saw Walton run toward it, panicked, drove away, and returned to find him gone. Five days later, he was found. Walton and his six witnesses have taken multiple polygraph tests with varying results, which says more about the unreliability of the polygraph as a lie detector than anything else. Travis Walton's aliens were essentially a

mix of grays and Nordics, which makes his account as suspect as both
the account of Barney Hill, whose gray aliens were clearly derived from
the *Outer Limits* episode "The Bellero Shield," and the account of George
Adamski's Venusian, Orthon.

As with the supposedly sober and conservative witnesses to George
Adamski's meeting with Orthon, who turned out to be UFO enthusiasts,
the presentation of Travis Walton as an ordinary person who just hap-
pened to be abducted by aliens turns out to be false. He was already a
UFO aficionado. The late UFO skeptic Philip Klass, in a report on the
Travis Walton case published by NICAP (the National Investigations
Committee on Aerial Phenomena), said the following (parentheses and
ellipsis in the original, bracketed material added):

> Dr. Howard Kandell, one of the two physicians who examined Walton at APRO's
> [Aerial Phenomena Research Organization's] request, was asked if the Waltons
> had indicated any prior interest in UFOs. Kendell replied: "They admitted to that
> freely, that he (Travis) was a 'UFO Freak' so to speak . . . . He had made remarks
> that if he ever saw one, he'd like to go aboard." Dr. Jean Rosenbaum, a psychiatrist
> who examined Walton, was asked whether he had mentioned any prior interest
> in UFOs. Dr. Rosenbaum replied, "Everybody in the family claimed that they
> had seen them (UFOs). Travis has been preoccupied with this almost all of his
> life . . . then he made the comment to his mother just prior to this incident that if
> he was ever abducted by a UFO she was not to worry because he'd be all right."
> Duane Walton has stated that he and Travis had often discussed the possibility of
> getting a ride on a UFO.[8]

Duane also claimed that when he was a child, a UFO had pursued him.

Klass suspected that Walton, in collusion with his boss, Mike Rogers,
had concocted his supposed abduction, rigging a light show to deceive
the other members of Rogers's crew as a means of getting Rogers out of
a financial bind (emphasis in the original):

> Mike Rogers had submitted a bid in the spring of 1974 to the U.S. Forest Service
> for a timber thinning operation of 1,277 acres of land in a National Forest, located
> in the Apache-Sigraves area. His bid was accepted and was 27% under the mid-
> figure submitted by the other companies. By the following summer (1975) it was
> clear to Rogers that he'd grossly underestimated the magnitude of the job and
> could not complete it on time. He applied for an extension, which was granted,
> but he was penalized $1.00 per acre for all work performed after the expiration of
> the original contract date. The new work completion deadline was November 10,
> 1975. As the new deadline approached, it became clear that, once again, they could

not possibly complete the work by that time and he would have to ask for another extension, that would result in another pay cut. More serious, the Forest Service was withholding 10 percent of the payments until the job was done. With winter at hand, Rogers could not finish until this next spring to collect these funds. The alleged UFO incident gave Rogers a legal basis for terminating his money-losing contract on the ground that his crew would not return to the work site out of fear, allowing Rogers to collect the withheld funds and pay his crew. . . .

On Nov. 18, 1975, Rogers wrote to inform the Forest Service that he could not complete this contract because of the UFO incident, "which caused me to lose my crew and will make it difficult to get any of them back to the job site." The Forest Service put the remaining work up for bid, handed it to another contractor and later released the funds it had been withholding to Rogers.[9]

Thus there was an excellent reason for Travis Walton and Mike Rogers to have rigged a fake UFO encounter.

In 1978, Travis Walton wrote a book about his alleged abduction titled *The Walton Experience*, which was the basis for the 1993 film *Fire in the Sky*. When the film was released, Walton reissued the book, now under the film's title. Thus he has benefited from his narrative both in terms of notoriety and financial gain.

BILLY MEIER

Eduard Albert Meier is a Swiss farmer whose penchant for Old Western attire prompted an American friend of his to call him "Billy" after Billy the Kid. He has claimed to have hundreds of contacts with extraterrestrials of the Nordic alien type, who come from the Pleiades and whom he calls the Plejaren.

Meier claims that the Plejaren initiated contact beginning in 1942, when he was five years old. He says that the being who contacted him was an elderly individual named Svath and that these contacts lasted until 1953, when Svath died. From 1953 to 1964, he says, he had contacts with an attractive female alien named Asket. After 1964, the alleged contacts with the Plejaren ceased for 11 years. Then in 1975, when Meier was 38, he claims, another attractive woman of the Plejaren contacted him. Her name was Semjase (pronounced sem-YAH-see) and Meier says that she is Svath's granddaughter.

Meier says that the Plejaren, or Pleiadians, look so much like northern Europeans because they are related to humans. Supposedly, common ancestors of Earth people and the Plejaren inhabited Earth thousands of years ago. However, Meier also asserts that the Plejaren are the angels of the Bible and that Jesus was the child of Mary and one of the Plejaren. He says that the Plejaren now come from an earthlike planet slightly smaller than our own, called Erra, which has a population of 500 million people. By comparison, the population of Earth is about 7 billion. Plejaren is one of 10 planets orbiting their sun, a star they call Tayget. Four of these planets are inhabited. The Plejaren lifespan is much longer than ours, being about 1,000 years. Until 1995, he claims, the Plejaren had three secret bases on Earth, located in Switzerland, North America, and Asia. Between 1975 and 1988 there were as many as 2,800 extraterrestrials stationed on our planet; however, there are now (conveniently for Meier) no Plejaren on Earth.

As did Adamski before him, Meier claims to have ridden in flying saucers. In Meier's case these are Pleiadian "beam ships." However, although Adamski's claims were limited to visits to other planets in the solar system, Meier claims to have visited other star systems, other galaxies, and even other universes. Furthermore, he claims to have even gone back in time on board a Pleiadian beam ship. To back up his claims, he has produced films of the beam ships flying in the vicinity of his farm, remarkably clear still photos of the ships, and even photos supposedly taken when he went back in time. He also claims to have photographed two of the Pleiadian women, Asket and Nera. In addition he has produced metal samples that he says are of Pleiadian origin, and he claims to have received prophetic announcements from these extraterrestrial visitors that came true.

Concerning the scientific plausibility of Meier's claims, there is a great problem with the age of the Pleiades star cluster. Using what is called the Hertzprung–Russel Diagram (HRD), formulated in 1910 by Ejnar Hertzprung and Henry Norris Russel, astronomers are able to tell the age of stars based on what we know about stellar evolution and the brightness of a star measured against its color and emission spectrum. Using the HRD, astronomers can date the stars in the Pleiades cluster as being between 70 million and 150 million years old. Any planets orbiting any of those stars would, likewise, be of that age. By comparison, using such radiometric dating techniques as uranium–lead dating, we know that there are rocks from Quebec that are 4.28 *billion* years old and that certain minerals in zircon crystals from western Australia are between 4.3 billion and 4.4 billion years old. Samples from meteorites have been dated at between 4.4 billion and 4.6 billion years old, and the solar system is, based on this material, dated as being about 4.6 billion years old. The earliest era in the geological history of the Earth, the Hadean, was from 4.5 billion to 4.0 billion years ago, a period of 500 million years. This was the period before there was any life on Earth. Thus, assuming geologic processes similar to those on Earth, the oldest planets in the Pleiades star cluster, being only about 150 million years old, would still be in the earlier phases of their own equivalent of our Hadean period and would be lifeless. No life on these planets means—assuming geologic processes akin to those on Earth—no Plejaren. However, let us, for the sake of argument, speculate that measurements of the ages of the stars in the Pleiades, based on the HRD, are way off and that those stars are in fact not 150 million but 1 billion years old. Again, using Earth's geological time scale as a guide, that would put the oldest planets in the Pleiades in the Archean Era, when there were little more than blue-green bacteria on Earth. Again, this would mean no Plejaren.

To some degree Meier has a way out of this problem: he claims that the Plejaren originally came from Earth and settled in the Pleiades thousands of years ago. However, had they landed on one of the planets in that star cluster, one that was at least into an equivalent of our Archean Era, they would have found a planet with an atmosphere devoid of free oxygen and thus without breathable air. Also, although Meier has seemingly covered his bases by the assertion the Pleiadians were originally from Earth, separation of their population from ours over thousands of years should have produced a significant amount of genetic divergence through genetic drift. Thus there should be considerable racial differences between the Plejaren and us, and they certainly should not closely resemble northern Europeans. So the scientific plausibility of very human-looking—so much as to appear to be northern European—aliens' in any way inhabiting planets in the Pleiades star cluster is nil.

One also has to wonder why the Plejaren, who have supposedly given Meier prophetic warnings that the human race must abandon its propensity for violence and war and ill treatment of the planet, contacted a Swiss farmer living in a remote part of his country rather than boldly announcing these warnings in a presentation to the United Nations. For that matter, why haven't they used their superior technology to simply commandeer every television station on Earth to announce their warnings to all of humanity? Not only is their biology implausible—so also is their sense of logic.

Meier's evidence of contact also disintegrates on examination. As we will soon see, his photo purporting to be of the Pleiadian women Asket and Nera is a fake. There is also strong evidence that his photos and films of Pleiadian beam ships flying over his farm are fakes. However, perhaps the most egregious example of falsehood is to be found in an image of a *Pteranodon*, a Mesozoic flying reptile, which Meier originally claimed was a photo he took when he went back in time aboard a Plejaren beam ship. It turns out that the supposed photo was actually a cropped portion of a painting by Czech artist Zdenik Burian of a *Pteranodon* dropping fish for its babies to eat, as featured in the book *Life before Man*, originally published in Prague in 1972 before being picked up by English publishing house Thames & Hudson as a coffee-table book because of the outstanding quality of Burian's paintings. In July 2005, Billy Meier claimed that this image was not his original but a forgery perpetrated by the "Men In Black." He said that his Pleiadian contact in this case, whose name is Quetzal (possibly derived from the Aztec god Quetzalcoatl), had destroyed all the original photographic materials in response to the actions of the Men in Black.[10]

Meier used the same tactic when his supposed photo of Asket and Nera was shown to be a cropped photo of two dancers from the Golddiggers dance troupe, which regularly performed on the *Dean Martin Variety Show* from 1968 to 1974. Once his photo was exposed as a fake, he said the following:

> Unfortunately, during my conversation with Ptaah he revealed that the two women depicted on the photos are not Asket and Nera from the DAL Universe, but two American look-alikes. These photos are malicious hoaxes and were switched upon the order of and in collaboration with the "Men in Black."[11]

At least Ptaah—whose name would seem to be a variant of Ptah, an ancient Egyptian god—got one thing right: the photo is fake.

Meier also left himself an out—along with giving himself consider-able latitude in terms of fulfillment—when it came to prophecy. In his *Contact Notes*, under *Contact 251, Part 2*, he (or, supposedly, the Plejaren) said the following:

> This will signal the first threat of a looming third world war, as foretold by a prophecy, unless terrestrial Man strives to avert this danger through reasoning and appropriate thoughts and actions. Should Man fail to act against the fulfillment of this prophecy, a new and extremely destructive weapon will be built that will produce disastrous consequences in the next world war. One important factor in this scenario is the criminal neglect to monitor the Earth from space. New weapons will once again create quite a stir, and so will the death of 4 heads of state who will die within 7 days from each other. These then are the last danger signs, which foretell that within merely 2 years of these events the long-feared world war will indeed erupt, unless terrestrials finally gain mastery over their reasoning to stop all these ills. Should this not be done, mankind will fail in its attempt to protest and boycott the new deadly weapons, because by this time the armories of many nations will be full to their capacity. Passing laws to prohibit the use of these weapons will be ineffective at this late stage. World War III cannot be averted if Man fails to finally become reasonable![12]

In December 2006 four heads of state did indeed die, though not all within seven days of each other. Also, one has to stretch the term *head of state* considerably to make the prophecy work. The heads of state, listed with the dates of their terms as head of state, followed by the dates of their deaths, were:

Augusto Pinochet: dictator of Chile 1972–88, died of a heart attack December 10, 2006

Sapamurat Niyazov: dictator of Turkmenistan 1985–2006, died of a heart attack December 21, 2006

Gerald Ford: president of the United States 1974–77, died of a stroke December 26, 2006

Saddam Hussein: dictator of Iraq 1979–2003, executed December 30, 2006

Even if we give some latitude in the case of the 11 days, rather than seven, that passed between the deaths of Pinochet and Niyazov, the lat-ter was the only one acting as head of state at the time of his death.

Pinochet lost power as dictator when Chile returned to democracy in 1988, Ford's three-year presidency ended in 1977, and Saddam Hussein was overthrown in 2003. Also, new weapons didn't make a great stir, and although the world still has its share of international tensions, World War III doesn't seem to be nearly as much in the offing as it did prior to the end of the Warsaw Pact and the dissolution of the Soviet Union in 1991. Also, although significant changes in world regimes or their social and environmental policies didn't take place in the two years following these deaths, World War III did not begin in 2008.

One of Meier's websites relates a number of prophecies showing a mix of some seemingly startling predictions and others that are totally absurd. The predictions are from a contact report dated February 28, 1987, and are sequentially numbered. In number 232, the Pleiadian Quetzal tells Meier the following:

> Also an epidemic known as Ebola will cause many deaths, as well as other unknown diseases which will sporadically arise in epidemic proportions and will be new to the human being, causing great concern.[13.]

This is initially quite startling until one looks up the history of the Ebola virus and finds that one of the earliest outbreaks of Ebola occurred in Zaire in August 1976, killing 280 of the 318 people infected, for a greater than 88% mortality rate. Thus Meier's prediction nearly 11 years later that Ebola will cause many deaths is hardly that startling, and one would not need the help of an extraterrestrial sentient being to predict that Ebola would be a problem.

In prophecy 242, Meier predicts the dissolution of the Soviet Union, chiefly caused by Mikhail Gorbachev. Again, because this prediction was made in 1987 of an event that took place in 1991, the signs of this drastic change were readily apparent. For example, agitation for the break-up of the Warsaw Pact showed itself in the late 1980s, resulting in the destruction of the Berlin Wall and calls for German reunification in 1989, which was enacted in 1990. In the next prophecy, 243, Meier predicts a war between Russia and China over Inner Mongolia, resulting in Russia's losing it to China. Actually, Inner Mongolia is already part of China and hasn't been claimed by Russia. In fact, there hasn't been any significant border dispute between Russia and China since the break-up of the Soviet Union.

Perhaps the most bizarre predictions are 269 and 270:

#269: Many lives will be lost during a civil war in Wales, where differences between various parties will arise before the Third World War.

270: Welsh and English forces will clash especially near Cymru, and claim many lives and cause great devastation.[14]

It's hard to imagine either a Welsh civil war or an armed conflict between Wales and England. Wales has been an integral part of the United Kingdom since its conquest by Edward I in 1283. Cymru is actually the Welsh word for Wales, so it's hard figure out what Meier is talking about in prophecy 270.

Despite all this, supporters of Billy Meier still point to the very clear photos of Pleiadian beam ships flying over his farm. They argue that for the photos to be fakes, we have to believe that a one-armed man who had no knowledge of Photoshop or other digital photography programs could have made such realistic photos and films. Although Meier did lose an arm in a bus accident in his youth, he is still quite active, does well at farming, and can still build models. As to his requiring computer skills in digital photography, stop-action animator and movie model builder Alan Friswell was able to reproduce Meier's UFO photos, building models with materials he obtained at a local hardware store and then cleverly staging them.[15] One of the most spectacular of the Pleiadian space ships has been dubbed the "wedding cake ship" because its elaborate tiered structure, sporting rows of silver spheres, does indeed somewhat resemble a wedding cake (Fig. 6.1). Again, model makers were able to locate all the components of the wedding cake UFO. Meier seems to have used a particular brand of container lid for its base, specifically the lid of a Harcoster Universal Container. Upon this he seems to have mounted a pedestal stand to which he attached silver spherical Christmas tree ornaments. He seems to have made the underside of this model from a pressure cooker lid.[16] It's also noteworthy that in Meier's films, the Pleiadian ships often appear and disappear erratically, which Meier explains as them materializing and dematerializing in our skies. Although it cannot necessarily be proven that Meier faked his photos and films with

FIGURE 6.1. (A) Drawing of George Adamski's "Venusian scout ship" based both on
Adamski's photographs and a model made from a Chrysler hubcap, a coffee can, and
three ping-pong balls. (Courtesy D. Loxton.) (B) A model of Bill Meier's "wedding cake
UFO by Phil Langdon made of the lid of a Harcoster Universal Container, a pedestal
stand, silver spherical Christmas tree ornaments, and a pressure cooker lid. (Courtesy
P. Langdon.)

homemade models, one must exercise Ockham's Razor, also called the
Law of Parsimony, in cases such as these: What is the simplest explanation for the photos—that they are of easily fabricated models or that
they are starships from the Pleiades? The fraudulent nature of the alleged
photos of a *Pteranodon* and of Asket and Nera, followed by Meier's backtracking and accusations of conspiracy to explain the hoaxes, strongly
indicates that his "beam ship" photos and films are likewise faked.

There is one last piece of evidence we must consider regarding Meier's claims. He asserts that during his 105th contact with Semjase, she gave him a number of specimens to use as proof of his Pleiadian contacts. Four of these were metallic, one was biological (a bit of blonde hair, supposedly Semjase's, with split ends), and nine were minerals or crystals. The metal samples were analyzed multiple times. The first analysis, performed in Zurich in March 1978 at the Federal Material Examination Experimental Laboratory for Construction, Industry and Business, merely resulted in a list of the elements found in them: aluminum, calcium, chlorine, copper, lead, phosphorus, potassium, sulfur, silicon, silver, and titanium. On May 12 of that same year, a second analysis, done by Dr. Walter W. Walker, a metallurgist at the University of Arizona at Tucson, yielded bismuth, calcium, copper, and silver. Walker commented that although the samples were unusual, there was nothing in them to indicate that they were of extraterrestrial origin. He also pointed out that they were "unsuitable for high performance structural use due to low strength and tending to react with the atmosphere."[17] At the request of Meier supporter Wendelle Stevens, Dr. Marcel Vogel, a senior research chemist at IBM in San Jose, California, analyzed the samples a third time. Meier's supporters seized on the result of this analysis, because Vogel said of the sample he examined that it "could not have been manufactured by any technology here on Earth."[18] However, when later questioned by Kal Korff, Vogel said that he found nothing about the sample he tested to indicate extraterrestrial origin. He was puzzled to find that it contained aluminum, silver, and thallium in high states of purity. The main riddle for Vogel was the presence of thallium, a very rare metal. A fourth analysis, done by Dr. Robert Ogilvie of the Massachusetts Institute of Technology (MIT) solved the mystery of the thallium: It wasn't there. Vogel had misinterpreted the graphs on the specranalyzer.[19] The odd mix of metals in Meier's samples puts one in mind of the mix of metals in Bob White's alleged alien artifact, which turned out to be a stalagmite made of various different metals that had accreted over time on the inside of a foundry grinder (see chapter 5).

We can see in the multiple tests performed on the metal samples desperation on the part of true believers trying to find validation for their belief in the face of strongly disconfirming evidence. Although obtaining a second opinion might be reasonable, Stevens responded to the findings of *two* metallurgists from different labs on different continents not by accepting their expert opinions but rather by eliciting the opinion of Dr. Marcel Vogel, a chemist but not a metallurgist. Once Vogel found the anomalous presence of thallium in the metal sample, it was reasonable to have the sample tested again by a metallurgist—who found that Vogel was in error for having claimed that the metal sample contained thallium. Not only was Vogel not a metallurgist, and hence not really competent to examine the sample, but his objectivity is suspect: Marcel Vogel also claimed to have psychic powers, including the ability to break iron bars using the power of his mind—an alleged feat that gained the attention not of refereed scientific journals but rather of the *National Enquirer*.[20] Thus, in their search for validation of the extraterrestrial origin of the metallic samples supposedly given to him by the Pleiadian woman Semjase, the supporters of Billy Meier have ignored the expert opinions of three metallurgists from reputable labs, including MIT, and have instead seized on the far less expert opinion of a chemist whose scientific objectivity is highly questionable. In other words, they kept searching until they got the answer they wanted to hear—and ignored all other findings.

When looking into the background of Erich von Daniken (see chapter 8), we found that his past history of having been convicted of embezzlement and forgery cast doubt on his honesty with respect to his claims regarding ancient aliens, particularly in regard to ancient artifacts he claimed to have found in a South American cavern. Thus it is reasonable to look into the past of Billy Meier as a test of his reliability as a witness. As Kal K. Korff points out in the first chapter of his 1995 book *Spaceships of the Pleiades: The Billy Meier Story*, Meier's past wasn't one to inspire confidence in his moral character. Having dropped out of school before completing the sixth grade, he dodged truant officers while working as a manual laborer and was sentenced for theft to six months at the Aarburg Correctional Facility, from which he escaped. He crossed into France and there joined the French Foreign Legion and was sent to Algeria. This, however, didn't suit him, so he went AWOL and returned to Switzerland to serve the rest of his six-month sentence. After this, he spent 12 years traveling and working at odd jobs. In 1965 he lost most of his left arm in a bus accident. In 1967, at a Christmas party in Thessaloniki, Greece, Meier, now 30, met 17-year-old Kalliope Zafiriou. The two fell in love and eloped to Korinthos because her parents didn't approve of him as a suitor. Their first child, born in 1967, was a daughter they named Gilgamesha. They moved to Switzerland in 1970, where they had their next child, a boy they named Atlantis Socrates. For a time they lived on welfare that Billy got for his disability. In 1973 they had a third child, a boy they named Methusalem, and moved to a farm they leased in the town of Hinwill. So far, Meier's life merely shows a person who, perhaps, had difficulty finding himself. However, the names that the couple gave their children—Gilgamesha, Atlantis Socrates, and Methusalem (Methuselah)—might presage Meier's future metaphysical turn.

In 1974, Billy borrowed money to place an ad in a German magazine called *Esoterica* to set up a metaphysical group. On January 28, 1975, he announced to the group that he had made contact with aliens from the Pleiades. Most of those in his group were skeptical of the story. However, some became his loyal supporters. Thus when the town government of Hinwill exercised eminent domain to build apartments on the farm he was leasing, Billy's friends were able to raise the equivalent of $240,000 for him to buy a 50-acre farm near the village of Hinterschmidruti, where he set up the Semjase Silver Center, named after his favorite Pleiadian.

In addition to claiming to have made more than 700 contacts with the Pleiadians/Plejaren, Meier has claimed to be the reincarnation of Jesus Christ. Meier also wrote a book titled the *Talmud Immanuel*, a celestial teaching supposedly handed down to him by the Pleiadians. It purports to be the true testament of Jesus Christ, written after his crucifixion. It is in this work that Meier claims that Jesus was sired on Mary by a Pleiadian named Plejos. The book also claims Jesus survived the Crucifixion. This is a fairly common alternative theory about Jesus, espoused by various people and groups ranging from the second century gnostic Basilides, whose teaching that someone else was substituted for Jesus on the cross was repeated in the *Qur'an* (*Surah An-Nisa*, "The Women," Q 4:157, 158), to those trying give a "scientific" explanation for the Resurrection (a way of denying the Resurrection without being honest enough to simply claim that it just didn't happen). Had Meier's contact with the Plejaren merely involved their urging the human race to avoid nuclear war and care for the environment, perhaps the messages would be at least potentially believable, though one would still wonder why the Pleiadians hadn't come to the United Nations or at least sent embassies to Washington and Moscow in the 1970s. However, the alternate story of Jesus, plus Meier's claims that *he* is in fact Jesus reincarnated, bespeaks such self-aggrandizement as to make one doubt any spark of sincerity in the man's narrative.

Another indication of dishonesty on the part of Meier and his followers is an antagonism toward critics and skeptics and an unwillingness to engage them openly and directly. When Kal K. Korff went undercover to the Semjase Silver Center, disguised as a Meier devotee named Steve Thomas, he asked Bernadette Brand, one of the senior members of Meier's group, what they would do if Kal Korff or someone of that sort came to the Center. She answered him, "We were warned about him by Lee (Elders). We would not let him in or we would never let him leave."[21] The last part of her statement sounds particularly ominous.

What have Meier's claims of Pleiadian contact gotten him? Notoriety? Certainly. Wealth? Definitely. According to Derek Bartholomaus, who runs the website BillyMeierUFOcase.com, membership in Meier's organization, FIGU (Freie Interessengemeinschaft für Grenz und Geisteswissen-schaften und Ufologiestudien, or "Free Community of Interests for the Fringe and Spiritual Sciences and Ufological Studies"), which is supposedly dedicated to the teachings of the Plejaren, is rather pricey. First, one must be a passive member for two years. This requires subscription to his quarterly newsletter, costing SFr. (Swiss Francs) 75 ($98) and an annual membership fee of SFr 30 ($39), a one-time application fee of SFr 30 ($39), a mandatory annual contribution of 7% of one month's salary, and an obligation to work three days a year at the FIGU Center in Switzerland. If one does not contribute the three days labor each year, one must pay a further contribution of SFr 70 ($91) for each day one hasn't worked, with a total of SFr 210 ($274) for three days, which would probably be the cost to any American who couldn't make the annual trip to Switzerland. Not counting the variable of 7% of one month's salary and assuming that one must pay in lieu of three days free labor, this comes to at least SFr 660, or $861 for the two years of *passive* membership. For this, you get little more than eight quarterly newsletters. Finally, one must attend at least one Annual Passive Group General Assembly at FIGU in Switzerland. Once one has been a passive member for two years, one can become a full member for a minimum annual fee of SFr 250 ($326), a registration ID of SFr. 30 ($39), and a subscription to the quarterly magazine at SFr 204 ($266), for a grand total of SFr 484, or $631, per year. Full membership in FIGU does not buy one any say whatsoever in the organization's affairs.

If FIGU has only 100 full members, then Meier grosses SFr 48,400, or $63,100 per year on memberships alone. Have Meier's claims given him power? In some ways, the Semjase Silver Center is a cultic outpost, and certainly Meier's claim that he is the reincarnation of Jesus Christ implies a will to power. Thus the checkered past, the gross fabrication of evidence, and the accrual of wealth and power are all indices of fraud.

DWIGHT YORK

Considering that he was convicted of more than 100 counts of child molestation, facilitated by his position as a cult leader, there is no question that Dwight York's main motive for his wholly dishonest assertion that he is a being from another planet was one of power. By the time he evolved his mythos of extraterrestrial origin, he had already purged his cult of potential unbelievers, first by driving out those supposedly not sufficiently devoted to Islam, then by changing his cult's focus from Islam to ancient Egyptian worship, then by physically relocating it from New York to Georgia. At this point, he was able to make his extraordinary claim that he was from the planet Rizq and had come to Earth aboard Comet Bennet without any of his followers' daring to ask for evidence of this claim.

York's deception was also helped by the exclusionary nature of his message. Although Vorilhon/Rael is trying to reach people across racial, national, and cultural divides, York's UFO cult specifically excluded all but African Americans. Furthermore, he initially set up his supposedly Islamic sect in opposition to the Nation of Islam. Accordingly, he was not interested in providing a story that could even vaguely be asserted to be logical or potentially provable. Indeed, his story seemed so deliberately outrageous as to, perhaps intentionally, exclude skeptics and prove the devotion of his followers.

RAEL

Claude Verilhon—who, as we note in chapter 8 on ancient aliens, has created a religion and renamed himself "Rael"—could be dismissed as an eccentric were it not for strong indications that he has committed plagiarism. The truthfulness of Verilhon/Rael's story of his encounter with a space alien has been questioned, and he has been accused of plagiarizing the work of French author Jean Sendy (1910–78). In 1968 Sendy wrote *The Moon: The Key to the Bible*. In that book he claimed that the word *God* in the biblical book of Genesis should be translated in the plural as *gods*. These gods, he said, were actually spacefaring alien humanoids, and Genesis was the factual history of these aliens' colonization of Earth. In 1969 Sendy published *Those Gods Who Made Heaven and Earth*, in which he claimed that the space travelers arrived here 23,500 years ago and created human beings. The similarities between his theory and Verilhon's appear too great to be coincidental.

The scientific plausibility of Vorilhon's assertion that benign space aliens created life on Earth a mere 25,000 years ago fails on a number of points. A somewhat apocryphal statement is attributed to famed biologist J. B. S. Haldane, once asked by theologians what his study of God's creation had told him about the nature of the Almighty: "He is inordinately fond of beetles." There are about 300,000 species of beetles in the world but fewer than 10,000 species of mammals, a rather odd portioning for a creator considering that we are supposed to be his prime concern. Skeptic Robert Todd Caroll applies this same failure of logic—one that plagues other creationists—to the Raelian assertion of the special creation of life on Earth:

> Apparently, the Raelians are not bothered by the rather absurd image of a race of superior beings working for thousands of years in a laboratory to create all our insects, fungi, bacteria, viruses, etc., not to mention all their lovelies that have gone extinct. Why would any beings do such a thing?[22]

Indeed, why would Vorilhon's Elohim create the HIV virus, particularly when his religion advocates complete sexual liberty and promiscuity?

Of course, this creation of all life on Earth 25,000 years ago also runs afoul of various forms of radiometric dating, such as potassium–argon dating and uranium–lead dating, which have dated some trilobites, for example, as having lived 500 million years ago. The only refutation of radiometric dating the Raelians have come up with is an attack on the accuracy of carbon-14 dating.[23] This is a tiresomely familiar and pointless creationist tactic. Because carbon-14 has a half-life of 5,730 years, and because after 10 half-lives only a bit more than 0.09% of a radioisotope present at a given time hasn't decayed, the carbon-14 present in any life form at its death will be almost entirely absent from its remains after 57,300 years. The absolute oldest possible carbon-14 date for any form of organic remains is about 100,000 years of age. Hence, any critique of the accuracy of carbon-14 dating is irrelevant to the dating most fossils, particularly those of 500-million-year-old trilobites.

The Raelian assault on evolution also shares with those of fundamentalist Christian creationists a certain level of vituperation. Consider this Raelian characterization of evolutionary scientists:

> The evolutionists are also false prophets, false informers, people who lead the majority of the population away from the truth about our creators, the Elohim. This population, which easily swallows and dumbly believes in everything said by these narrow-minded high priests in white coats . . . is purposely kept ignorant and so inevitably believes that which the officialdom says is true. Can you imagine what the Elohim feel when they see that humans attribute their masterpiece to random chance?[24]

Here we see not only the characterization of scientists as deceivers but also, in the accusation that they deliberately keep the majority of the population ignorant, the assertion of a conspiracy on their part. Finally, the statement includes the usual creationist mischaracterization of evolution as being solely the product of random chance, an assertion that ignores the disposition of random mutations by the rigorous and unforgiving force of natural selection.

The Raelians further claim that because living things can repair damages to the sequence of bases in strands of DNA, mutation—and with it evolution—is impossible. Of course, if this is the case, the HIV virus, among others, could not have arisen through mutation and must, again, have been deliberately created by the Elohim. Also, although such repair mechanisms are present in living organisms, they are far from perfect, as witnessed by the many hereditary diseases from which human beings suffer. Among these are hemophilia, sickle-cell anemia, color blindness, Tay–Sachs disease, muscular dystrophy, cystic fibrosis, and achondroplastic dwarfism. This last genetic defect stunningly refutes the Raelian assertion that mutation is impossible. Because the gene for achondroplastic dwarfism is dominant, it is expressed in all who carry it. On occasion two normal parents, who thus cannot carry that gene, will give birth to a child who is an achondroplastic dwarf. This can occur only as the result of a mutation. Thus the Raelian assertion that mutation is impossible is refuted, Q.E.D.

Concerning Vorilhon's evidence of the truth of his claim of contact with space aliens, we not only have nothing but his word but also have evidence that his word is not to be trusted. For example, it is likely that Vorilhon plagiarized material from Jean Sendy (1910–78), a French writer of esoterica. On page 13 of his book *Those Gods Who Made Heaven and Earth*, Sendy claimed that the Hebrew word *elohim* wasn't really the plural form of *el*. Rather, it meant "Those Who Came From the Sky"[25]—as Vorilhon claimed the space alien Yahweh told him. This is only one of many strong similarities between Sendy's writings and Vorilhon's.

As to motive for perpetrating falsehoods, consider that Vorilhon, as Rael, not only has worldwide notoriety but also is now supported by tithes from his followers—10% of one's income being suggested as a norm. Thus the motives of wealth, fame, and power apply. In the absence of either supporting evidence for his extraterrestrial contacts or evidence that these alleged contacts are hallucinatory, the likelihood of fraud on his part would seem high.

L. RON HUBBARD

Of all the strange characters reviewed in this book, one of the strangest was L. Ron Hubbard (1911–86), founder of the cult known as Scientology. In the 1968 recording in which L. Ron Hubbard relates the story of the tyrant and intergalactic overlord Xenu (see chapter 8 on ancient aliens) he does not tell how he supposedly came by this history. Thus we can do no more than conjecture about whether he picked it up in a trance state or deliberately and deceptively made it up out of whole cloth. The scientific implausibility of the narrative is underscored by Hubbard's assertion that the DC-8 passenger jet is the exact copy of the spaceships in which Xenu and his henchmen flew the unfortunate Thetans to the planet Teegeeack (Earth).[26] As with Dwight York, who delivered his message that he was from another planet to the already faithful, Hubbard's disclosure to the inner circle of Scientologists of the whole Xenu/Teegeeack myth did not require that he give any evidence to support the veracity of the narrative.

Although the Church of Scientology has hotly denied having an ancient alien mythos, a 1968 tape of their founder, L. Ron Hubbard, expounding just such a story has surfaced, though the church asserts it is false. Actor Jason Beghe, a former Scientologist who has left the church, has testified that the tape is real and that its ancient alien story lies at the core of the belief system of Scientology. The story has been publicized a number of times. Here is a summary: 75 million years ago, a galactic overlord named Xenu ruled 76 planets, all of which had a severe population problem. To solve this, Xenu rounded up 13.5 trillion of his subjects and flew them to the planet Teegeeack—which we know as Earth. Interestingly, on the tape Hubbard says of the vessels that Xenu used to transport these trillions of his subjects that they were remarkably similar to a certain 20th-century aircraft. He says specifically, "The DC-8 is the exact copy of the space plane of that day." The DC-8 was the chief passenger jetliner produced by Douglas from 1958 to 1972. Suffice it to say that none of these aircraft ever achieved escape velocity, nor were they rivals of either Saturn rockets or the space shuttle. Once these subjects had been flown to Teegeeack/Earth, they were dumped into volcanoes over which Xenu detonated hydrogen bombs. Having destroyed their bodies, Xenu captured their souls, called Thetans, in electromagnetic traps and implanted them with false ideas. These included concepts of God and organized religion. The souls of these Thetans, who are, of course, traumatized, attach themselves to human beings and are the source of all our neuroses.

Although this mythos is atypical and rather tangential to ancient alien theories, it is worth mentioning in passing because Hubbard was originally a science fiction author. Ancient alien theories owe much to science fiction. *The Morning of the Magicians* by Louis Pauwels and Jacques Bergier, a book originally published in French in 1960 and that was published in English in 1964, expounded a theory that Earth was once ruled by giants. Although this work is based more on esoterica than on any appeal to science, it does expound theories that turn up later in *Chariots of the Gods?*, which appeared in 1968. In his book *The Cult of Alien Gods: H. P. Lovecraft and Extraterrestrial Pop Culture*, Jason Colavito points out that ancient astronaut theories, beginning with *The Morning of the Magicians*, owe much to the work of science fiction and horror writer H. P. Lovecraft (1890–1937), particularly his novelette *At the Mountains of Madness* (1931). Lovecraft's Cthulhu mythos was the basis of this story and many others he wrote. In it ancient beings called the elder gods traversed space to come to the Earth in prehistoric times and still lurk, in other dimensions, under the sea, or on other planets, waiting for their chance to reclaim the Earth.

Scientology-based biographies of L. Ron Hubbard paint him as a hero of World War II, claiming, among other things, that he was awarded 21 medals, that he was wounded and given the Purple Heart, and that, as a U.S. Navy officer, he skillfully commanded a ship in battle. In point of fact, he was never wounded, he spent most of his time as a junior officer in the U.S. Navy—onshore—and, while in command of a coastal patrol craft, *PC-815*, he blundered into Mexican territorial waters and held gunnery practice, shelling an island he erroneously thought to be uninhabited. The Mexican government complained about this incident, and the U.S. Navy removed Hubbard from command of the ship.[27]

Before creating Dianetics, the forerunner of Scientology, Hubbard dabbled in the occult and wrote pulp science fiction. That he was a glib con artist was well known among his fellow science fiction authors. On August 27, 1946, L. Sprague de Camp wrote Isaac Asimov the following:

> The more complete story of Hubbard is that he is now in Fla. living on his yacht with a man-eating tigress named Betty-alias-Sarah, another of the same kind . . . . He will probably soon thereafter arrive in these parts with Betty-Sarah, broke, working the poor-wounded-veteran racket for all its worth, and looking for another easy mark. Don't say you haven't been warned. Bob [Robert Heinlein] thinks Ron went to pieces morally as a result of the war. I think that's fertilizer, that he always was that way, but when he wanted to conciliate or get something from somebody he could put on a good charm act. What the war did was to wear him down to where he no longer bothers with the act.[28]

Given Hubbard's history of deceptively inflating his military record and other absurd and inflated claims Scientologists have made on his behalf, the greatest likelihood is that he fabricated the Xenu myth out of whole cloth, as opposed to the story's being the product of a mental aberration.

Scientology seems to have suffered some serious blows lately, particularly with the 2013 Lawrence Wright book *Going Clear: Scientology and the Prison of Belief* and the 2015 HBO Alex Gibney documentary based on that book. Gibney read the book, then enlisted Wright as a coproducer and managed to get extraordinary access not only to damning footage about Scientology but also, and especially, to many high-ranking defectors who told their stories to the camera. The film originally aired at the Sundance Film Festival in early 2015, where it was the hit of the festival and received standing ovations from SRO audiences.[29] Now it has been released in theaters and on HBO—perhaps the only network able to tackle such a dangerous topic. (All the major broadcast networks were reportedly[30] too frightened of the Church of Scientology even to license material for inclusion in the film—perhaps unsurprising, considering its chilling tactics.[31]) Yet thanks to this wide release, the genie is now well and truly out of the bottle. The film's reach and effect will only continue to expand online.

Although most of the material in the documentary has been written about before and covered in numerous books (including Wright's, on which the documentary is based), there is a big difference between written exposés and a much hotter, more visceral medium such as video, as described on the pop culture site *Vulture*:

> The cult is right to be scared. Going Clear is spectacular stuff. Literally. Scientology thrives on spectacle, and one big advantage the film has over the book is that it can show us footage of Scientology's big "events," which look like Nazi rallies crossed with Hollywood awards shows. It can also show us video footage of the church's leaders dissembling and evading when confronted by reporters. Seeing is disbelieving: Watching this stuff, instead of merely reading about it, somehow makes Scientology both more ridiculous and more chilling.[32]

The documentary revisits many of the numerous shocking events that have already been revealed about the cult. These start with the way in which the cult manipulates members with their "auditing process." They hook them up to an ordinary galvanometer (they call it the "e-meter") to measure the involuntary galvanic skin response and then put them through intense interrogation, pressuring subjects to confess to things they did or didn't do and to "recover memories" of things that didn't happen. Once you are hooked, they soak you for money over and over again, making you spend money on more and more courses and auditing sessions to reach higher levels of the church ladder. Only after years of auditing (and thousands of dollars) do they reveal the bizarre mythology that their founder, pulp science fiction writer L. Ron Hubbard, cooked up to explain their belief system, as summarized by *Newsweek*:

> Seventy-five million years ago, according to Hubbard, people lived in a world very much like 1950s America, save for Xenu, the tyrannical overlord of the 76-planet Galactic Confederacy. Xenu sought to solve a burgeoning overpopulation problem by freezing people with glycol injections after luring them in under the auspices of "tax audits." The frozen bodies were then shipped in boxes via space planes to the prison planet Teegeeack (Earth), where they were dropped into volcanoes and blown up with hydrogen bombs. Being immortal, these disembodied spirits ("Thetans") were then trapped in an electromagnetic ribbon and placed in front of a "three-D, super colossal motion picture" for 36 days, where they were forced to look at images called R6 implants. "These pictures contain God, the Devil, angels, space opera, theaters, helicopters, a constant spinning, a spinning dancer, trains and various scenes very like modern England," Hubbard wrote. "You name it, it's in this implant."[33]

As Paul Haggis comments in the documentary, his first reaction on learning this story was "What the fuck are you talking about?"[34] He then wondered whether they were testing him to see whether he would believe this silly 1950s-style bad science fiction. But they are dead serious about this crazy myth, although they deliberately don't tell it to you until you are high in their system and deeply committed. They tried to keep it secret for many years lest it scare away the converts.

As Jim Lippard wrote about L. Ron Hubbard:

> There once was a man who considered himself an explorer, a military hero, a mystic, a philosopher, a nuclear physicist, and an expert in human nature. In fact, he was none of these things. He was an adventurer, a writer of pulp fiction, and a teller of tall tales. He was a college dropout, a bigamist, convicted of petty theft and fraud, and named as an unindicted co-conspirator in a plot to infiltrate and steal information from U.S. government agencies. Despite his unimpressive physique, he was a larger-than-life, charismatic figure who persuaded thousands

of people to believe in and pay large sums of money to learn more about a view of the world he constructed from a foundation of pseudoscience, bad philosophy, science fiction, and space opera. He came to believe his own claims of developing the power to shape the world to his tastes and improve one's physical and especially mental states through specific techniques he invented that precluded all psychiatric drugs, and yet he died alone with matted hair and rotting teeth, with the anti-anxiety drug Vistaril in his system.[35]

After a spectacular introduction, the documentary does a good job of summarizing the unsavory and dishonest past of L. Ron Hubbard and how he dreamed up the entire Scientology scam to get rich. It is full of damning clips and quotes, many in his own words or in the words of his wife at the time, Sara Northrup, who helped him. It then goes through the recent years of Scientology since Hubbard's death in 1986, during which it has been ruled with an iron fist by his successor, David Miscavige. All the well-documented scandals are there: the widespread abuse of its members, who have gone through slave labor and bizarre tortures when Miscavige demands it; the way the Church of Scientology harassed the IRS until it got tax-exempt status as a religion;[36] its cozying up to celebrities such as Tom Cruise and John Travolta to gain positive media attention; and its brutal secret police, who will use any tactic in the book, from lawsuits and publicity attacks to spying to physical intimidation, to protect the organization. There are also some shocking stories that have not received widespread attention, including how the Church of Scientology actively undermined Cruise's marriage to Nicole Kidman to pull him back into the Church and even groomed a "girlfriend" for him who was a Church loyalist, how it uses the slave labor of its members to trick out Cruise's many cars and motorcycles and airplanes and give him anything he wants (but there is no mention of how Katie Holmes has spoken out about the Church's effect on their marriage), and how it uses dirty secrets to keep John Travolta in line. All these allegations are made by highest-level defectors with many years inside the Church, including the Church's former number-two man and enforcer, Marty Rathbun, who was a direct witness to or agent in some of the most shocking allegations; longtime spokesperson Mike Rinder; Oscar-winning writer–director Paul Haggis (famous for writing *Million Dollar Baby, Crash, Flags of our Fathers, Letters from Iwo Jima, Casino Royale,* and *Quantum of Solace*), a member for 35 years; actor Jason Beghe, once one of Scientology's most publicized figures and a member for 13 years; Spanky Taylor, who was the liaison to John Travolta; SeaOrg cofounder Hana Eltringham, who breaks down on camera because the Church will no longer let her see her family; and several others.

After two hours of this unremitting dirty laundry, you find yourself stunned that this kind of organization could get away with so much for so long—and puzzled about what keeps believers locked into an abusive relationship with the Church of Scientology. The answer to the latter is standard across all cults: they hook you with something innocuous at the beginning (you feel good after a light auditing session, just as you would after spilling your problems to your therapist); then, once you become part of their tight-knit community, it's very hard to get out—in part because they have so much dirt on you and also because they will intimidate and harass you and cut you off from your loved ones still in the Church.

But how did they get away with it for so long? For decades, the Church of Scientology's use of ruthless tactics against anyone who would reveal its secrets or defect kept everyone in a conspiracy of silence about its Dark Side even as it generated positive PR with its appealing ad campaigns and celebrity endorsements from Cruise, Travolta, Kirstie Alley, Beck, and others. In the days of print media, a book or a magazine or newspaper article might criticize the Church but had a limited readership. But the biggest factor changing this code of silence has been the internet. No government can control it or completely silence it (even in China or North Korea or other countries under oppressive regimes), and no religion can censor it, either. For almost 15 years now, one secret after another has been revealed and gone viral, so Scientology has a huge media trail that is only an internet search away. As the secrets have begun to come out, more and more people have come forward, and the mainstream media have become bolder. In one highly damaging example, the cartoon *South Park* dedicated an entire episode to mocking Scientology. (The Church attempted to dig up dirt[37] on creators Trey Parker and Matt Stone in retaliation but without success.) *Going Clear* raises the stakes by putting all the Dark Side of Scientology in one powerful documentary, and it will live in cyberspace forever.

Not only is the internet more powerful than the Church's secrecy, but also lately the Church's PR efforts have been a disaster. Cruise's antics, from jumping on Oprah Winfrey's couch to attacking psychology in many interviews, have made him look more and more ridiculous in the public eye, and they reflect badly on the Church. So does John Travolta's recent behavior. These actors are still working and getting decent movie roles, but many people no longer admire them or want to emulate them. The Church of Scientology no longer even gives press conferences, because it doesn't want reporters asking hard questions. For some cults and churches that recruit among the poor and less literate, a media scandal might not be important. But for Scientology, publicity is crucial, because its recruiting depends on hooking relatively affluent young members who are searching for spiritual answers. These people are internet-savvy and are likely to read about the Church online and watch damning videos before they have the chance to get hooked. Indeed, there is much evidence that even as the Church gets richer (more than $3 billion in assets, mostly invested in pricey real estate,[38] from milking its membership while paying no taxes), it is also losing members at an alarming rate. Even though the Church claims it has millions of members, more reliable sources indicate only about 40,000 to 50,000 worldwide:[39] maybe 25,000 in the United States[40] and the remainder in a few other countries.[41] If this wave of bad publicity on the internet continues, especially with the widespread release of *Going Clear*, expect the religion to decline to just a lot of expensive buildings with almost no one in them.

In many ways, the story of Scientology is similar to that of another cult that misappropriated the cachet of "Science" in its name: "Christian Science" (which is neither Christian nor scientific). When it was founded by Mary Baker Eddy in 1879, its message of faith healing was powerful, and it became a large organization with hundreds of thousands of members. But today it is virtually extinct, with lots of property ("Christian Science Reading Rooms") but almost no one occupying that property. Estimates made in 2009 suggested that there were fewer than 50,000 remaining members[42] worldwide. At a plausible attrition rate of 4,000 lost members per year (extrapolating from declining *Christian Science Sentinel* subscription rates), Mary Baker Eddy's once powerful church could well be extinct within a decade. What killed it? Modern medicine. When Christian Science was founded, medicine was primitive and was often detrimental to a patient's health. Just as in the case of homeopathy, harmless placebos such as faith healing or homeopathic "cures" often were just as effective as real medicine. But a century later, real medicine has made enormous advances, and Christian Scientists keep dying off by refusing it. They are also very stodgy and conservative (offering "reading rooms" rather than an internet presence) and are not very active in recruiting. Will Scientology follow Christian Science into extinction? If present trends continue, it's very likely.

BOB LAZAR

As with that of L. Ron Hubbard, the story of the alien contact claimed by Robert Scott Lazar is indirect, though much more physical. We briefly met Lazar in chapter 3, but his case is worth examining in greater detail. In November 1989, Lazar, who had worked at the Las Alamos Laboratory, was interviewed by George Knapp on the Las Vegas television station KLAS. In this interview and others, Lazar claimed to be a physicist holding degrees from the Massachusetts Institute of Technology (MIT) and the California Institute of Technology (Caltech), who had worked at the S-4 section of Area 51. While he was there, he claimed, he had seen a number of alien spacecraft that had crashed at Roswell, New Mexico (see chapter 3). These were being reverse-engineered by U.S. scientists to create new technologies. He further claimed that these craft used an antigravity propulsion system that used element 115 as its fuel. He said that each spaceship carried two kilograms of element 115 and that the two kilograms would last several years before needing to be replenished. He asserted that the U.S. government had, courtesy of aliens from Zeta Reticuli (the grays), 500 pounds of element 115 at Area 51 and that it was of extraterrestrial origin, because element 115 wasn't found on Earth. He also said it was comparatively stable, because that particular heavy radioactive element was in what is called an island of stability. Lazar further asserted that the thermionic generators of these spacecraft had an efficiency of 100%, though he freely acknowledged that such an engine would violate the second law of thermodynamics.

To understand Lazar's claim regarding element 115, we need to review some basic science having to do with elements at the upper end of the periodic table. In nature, the heaviest element, the one with the greatest number of positively charged protons in its nucleus, is uranium, element 92—it has 92 protons in its nucleus. By bombarding uranium with various subatomic particles, nuclear physicists have created elements beyond those that occur in nature, the most important of these being plutonium, element 94. These artificially created elements are known collectively as transuranic elements. Although plutonium-239 has a half-life of 24,100 years, most transuranic elements have extremely short half-lives, often measured in fractions of a second. Paradoxically, however, some isotopes of these elements, at least theoretically, have longer life spans, existing as they do in the aforementioned "island of stability."

In 1989, when Bob Lazar first made his claims, element 115, provisionally named ununpentium (abbreviated Uup), had not yet been synthesized on Earth. Physicists didn't manage to create it until 2003. Thus Lazar's assertions about its properties couldn't be checked. All known isotopes of ununpentium are highly radioactive and have extremely short life spans. (Differing isotopes of the same element have the same number of protons in their nuclei but differing numbers of neutrons.) The most stable isotope of element 115 is Uup-289, which has a half-life of 220 milliseconds. Generally speaking, after 10 half-lives, a moderate amount of a radioactive isotope created at the same time is considered to have virtually entirely decayed away. As already noted, after 10 half-lives, only a bit more than 0.09% of the original sample remains. Ten half-lives of Uup-289 adds up to 2.2 seconds, at which point there would only be 0.45 pounds of the original 500 pounds. For greater concentrations of a given radioisotope, 100 half-lives might be required for all of it to have decayed. Certainly 500 pounds of element 115 would fit into that category. So 500 pounds of Uup-289 concentrated in one spot might last for a stunning 22 seconds!

This isn't the only problem with Uup-289, however. Its decay chain leads through a number of transuranic elements, all with half-lives measured in seconds or fractions of a second, to isotopes that decay by spontaneous fission rather than alpha decay. For the very small quantities of Uup-289 created in laboratories, this is not a problem. However, 500 pounds of Uup-289 concentrated in one spot decaying in a matter of seconds into material that spontaneously fissions could result in a sustained, explosive chain reaction, possibly resulting in something such as an atomic bomb detonating in Area 51.

To be fair to Lazar, however, physicists speculate that Ununpentium-291, an isotope they have yet to create, might have a half-life of several seconds. Such a hypothetical isotope of element 115, it is believed, could be bombarded to create a heavy isotope of element 112, copernicum (abbreviated Cn). This isotope, Cn-291, it is predicted, would have a half-life of 1,200 years, being in the middle of the island of stability. However, to date, all isotopes of copernicum that physicists have created are highly unstable. The most stable known isotope of copernicum, Cn-285, has a half-life of 29 seconds. So 500 pounds of Cn-285 would, at best, have decayed away in a bit over 48 minutes; and Cn-285 quickly decays into another very unstable transuranic isotope, hassium-277, most of which undergoes spontaneous fission. So again we have the possibility of 500 pounds of the stuff spontaneously detonating. Had Lazar said that what was being stored at Area 51 was Cn-291, the hypothetical long-lived isotope of element 112, his assertion would be somewhat viable. But because he said that the 500 pounds of extraterrestrial anti-gravity fuel was element 115, his story fails the criterion of scientific plausibility.

To the further detriment of his story, Lazar has falsely claimed scientific credentials he doesn't have. Physicist Stanton Friedman, a strong proponent of the theory that the Roswell crash was that of an extraterrestrial spacecraft, said the following of Lazar's credentials:

> He was publicly asked when he got his MS from MIT. He said, "Let me see now, I think it was probably 1982." Nobody getting an MS from MIT would not know the year immediately. He was asked to name some of his profs, He said: "Let's see now, Bill Duxler will remember me from the physics department at Caltech." I located Dr. Duxler. He's a Pierce Junior College physics prof, and never taught at Caltech. Lazar was registered in one of his courses at the same time Lazar was supposedly at MIT! Nobody who can go to MIT goes to Pierce JC, not to mention the rather long commute between LA and Cambridge, Mass. . . . I checked his High School in New York State. He graduated in August, not with his class. The only science course he took was chemistry. He ranked 261 out of 369, which is in the bottom third. There is no way he would have been admitted by MIT or Caltech. An MS in Physics from MIT requires a thesis. No such thesis exists at MIT, and he is not

on a commencement list. The notion that the government wiped his CIVILIAN records clean is absurd. I checked with the Legal Counsel at MIT—no way to wipe all his records clean. The Physics department never heard of him and he is not a member of the American Physical Society.[43]

In fact, Lazar didn't even work at the S-4 section of Area 51. He was employed not by the government but rather as a technician working for a private company that contracted work at Los Alamos.

Thus Lazar's science doesn't hold up, his credentials are false, and when the insufficiency of his evidence is pointed out, he employs the same tired tactic used by Billy Meier and others—claims of destruction of evidence perpetrated by powerful and sinister conspirators determined to keep the truth from the public. Lazar, a technician working for a company contracting work at a federal lab—in short, rather a minor player—gained considerable attention and notoriety, becoming a hero to UFO enthusiasts and conspiracy buffs. The failed science, the fabricated degrees from MIT and Caltech, and the alluring motive of posing as a whistleblower exposing a government cover-up all strongly indicate fraud. It should be noted that when interviewed 25 years later, Lazar claimed that he wished he hadn't made his alleged revelations—they had brought him, he said, nothing but trouble. Travis Walton has also expressed such sentiments. However, Walton's story did save his friend, Mike Rogers, from a financial debacle and resulted in a book deal as well as his story's being made into a motion picture. Perhaps Lazar's regrets stem from the fact that he did not enjoy similar financial success.

SUMMARY

All or most of the eight narratives examined in this chapter share a number of striking common attributes. In most of the narratives, the space aliens are fully human, though in Walton's case, they also included grays. The grays are also obliquely alluded to in Lazar's scenario. Adamski's Venusians and Meier's Pleiadians were or are essentially Nordic-type aliens from the basic UFO mythos, as were some of the beings in Walton's account. Frank Stranges's Valiant Thor was indistinguishable from any European-derived American. Claude Vorilhon's Yahweh was slightly different from most humans, being about four feet tall, but would certainly pass for human. Graphic representations of Xenu also show him as a human being, and Dwight York claimed that he, himself, was an extraterrestrial being. Another common attribute to these narratives is that all of them resulted in notoriety, wealth, power as a cult leader, or a combination of these for their authors.

With the exception of Lazar's claim of an alien anti-gravity generator using element 115, these narratives also demonstrate a lack of imagination that would seem to be a part of their overall absurdity. Listening to the recording of L. Ron Hubbard saying the DC-8 passenger jet is an exact copy of the spaceships in which Xenu and his henchmen transported the Thetans to Earth, or hearing Frank Stranges say that the Venusian visitor to the Pentagon was named "Valiant Thor," one wonders who could possibly believe such nonsense. Perhaps, as with the repeated acts of self selection among followers of UFO religions, the absurdity of these claims is deliberate: they act as filters, screening out any who are likely to be skeptical, thereby ensuring that those who initially support a fraudulent narrative won't later come to their senses and cease financially supporting the narrative's perpetrators. Put another way, those who initially accept an absurd story are more likely to drink the Kool-Aid.

7

Close Encounters of the Fourth Kind: Alien Abduction

"I WAS CAPTURED BY ALIENS"

On February 10, 1964, the popular science fiction television show *The Outer Limits* aired an episode titled "The Bellero Shield." Directed by John Brahm and written by Joseph Stefano and Lou Morheim, the episode starred Martin Landau as Richard Bellero, a scientist working on lasers; Sally Kellerman as his ambitious wife; and, most significant, John Hoyt as an alien from another dimension, reeled into the Belleros's home when the scientist fires a laser out into space. The alien depicted in "The Bellero Shield" was humanoid, stood somewhat below average human height, wore a gray jumpsuitlike garment, had a bulbous bald head—also gray—and had curious, slightly slanted eye sockets that somewhat wrapped around to the sides of his head. He had at his disposal a technology more advanced than ours, and he was benign and altruistic. He was, in fact, the prototype of the classic "gray alien" so common in UFO encounters. Thus the source of the grays of modern UFO myth was bequeathed to us via modern medium of television. The "grays" were firmly imprinted into the annals of UFO encounters via the "recovered" memory of Barney Hill, revealed through hypnosis only days after "The Bellero Shield" aired.

To understand how this happened, we have to know the details of this narrative. Beginning in the latter half of the 20th century, stories of aliens from beyond our solar system abducting people, sometimes experimenting on them and inserting probes and other implants into their bodies, as well as often giving them dire prophetic warnings of an impending apocalypse, have become a staple of the UFO community. One of the earliest of these, which became the classic tale of alien abduction, is that of Barney and Betty Hill. Here's their story.

THE ORIGINAL ABDUCTION STORY:
BARNEY AND BETTY HILL

On the evening of September 19, 1961, around 10:30 p.m., the Hills, who were driving home from a vacation in Canada, encountered a UFO on a lonely stretch of highway near Indian Head, New Hampshire. They observed a bright point of light in the sky that at first seemed to be a meteorite but that then began to move erratically—and that then seemed to follow them. Barney stopped the car to observe the light, which, as it came closer, resolved into an oddly shaped craft with flashing lights. The craft rapidly descended toward their parked car. Barney started the car up and drove toward Franconia Notch along a narrow, mountainous stretch of road. The craft appeared to be following them, and Betty testified that it was at least one and a half times the length of the Old Man of the Mountain, a natural rock face 40 feet long. About one mile south of Indian Head, the object descended rapidly and silently to about 100 feet above the Hills' car. It now blocked their path. According to Barney, it was pancake-shaped.

Barney stopped the car and stepped out, carrying a pistol he had previously retrieved from the trunk of his car. He looked through binoculars he had also taken from his car and saw 8 to 11 humanoid figures, dressed in glossy black uniforms, peering out of the craft's windows. All but one of the figures moved away. The one remaining communicated to Barney, seemingly through telepathy, telling him to stay where he was and to keep looking. Bat-wing shaped fins covered with red lights began to telescope out of the sides of the craft, and a long structure descended from the bottom of it. The silent craft approached to what Barney estimated was within 50 to 80 feet overhead and 300 feet away from him.

Barney ran back to his car. In a near hysterical state, he told Betty, "They're going to capture us!" Barney reported that the binocular strap had broken when he ran from the UFO back to his car. He recalled driving the car away from the UFO but noted that afterward he felt irresistibly compelled to pull off the road and drive into the woods. He eventually sighted six men standing in the dirt road. The car stalled, and three of the men approached the car. They told Barney not to fear them. He was still anxious, however, and he reported that the leader told him to close his eyes.

Barney said he and Betty were taken onto the disc-shaped craft, where they were separated. He was escorted to a room by three of the men and told to lie on a small rectangular exam table. A cuplike device was placed over his genitals. He did not experience an orgasm, though he thought that a sperm sample had been taken. The men scraped his skin and peered into his ears and mouth. A thin tube or cylinder was inserted into his anus and quickly removed. Someone felt his spine, seemingly counting his vertebrae.

As the couple was being separated, Betty protested, but a man she called "the leader" told her that if she and Barney were examined together, it would take much longer to conduct the exams. A new man, similar to the others, entered to conduct her exam. Betty referred to him as "the examiner." He had a pleasant, calm manner; but when he spoke to her, his command of English was poor and Betty had difficulty understanding him. (In contrast to this, the English spoken by "the leader" had been polished.) "The examiner" told Betty that he would conduct tests to note the differences between humans and the aliens. He first seated her on a chair and shined a bright light on her. He cut off a lock of her hair, then examined her eyes, ears, mouth, teeth, throat, and hands and collected trimmings from her fingernails. After examining her legs and feet, he used a dull knife, similar to a letter opener, to scrape some of her skin onto what resembled cellophane. He then tested her nervous system and finally thrust the needle into her navel, which caused Betty agonizing pain. But the leader waved his hand in front of her eyes, and the pain vanished. The examiner left the room, and Betty engaged in conversation with the leader. She picked up a book with rows of strange symbols that the leader said she could take home with her, though later the aliens took the book away from her. She also asked where he was from, and he pulled down an instructional map dotted with stars, which she copied.

Later, the Hills found themselves back in their car. Barney saw the craft again shift its location to directly above their car and drove away at high speed, telling Betty to look for the object. She rolled down the window and looked up but saw only darkness above them, even though it was a bright, starry night. Almost immediately, the Hills heard a rhythmic series of beeping or buzzing sounds that seemed to bounce off the trunk of their vehicle. The car vibrated, and a tingling sensation passed through the Hills' bodies. Betty touched the metal on the passenger door expecting to feel an electric shock but felt only the vibration. The Hills said that at this point, they experienced the onset of an altered state of consciousness that left their minds dulled. A second series of beeping or buzzing sounds returned the couple to full consciousness. They found that they had traveled nearly 35 miles south but had only vague, spotty memories of this section of road. They recalled having made a sudden unplanned turn, encountered a roadblock, and observed a fiery orb in the road.

Once they arrived home, about dawn, the Hills had some odd sensations and impulses they could not readily explain: Betty insisted that their luggage be kept near the back door rather than in the main part of the house. Their watches had stopped and would never run again. Barney noted that the leather strap for the binoculars was torn, though he could not recall its tearing. The toes of his best dress shoes were inexplicably scraped. They took long showers to remove possible contamination, and each drew a picture of what they had observed. Their drawings were strikingly similar.

After sleeping for a few hours, Betty woke and placed the shoes and clothing she had worn during the drive into her closet, observing that the dress was torn at the hem, zipper, and lining. Later, she noted a pinkish powder on her dress. She hung the dress on her clothesline, and the pink powder blew away; but the dress was irreparably damaged. She threw it away but then changed her mind, retrieving it and hanging it in her closet. Over the years, five laboratories have conducted chemical and forensic analyses on the dress. There were shiny, concentric circles on their car's trunk that had not been there the previous day. Betty and Barney experimented with a compass, noting that when they moved it close to the spots, the needle would whirl rapidly.

But when they moved it a few inches away from the shiny spots, it would drop down.

On September 21, Betty telephoned Pease Air Force Base to report their UFO encounter, though for fear of being labeled eccentric, she withheld some of the details. On September 22, Major Paul W. Henderson telephoned the Hills for a more detailed interview. Henderson's report, dated September 26, determined that the Hills had probably misidentified the planet Jupiter. This was later changed to "optical condition," "inversion," and "insufficient data" (Report 100-1-61, Air Intelligence Information Record). His report was forwarded to Project Blue Book, the U.S. Air Force's UFO research project.

THE GENESIS OF AN ABDUCTION

That's the story—that is, that is the *full* story, developed some time after the supposed abduction and after the Hills had undergone hypnosis to retrieve their "repressed memories" of the abduction. Still, how can one argue with such detail, independently related by both Barney and Betty under hypnosis? The problem here is that the Hills' shared hypnotic memories had indeed been shared—well before they underwent hypnosis. What the Hills actually experienced was that they saw some lights in the sky they couldn't identify and arrived home much later than they had expected to, having little memory of the last 35 miles of their late-night journey.

Ten days later, Betty began having a series of nightmares in which much of what was later reported after hypnosis occurred. She related the details of these dreams to Barney. In November 1961, Betty began writing down the details of her dreams. In one dream, she and Barney encountered a roadblock and men who surrounded their car. She lost consciousness but struggled to regain it. She then realized that she was being forced by two small men to walk in a forest in the nighttime, and she saw Barney walking behind her, though when she called to him, he seemed to be in a trance or sleepwalking. The men stood about five feet to five feet four inches tall and wore matching blue uniforms, with caps similar to those worn by military cadets. They appeared nearly human, with black hair, dark eyes, prominent noses, and bluish lips. Their skin was a grayish color. In the dreams, Betty, Barney, and the men walked up a ramp into a disc-shaped craft of metallic appearance.

Also, within days of the "encounter," Betty borrowed a book from a local library on UFOs, written by Donald Keyhoe, Maj. USMC (Ret.). On September 26, Betty wrote to Keyhoe. She related the story, including the details about the humanoid figures that Barney had observed through binoculars. *Betty wrote that she and Barney were considering hypnosis to help recall what had happened.* Her letter was eventually passed on to Walter N. Webb, a Boston astronomer and NICAP member. NICAP, the National Investigations Committee on Aerial Phenomena, was a civilian unidentified flying object research group active in the United States from the 1950s to the 1980s. Webb met with the Hills on October 21, 1961. In a six-hour interview, the Hills related all that they could remember of the UFO encounter. Barney asserted that he had developed a sort of "mental block" and that he suspected there were some portions of the event that he did not wish to remember. He described in detail all that he could remember about the craft and the appearance of the "somehow not human" figures aboard the craft. Webb stated that they were telling the truth and that the incident probably occurred exactly as reported, except for some minor uncertainties and technicalities that must be tolerated in any such observations when human judgment is involved (e.g., exact time and length of visibility, apparent sizes of object and occupants, distance and height of object).

On November 25, 1961, NICAP members C. D. Jackson and Robert E. Hohmann interviewed the Hills at length by. Having read Webb's initial report, Jackson and Hohmann had many questions for the Hills. One of their main questions was about the length of the trip. Although the Hills had noted that they had arrived home later than anticipated (the drive, 178 miles, should have taken about four hours), they did not realize that they arrived home seven hours after their departure from Colebrook. When Hohmann and Jackson noted this discrepancy to the Hills, the couple had no explanation. Despite all their efforts, the Hills could recall almost nothing of the 35 miles between Indian Head and Ashland. Although Betty's recall was somewhat fuller than Barney's, both were able to recall an image of a fiery orb sitting on the ground. Betty and

Barney reasoned that it must have been the moon, but Hohmann and Jackson informed them that the moon had set earlier in the evening.

On November 23, 1962, the Hills attended a meeting at the parsonage of their church where the invited guest speaker was Captain Ben H. Swett of the U.S. Air Force, who had recently published a book of his poetry. After he read selections of his poetry, the pastor asked him to discuss his personal interest in hypnosis. After the meeting broke up, the Hills approached Captain Swett privately and told him what they could remember of their strange encounter. He was particularly interested in the "missing time" of the Hills' account. The Hills asked Swett whether he would hypnotize them to recover their memories, but Swett said he was not qualified to do that and cautioned them against going to an amateur hypnotist, such as himself.

On March 3, 1963, the Hills first publicly discussed the UFO encounter with a group at their church. On September 7, 1963, Captain Swett gave a formal lecture on hypnosis to a meeting at the Unitarian Church. After the lecture, the Hills told him that Barney was going to a psychiatrist named Stephens, whom he liked and trusted. Captain Swett suggested that Barney ask Stephens about the use of hypnosis in his case. When Barney next met with Stephens, he asked about hypnosis. Stephens referred the Hills to Benjamin Simon of Boston. On November 3, 1963, the Hills spoke before an amateur UFO study group, the Two State UFO Study Group, in Quincy Center, Massachusetts.

The Hills first met Simon on December 14, 1963. Early in their discussions, Simon determined that the UFO encounter was causing Barney far more worry and anxiety than he was willing to admit. Though Simon dismissed the popular extraterrestrial hypothesis as impossible, it seemed obvious to him that the Hills genuinely *thought* they had witnessed a UFO with humanlike occupants. Simon hoped to uncover more about the experience through hypnosis. Simon began hypnotizing the Hills on January 4, 1964. He hypnotized Betty and Barney several times each, and the sessions lasted until June 6, 1964. Simon conducted the sessions on Barney and Betty separately so that they could not overhear one another's recollections. At the end of each session, he told them that they would not remember what they had told him.

Under hypnosis, Betty's account was very similar to the events of her five dreams about the UFO abduction, but there were also notable differences. Details pertaining to her capture and release were different. The technology on the craft was different. The short men had a significantly different physical appearance from the ones in her dreams. The sequential order of the abduction event was also different from that in Betty's dream account. She filled in many details that were not in her dreams and contradicted some of her dream content. It is interesting that Barney's and Betty's memories in hypnotic regression were consistent but contradicted some of the information in Betty's dreams.

Simon gave Betty the posthypnotic suggestion that she could sketch a copy of the "star map" that she later described as being a three-dimensional projection similar to a hologram. She hesitated, thinking that she would be unable to accurately depict the three-dimensional quality of the map she says she saw on the ship. Eventually, however, she did what Simon suggested. Although she said the map had many stars, she drew only those that stood out in her memory. Her map consisted of 12 prominent stars connected by lines and three lesser ones that formed a distinctive triangle (Fig. 5.1). She said she was told that the stars connected by solid lines formed "trade routes," whereas dashed lines were to less traveled stars.

After extensive hypnosis sessions, Simon speculated that Barney's recall of the UFO encounter was possibly a fantasy inspired by Betty's dreams. Though Simon admitted that this hypothesis did not explain every aspect of the experience, he thought it the most plausible and consistent explanation. Barney rejected this idea, noting that although their memories were in some regards interlocking, there were also portions of both their narratives that were unique to each. Barney was now ready to accept that they had been abducted by the occupants of a UFO, though he never embraced it as fully as Betty did.

Thus the Hills' shared memory of the event was *indeed* shared, based on Betty's dreams. The full story, based on the Hills' hypnosis sessions, didn't come together until about two years after their UFO sighting. During the intervening period, Betty had related the content of her dreams to Barney, she had taken a book out of the library on UFOs, and the Hills had been interviewed twice by members of NICAP—people looking for evidence of UFO sightings and encounters.

Brian Dunning, in his article on the subject on *Skeptoid.com*, notes that not only had Betty related her dreams to Barney, but she had also been writing them down:

> All of the significant details you may have heard about the Hills' medical experiments came from Betty's two years of writings: A long needle inserted into her navel; the star map; the aliens' fascination with Barney's dentures; the examination of both Betty and Barney's genitals; and the overall chronology of the episode, including being met on the ground by the aliens, a leader coming forward and escorting them to exam rooms, the aliens' general demeanor and individual personalities, and the way they spoke to Betty in English but to Barney via telepathy. Betty wrote all of this based only on what she claims were her dreams, and probably told the story to Barney over and over again until his ears fell off over a period of two years, before they ever had any hypnosis.[1]

Dunning also notes that Betty's original "memory" of the aliens was altered after an interesting media event:

> Betty's written description of the characters in her nightmare depicted short guys with black hair and "Jimmy Durante" noses [Fig. 7.1A]. It was only in Barney Hill's hypnosis sessions that we got the first description of skinny figures with gray skin, large bald heads, and huge black eyes. After Betty Hill heard these sessions, suddenly her own hypnosis accounts began to describe the same type of character, and from that moment on, she never again mentioned her original Jimmy Durante guys. Many modern accounts wrongly state that her original nightmares also described grays.[2]

So where did Barney's—now classic—description of the alien "grays" come from? Barney first described and drew the "gray" type alien during a hypnosis session on February 22, 1964. Remember that the *Outer Limits* episode "The Bellero Shield," featuring its prototypical classic gray alien (Fig. 7.1B), aired on February 10, 1964, less than two weeks before the hypnosis session in which Barney described the aliens who abducted him and Betty as small men with gray skin, large, bald heads, and huge black eyes.

Further evidence that Barney's gray aliens derived from "The Bellero Shield" is his reference to their telepathic communication. He said that

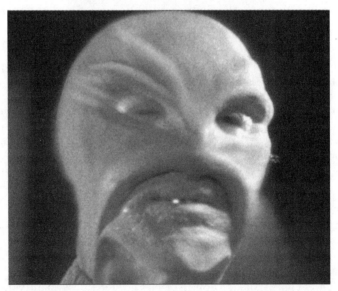

FIGURE 7.1. Who's the real alien? (A) Jimmy Durante: Betty Hill's initial description of her extraterrestrial abductors was of short men with bluish lips, black hair, and large "Jimmy Durante" noses. (© Pictorial Press Ltd/Alamy stock photo.) (B) Only after the airing of *The Outer Limits* episode "The Bellero Shield" did Barney Hill "remember" his abductors as grey aliens. Pictured is John Hoyt as the "Bifrost alien." (*The Outer Limits*, season 1, episode 20, "The Bellero Shield.")

the aliens seemed to speak with their eyes. In "The Bellero Shield," when Judith, Richard Bellero's ambitious wife (played by Sally Kellerman), asks the alien whether he can read her mind and how he can understand English, he tells her that he understands her by reading her eyes and adds, "In all the universes, in all the unities beyond all the universes, all who have eyes have eyes that speak."

Betty's description of the aliens inserting a needle into her navel might have an older ancestor. Here, from the 2013 episode of *Unexplained Mysteries* titled "Touched by an Alien," is what Betty said in a tape recorded while she was under hypnosis:

> The examiner has a long needle. It's bigger than any needle I've ever seen. And he sticks it into my navel. And I'm crying and I tell him, "It's hurting! It's hurting! It's hurting! Take it out!"[3]

This is strikingly similar to the testimony of the ecstatic vision experienced by the Spanish Carmelite nun St. Teresa of Avila (1515–82), in which she encountered a beautiful angel:

> I saw in his hand a long spear of gold, and at the iron's point there seemed to be a little fire. He appeared to me to be thrusting it at times into my heart, and to pierce my very entrails; when he drew it out, he seemed to draw them out also, and to leave me all on fire with a great love of God. The pain was so great, that it made me moan; and yet so surpassing was the sweetness of this excessive pain, that I could not wish to be rid of it.[4]

St. Teresa's ecstatic vision, as captured in the famous statue (Fig. 7.2) by Gian Lorenzo Bernini (1598–1680), shows her, as the angel thrusts his spear toward her body, in a state of ecstasy not unlike that of a woman in the throes of an orgasm. Certainly the imagery of her body being pierced, undergoing a torment that made her moan but that was of a "surpassing sweetness," calls to mind orgasmic ecstasy. In St. Teresa's day, people saw visions of angels and demons. Today, in our scientifically oriented and secularized society, people prone to see visions experience extraterrestrial aliens. The neurophysiology of visionary experiences and nightmares is the source of not only much of the imagery of these visionary experiences but also the conviction of their reality by those experiencing them.

FIGURE 7.2. Bernini' Ecstasy of St. Teresa of Avila—like Betty Hill, pierced in the abdomen by a heavenly being. (By Alvesgaspar (Own work) [CC BY-SA 4.0 (http://creativecommons.org/licenses/by-sa/4.0)], via Wikimedia Commons.)

OTHER ABDUCTIONS

However, before considering the nature of such visionary experiences, let us look at some of the details common to many other alien abduction stories, as related by abductees in "Touched by an Alien." All these encounters take place at night, often on dark, lonely roads, sometimes in their bedrooms. Many times the alien examinations are intrusive, often involving the implantation of foreign objects into their bodies. Significantly, the abductees often report a strange paralysis during the encounter. Here is what "Judy" (a pseudonym) says about her first encounter with an alien:

> I woke up in the middle of the night and I felt that someone was beside the bed, and I reached out my right hand and I touched the shoulder of someone who was probably the size of a five year-old. I was paralyzed. I couldn't move. It was terrible; but I didn't know at the time what was going on, because I didn't know about alien abductions.[5]

"Judy" is not alone in experiencing this feeling of paralysis. Consider the report of the anonymous abductee in MUFON case #32262. After relating an incident in which he, then five years old, fell off an outer railing on a second story and experienced a soft landing, the 41-year-old witness tells of an event that happened when he was 24 (emphasis added):

> At the age of 24: I was away at college and living in an old c.1894 house with four other students. For about two months at least 2 mornings per week I had a series of waking up and being completely unable to move anything except for my eyes for approx 30 minutes. Another thing was the "weight" that heavily pressed my body into the mattress; it felt like someone was on top of me. That was a disturbing times [sic] but it ceased at two months. Not ever was I into "ghosts" but that made me serious about wondering if they really exist. I searched the old house at Humboldt State University's library and found that the 1894 house was built right over the ashes of a previous building. The wondering about a ghost faded as I was in an intensive program that needed my focus.
>
> 2 years ago: I woke up in my bed and looked at my watch seeing that three days had passed with no recollection of anything. I was extremely upset about it and went to my Doctor who took a blood test and everything was normal. I did not tell him about why I was so frazzled. In addition I had multiple dreams where I was unable to move except for turning my head. It was a very strange room with two other tables with people laying on them and again there was a being 8–9 feet tall wearing an outfit with a hood and shielding for the face working on the body to my right. I can recall turning my head to the left where small rectangle windows looking down at Clam Beach from I guess around 7000 feet, just a guess. During the two years I had multiple dreams like that.[6]

Like "Judy," the anonymous witness woke in the night unable to move. As we will see, this experience of feeling a weight pressing down on his body is highly significant.

Probably the most celebrated abduction, other than that of Barney and Betty Hill, is that of author Whitley Strieber, which he recounted in his best-selling 1987 book *Communion* (followed by several others in a series). For all the hype it has been given, his alien encounter is little more than a replay of the usual scenario. In 1985 Strieber, his wife and young son spent the Christmas holiday at their cabin in an isolated forest area in upper state New York. After retiring for the evening on the night of December 26, he was awakened by a strange sound and saw a shadowy being in his room. After that, he found himself sitting in the woods surrounding his cabin. His memories of what had occurred were fragmented. To reestablish memories of that night and understand what had happened, he underwent regressive hypnosis by Dr. Donald F. Klein. Under hypnosis, he recalled being paralyzed and levitated out of his bedroom into a craft of some sort. There he saw a number of different types of beings, the most notable of which was a slender, weak-looking one that had mesmerizing, slanted black eyes. A picture of this being appears on the front cover of *Communion*. It is a typical "gray." Under hypnosis Strieber remembered the aliens thrusting a needle into his brain, then inserting an object into his rectum. The aliens also took a blood sample from his finger. Dr. Klein diagnosed Strieber as suffering from temporal lobe epilepsy, a condition that can cause hallucinations, hypergraphia (excessive, compulsive writing or drawing), euphoria, and a sensation of sudden insight or revelation, often of a religious nature. Like the Hills, Strieber recalled the details of his encounter only under hypnosis.

Prophetic visions are another staple of alien abductions. Many times the abductees are shown horrific visions of the future involving either natural disasters, climatic or geologic, or nuclear holocausts. Whitley Strieber reported the aliens' having shown him apocalyptic visions of what appeared to be a nuclear war. Other abductees are warned that the human race has become estranged from nature and needs to drastically change its ways or it will face extinction. An abductee identified in the *Unexplained Mysteries* episode only as "Joe" tells of the stern message delivered by the aliens after they imparted to him a vision of a possible future apocalypse: "The message is: 'Wake up. Grow up. The sandbox you're littering is not your own, that you are not alone.'" Other abductees voice similar warnings, saying that the human race has become estranged from nature and needs to drastically change its ways or face extinction. One can certainly sympathize with such a message. However, if the benevolent space aliens really want us to change our ways, why relate the warning to ordinary people on lonely roads in the middle of the night? Wouldn't it be much more effective to hover over Washington, DC, at midday and rap the knuckles of recalcitrant congressmen and senators?

Sometimes, however, the prophecies are not quite so apocalyptic. Often the aliens only predict that great changes are coming. Consider a supposedly anonymous alien message, which we found on *YouTube*, a portion of which states:

Like all conscious races at this stage of progress, you may feel isolated on your planet. This impression makes you sure of your destiny. Yet, you are at the brink of big upheaval that only a minority is aware of.[7]

Here is a great psychological hook wherein believers can fill fulfilled be the quasi-religious nature of UFO cults: *You, being aware of the coming New Age, are a member of an elite cadre.*

MEMORIES: REPRESSED, IMAGINED, AND IMPLANTED

Common to the abduction narratives of both the Hills and Whitley Strieber is that they were initially repressed and only later recovered and reconstructed under hypnosis. Herein lies a problem. Although the doctrine of repressed memories of traumatic events, particularly those in childhood, is a mainstay of psychoanalysis, there is compelling evidence that such memories are not memories at all but rather are fabrications, fantasies often induced by leading questions from those "uncovering" the memories. Actual traumatic memories are usually not repressed. Dr. Daniel L Schacter says of them:

> Evidence concerning memory for real-life traumas in children and adults indicates that these events—such as the Chowchilla kidnappings, the sniper killing at an elementary school or the collapse of sky walks at a Kansas City hotel—are generally well remembered. . . . Complete amnesia for these terrifying episodes is virtually non-existent.[8]

The Chowchilla kidnapping occurred near Chowchilla, California, on July 15, 1976. The kidnappers stopped a school bus and abducted 26 children and their adult driver. They loaded their captives into two separate vans, drove them to a rock quarry in Livermore, California, then loaded them into a truck that they then buried. They left them with little food and water and a few mattresses. The bus driver, Frank Edward "Ed" Ray, piled up mattresses that were in the truck and held the oldest child, who was 14, up to the opening while the boy dug out the ceiling, allowing himself and the others to escape. Surely being imprisoned in a buried truck and left to potentially suffocate would be the sort of memory children would repress?

In her 2013 TED talk (TED stands for Technology, Entertainment, Design), "The Fiction of Memory," cognitive psychologist Elizabeth

Loftus deals with the faultiness of memory, beginning with the tragic case of Steve Titus. At the age of 31, Titus, a restaurant manager in Seattle, was engaged to be married with the love of his life. His future would seem to have been a happy one. However, he was pulled over by police because his car resembled that driven by a man who had raped a female hitchhiker. The police photographed him, and the victim, looking at his picture in a photo lineup, said he was the closest match to the rapist. Titus was put on trial for the rape, and the victim positively identified him as the rapist. He was convicted on her testimony. However, he managed to enlist the help of an investigative journalist, who eventually found the real rapist. Titus was freed but had lost his job, which he was unable to get back. He was also embittered to the point that his rage estranged his fiancée. He also lost all his savings. He sued the authorities but died of a stress-related heart attack days before his suit was due to go to court. He was only 35 years old. Loftus goes on to note a project in which 300 people convicted of crimes they didn't commit were freed based on DNA evidence. Unfortunately, many of them had been in prison for 10—30 years before being exonerated. In three-quarters of the cases, their convictions were the result of faulty eyewitness testimony. Explaining how eyewitness testimony could be so wrong, she says the following:

> Like the jurors who convicted those innocent people and the jurors who convicted Titus, many people believe memory works like a recording device. You just record the information, then call it up and play it back when you want to answer questions or identify images. But decades of work in psychology have shown that this just isn't true. Our memories are constructed. They're reconstructed. Memory works a little more like a Wikipedia page. You can go in there and change it; but so can other people.[9]

Loftus goes on to recount an experiment in which people viewed a simulated accident. Those conducting the experiment asked some of the subjects how fast the cars were going when they hit each other. They asked others how fast the cars were going when they *smashed* into each other. Those who heard the more graphic, emotionally charged word *smashed* not only estimated higher speeds on impact but also testified to seeing shattered glass on the pavement that simply wasn't there. A leading word, or an indication of what the investigator is looking for, can slant testimony. Investigating the sharp rise in cases in which people going into psychotherapy for problems such as depression seemingly recovered memories of, among other things, satanic ritual sexual abuse, Loftus found that the factors leading to the construction of false memories were imagination exercises, dream interpretation, hypnosis, and exposure to false information.

Probably the most egregious case of "recovered" memories that were, in fact, implanted was that of the McMartin Preschool trial. The accusations against members of the McMartin family of egregious, pro-tracted child molestation and Satanic ritual abuse were prompted by baseless accusation on the part of a parent of one of the children, Judy Johnson—who, it was later learned, turned out to be suffering from paranoid schizophrenia and who eventually died in 1986 of complica-tions arising from chronic alcoholism. The accusations were first made in 1983. Based on what turned out later to be entirely fabricated memories, prosecutors charged Virginia McMartin, Peggy Ann Buckey, Mary Ann Jackson, Betty Raidor, Babette Spitler, Peggy McMartin Buckey, and Ray Buckey with multiple counts of child abuse and molestation. The pretrial investigation went from 1984 to 1987. The trial went from 1987 to 1990. In the end, there were no convictions, and all charges were dropped. For all that, Ray Buckey spent five years behind bars, having been denied bail, and the McMartin preschool went out of business. One of those children who reported being sexually abused, Nathan D. Zirpolo, as an adult recanted his testimony and apologized to the McMartins in 2005. He said the following of those eliciting his story:

> Anytime I would give them an answer that they didn't like, they would ask again and encourage me to give them the answer they were looking for.[10]

So the lives of seven people were permanently damaged by memo-ries, elicited from children under pressure, of events that simply never happened.

Now let us consider the "recovered" memories of the Hills and Whit-ley Strieber in light of these studies. We have the exposure of Betty Hill to a book on UFOs, Barney's exposure to Betty's dreams and the gray alien of "The Bellero Shield." Whitely Strieber certainly could have been exposed to all the UFO abduction literature extant between the Hills' 1961 encounter and his own experience 24 years later, in 1985. Harvard psychologist Susan Clancy, author of *Abducted: How People Come to Believe They Were Abducted by Aliens*, refers to what she calls "socially available scripts." In the 24 years between 1961 and 1985, there were a plethora of socially available scripts. Susan Clancy notes that all of the abductees she interviewed remembered the details of their abduction experiences only either under hypnosis or in guided psychiatric sessions. In an interview on NPR's *Day to Day* show (November 9, 2005) she said the following:

> The modal subject I spoke with typically had a number of sleep experiences that they thought were strange. For example, maybe they've had an episode of sleep paralysis. Yeah, it's essentially common. You wake up in the middle of the night, you can't move. It's a very scary experience. It lasts for about a minute; many people have it. And for the alien abductees I spoke with, some of them had that episode and they woke up and they thought, "Oh, my God. What was that?" And then later on in their lives, they read something or saw something on TV and they said, "You know, what happened to me that night is a lot like what happened to Bud Hopkins' subjects or what I saw on 'Close Encounters of the Third Kind.'" Bud Hopkins is an abduction researcher who's written a number of books about the stories that his subjects remember under hypnosis.

Clancy says the following of the notion that hypnosis is a good tool for recovering lost memories:

> This belief is wrong. A wealth of solid research, conducted over four decades, has shown that hypnosis is a bad way to refresh your memories. Not only is it generally unhelpful when you're trying to retrieve memories of actual events, but it renders you susceptible to creating false memories—memories of things that never happened, things that were suggested to you or that you merely imagined. If you (or your therapist) have preexisting beliefs or expectations about "what might come up," you're liable to recall experiences that fit those beliefs, rather than events that actually happened. Worse, neither you nor your therapist will realize this, because the memories you do retrieve seem very, very real.[11]

A MASS ABDUCTION

One objection often raised against treating abduction narratives as the product of sleep paralysis is that mass abductions have occurred in which all of those abducted reported the same story. Of course, there is always the possibility of cross-contamination, as in the case of the Hills, where Betty told Barney details of her dreams. One striking tale of mass abduction occurred on the evening of Thursday, August 26, 1976. Four men—Chuck Rak, Charlie Foltz, and identical twins Jim and Jack Weiner—went camping and fishing along the Allagash River in Maine, setting up camp on Eagle Lake. Before going out in a canoe to fish, they built a large fire at their campsite to act as a beacon. When they were out in their canoe, they saw a ball of glowing, pulsating light that hovered about 200 to 300 feet above the lake. Foltz shined his flashlight at it, blinking an SOS code, whereupon the light began to follow the men. It shined a cone of light on the canoe. The next thing they knew, the four of them were standing on the shore, looking at the light ascend and disappear. They found their fire had burned down to embers. Since they had built the fire with heavy logs, it should have lasted for hours. The men could not account for the missing time.

It wasn't until 1988, 12 years later, that Jack Weiner began having nightmares. His identical twin, Jim, also began having the same dreams. It is noteworthy that Jim Weiner had, in the intervening period, suffered a head injury, which caused him to have temporo-limbic epilepsy. He attended a UFO conference in 1988, where he met Ray Fowler, another abductee and UFO investigator. Fowler urged Jim Weiner to undergo hypnosis with professional hypnotist Anthony Constantino, who had helped Fowler recover his own UFO abduction memories that same year. According to a report taken from a 1993 article in *North Shore Sunday Newspaper* (Salem, Massachusetts), the four men gave a common report (emphasis added):

> Under hypnosis, they described the diffusely lit, sterile interior of the spacecraft, *the spindly-fingered big-eyed bald-headed aliens that Whitley Strieber popularized with his non-fiction book "Communion,"* and strange medical experiments conducted on each man.[12]

Surely, the common testimony of these four men is hard to refute. However, from the reference in the quote to Whitley Strieber's book, *Communion*, we can see that it was not free of cultural contamination. Also, there are strong reasons to doubt the independence of their common testimony. Consider that, as is usually the case in abduction narratives, it comes out through the highly suspect medium of hypnotic regression. In this case, the hypnotic regression was done a full 12 years after the 1976 sighting. Also, Jim Weiner suffered from temporo-limbic epilepsy, a variant of temporal lobe epilepsy, which is characterized by hallucinations of a spiritual nature. Taken together, the 12-year time lapse between the incident and the recovered memory of it, the possible cross-contamination between the twin brothers, the suspect medium of hypnotic regression, the cultural contamination of Whitely Strieber's book *Communion*, and Jim Weiner's temporal lobe epilepsy all cast serious doubts not only on the narrative itself but also on the likelihood of the common narrative's really being the result of four independent corroborating testimonies.

THE ALIEN SEX AGENDA, THE REPTILIANS, AND THE NORDICS

Although the grays in the narratives seen so far seem to be basically benign, if intrusive and cavalier in their attitude toward the rights of those they abduct, other narratives show a less benign, more selfish motive on the part of the grays and other aliens. Consider the abduction of Antonio Villas Boas, a Brazilian farmer, on October 16, 1957, which antedates the Hills' abduction by nearly four years. Boas was plowing his fields at about 1:00 a.m. when he noticed what looked like a bright red star in the sky. It dropped down and Boas tried to drive away only to find his tractor's motor had gone dead. The "star" was an alien craft from which four humanoid aliens emerged. Boas tried to flee but was paralyzed. He was taken aboard the craft and stripped, then had a gel applied to his body. A beautiful naked woman then entered the room. She had an angular face, catlike eyes, and long white hair. However, her pubic hair was bright red. Boas felt himself getting aroused and willingly had sex with the woman. After they had finished, she indicated by signs that he had impregnated her. Boas was then allowed to leave the craft. He reported feeling indignant at being used as a stud stallion. Considering the erotic fantasy nature of this narrative, it's not surprising that it didn't spark any of the notoriety enjoyed by the Hills' abduction. Although not having been used as a stud, Barney Hill did report the aliens' having placed a cup over his genitals and suspected that they had taken a sperm sample from him. One of those interviewed by Susan Clancy, a man identified as Joe, gave a more graphic description of a similar procedure:

> In one of his first memories, he was strapped down on a black marble table. An alien standing to his right with "menacing black eyes," forced a device like a suction cup onto his genitals. Sperm were sucked painfully out of his body.[13]

Most abductions involving sex between aliens and humans do not involve men being used as studs, however. In a 2003 episode of *Unexplained Mysteries*, "Taken: The Abduction Phenomenon," a number of women testify to either having been impregnated by aliens or having had infants extracted from their wombs by aliens. Typical of these is the story of photographer Kim Carlsberg, who relates how, having gone to bed one evening, she awoke paralyzed and was taken aboard an alien craft. There she was impregnated; the embryo was then removed from her womb. UFO researchers speculate that the grays have lost their ability to procreate naturally and need human DNA to reinvigorate their dying race.

Although the grays are generally seen by abductees as benevolent, a more recent phenomenon is that of hostile aliens who are deceptive and inimical to humanity. These are the reptilians. The earliest report of extraterrestrials was that of Herbert Schirmer, a police officer in Ashland, Nebraska, who claimed in 1967 that he had been taken aboard a spacecraft crewed by humanoids with slightly reptilian features (see Fig. 10.5) whose uniforms bore a winged serpent crest. Although Schirmer's aliens were vaguely benign, other reptilians were reported by abductees as malicious.

According to tone of the websites under the umbrella of *Arcturi*, the reptilians exist in a different dimension form ours and are manipulating us through thought control:

> The Reptilians exist in the third and fourth dimension manipulating the minds of society through thought control. Reptilian abduction, therefore, is often a product of mind abduction rather than physically taking the abductee's body into a Starship and flying through space. These abductions can sometimes be experienced in the form of "nightmares."[14]

The website also says that the reptilians are capable of changing their shape for the purposes of deception:

> The shape-shifting nature of Reptilians makes them exceedingly dangerous when abducting human minds. The Reptilian will often use the appearance of other benevolent alien races to disguise themselves. A common example is the use of a Nordic alien body in order to seduce the abductee with a sexual fantasy. Reptilians often use sexually aggressive attacks as the emotion of sex is one of the main drives of human beings. Interestingly, many ancient legends tell of us of great leaders (Julius Caesar, Alexander the Great) being conceived by intercourse between a woman and a serpent. The sexual nature of these attacks can be physical (in the flesh), mental (as a sexual perversion), and in the dream state.
>
> The Reptilians use abduction for feeding purposes. These entities thrive off of human trauma, horror, fear, and submission. Additionally, Reptilian abduction has served the purpose of DNA experimentation, which through the years have developed the human race. The Reptilians enjoy the process of control and this is their purpose for abductions.[15]

The assertion that malign aliens control their victims through sexual fantasies can itself give rise to rather lurid abduction/seduction narratives. Consider that of "Amy," whose story is told by Donald Worley. After describing an angel of light to Worley, "Amy" revealed the shocking truth:

> The angel came back and started pretending like he was just there to hang out with me. Then I fell asleep for 15 minutes unexpectedly in mid-conversation with him. I wake up feeling nauseous and there is white stuff all over my underwear and you know where. The lying thing tells me I fell asleep and he had to have sex with me. He played it off great so then an hour later we were chatting and playing and his reptilian shield thing comes off and after that the jig was pretty much up but he still tried to play it off when I asked him, "Uh huh, you are an evil Reptilian."[16]

Amy goes on to say that, "The nasty reptile caused me to have sex with him."

While being reptilian, these aliens have a generally human form. "Amy" describes them as follows:

> They are very tall, muscular on top, kind of small average sized heads, a little different shaped than humans, eyes shaped like a human's but I think they are red and yellow colored, very long pointed fingers. He looked like a reptile, like a human lizard, alligator skin, snake-like face and features, a mix of them all.[17]

For all that, due to his shape-shifting abilities "Amy" still felt a sexual desire for the reptilian, which led to nearly deadly results, only averted by the arrival of the benevolent grays:

> After a while he laid beside me and I kept asking him to continue and he would not. I kept trying to make him kiss me and he wouldn't talk to me or look at me. After a few minutes of this he told me to run. He said I had to get away from him because he wanted to kill me. He said he is evil by nature and cannot control his feelings.
>
> I asked him to leave my house and he said he couldn't until he killed me. I got out of bed and ran naked into the living room and I begged for Jesus and the Nordics to save me. About 15 minutes of this and the room filled with shadow-like forms of energy that appeared to be the shape of ships. I know this sounds crazy. Out of the ships came little men who looked like they were in some kind of armor. Also, a beautiful fairy- looking lady in a pretty dress came off a ship. She was larger than the little men. One man holding my hand said, "Don't worry, we are all around you." I couldn't see clearly but I knew they came off the ships. The one holding my hand I could tell was a Gray-type alien. I could tell by its size. I was crying and very confused. He told me he lived within my heart and it's ok, I could go back to my room. When I went back in, the reptile was gone and as soon as I laid on my pillow I went right to sleep. I have not heard or seen or been bothered by the trickster energy forms since then.[18]

The "Nordics" from whom Amy sought salvation are described at another website as follows:

> Nordic Aliens (a.k.a. Pleiadeains) are a humanoid extraterrestrial race of beings that are said to look very similar to humans. They are most commonly reported in Europe and in some small cases in the United States. They are allegedly one of the high ranking species in the Galactic Federation of Light.[19]

Although UFO enthusiasts haven't yet proposed that there are evil Jewish aliens, the description of the Nordics seems to imply a certain racial bias among at least some of them.

Just as with the grays, we can probably look to the medium of television for the origin of the malign reptilians, specifically to the 1983 television miniseries *V*, written and directed by Kenneth Johnson. In *V*, a race of seemingly human aliens arrive on Earth in 50 huge ships, which hover over the major cities of the world. The "Visitors," as the aliens are called, ostensibly wish to trade knowledge for Earth's mineral resources. However, it turns out that the aliens are not really human beings. Beneath their humanlike façade—a thin, synthetic skin, and humanlike contact lenses worn in public—they are actually carnivorous reptilian humanoids preferring to eat live food such as rodents and birds. Of course, like all evil aliens, they seek world domination.

A subplot of *V* is that they also wish to interbreed with humans. One teenager, Robin Maxwell, has a sexual relationship with a male Visitor named Brian, who impregnates her during one of the reptilians' "medical experiments."

As in *V*, the reptilians seek more than merely to seduce women. According to *Arcturi*, they are also the secret and nefarious masters of our world. One of the leading proponents of the reptilian UFO theory is David Icke:

> David Vaughan Icke (pronounced /aɪk/; born April 29, 1952) is an English writer and public speaker who has devoted himself since 1990 to researching what he calls "who and what is really controlling the world." . . . One of his main focuses is the influence that extra-terrestrial beings have on our world order, one of these alien races being the Reptilian Aliens or also known as Reptilians, Reptoids, or Lizard people. In The Biggest Secret (1999), Icke introduced the "Reptoid Hypothesis." He identifies the Brotherhood as originating from reptilians from the constellation Draco, who walk on two legs and appear human, and who live in tunnels and caverns inside the earth. They are the same race of gods known as the Anunnaki in the Babylonian creation myth, Enûma Eliš. Lewis and Kahn write that Icke has taken his "ancient astronaut" narrative from Israeli-American writer Zecharia Sitchin.[20]

Icke even claims that certain powerful, well-known families are of mixed human and reptilian descent. Specifically, he asserts the Rothschilds are one such family:

> The Reptilian Alien bloodline streaming through the global elite cannot be seen more clearly than with the aristocratic family the Rothschilds. The Rothschilds family holds, by far, the largest private fortune in modern history. The Rothschilds rose to power under the leadership of Mayer Amschel Rothschild, a German money changer and banker (said to be of the same Reptilian bloodline of Herod the Great). He spread his empire out through his five sons which he spread throughout Europe to "spread the empire." The Rothschilds used reptilian aristocratic technique of successfully keeping fortunes and businesses within the family through arranged marriages and interbreeding. It is not uncommon for Reptilian bloodlines of this rank to keep the aristocratic ties by marrying first and second cousins.[21]

It is painfully obvious that the proponents of this human–reptilian hybrid theory know nothing of biology. Consider the mule. Moreover, consider the hinny, tiglions, and ligers. The mule is a hybrid between a mare and a jackass, and the hinny is a cross between a stallion and a jenny. The horse, *Equus caballus*, and the ass, *Equus assinus*, are two closely related species within the same genus; thus they can be bred together to produce offspring. However, those offspring, mules and hinnies, are sterile. The same is true of tiglions, the offspring of male tigers and lionesses, and ligers, the offspring of male lions and tigresses. The tiger, *Panthera tigris*, and the lion, *Panthera leo*, are, like the horse and the ass, different species in the same genus. (The fact that wolves and domestic dogs can produce fertile offspring has led to their reclassification. Initially, wolves were in the species *Canis lupus* and dogs were considered a separate species, *Canis familiaris*. Now dogs are classified as a subspecies of wolves, *Canis lupus familiaris*.) That separate species as close to each other as are horses and asses or lions and tigers cannot produce fertile offspring together should tell even a layman that humans of Earth and aliens from another star system cannot possibly produce hybrids together. In any case, cold-blooded reptiles aren't likely to produce an intelligent species.

Another biologically unlikely aspect of the space aliens, whether they are Nordics, grays, or reptilians is that they are all remarkably like us, being basically humanoid in structure. There appear to be no intelligent life forms that have, for example, four legs and two arms or multiple appendages or that are structured—even in the case of the reptilians—somewhat like the velociraptors of the movie *Jurassic Park*. It seems highly unlikely that beings from another solar system would be so human in structure and appearance.

Having placed the biologically unlikely (if not impossible) reptilians within the sphere of conspiracies running the world, Icke has no problem linking them with feared secret societies:

> The Rothschilds were said to be major players in the Bavarian Illuminati, and one of Amschel's sons is confirmed to be a member. When the Bavarian Illuminati was [*sic*] exposed by several government interest groups the power shifted to a Masonic group called the Alta Vendita, which was led by Karl Rothschild another of Amschel's sons. Letters from this group are confirmed to have been sent out

> by this powerful Reptilian leader to the headquarters of the Masonic order of
> Europe.... The occult powers of both white and black magic encouraged by
> Reptilians are said to be the root of the Rothschilds' great wealth. Additionally,
> numerous ties to the illuminati, the Masonic Order, and other Occult groups
> combined with the interbreeding gives us clear indication of their Reptilian
> Agenda.[22]

The Illuminati and the Masons have long been stock bogeys of fundamentalist end-times prognosticators. This leads us naturally to the fundamentalist Christian take on UFOs and alien contacts.

THE ALIENS AS THE NEPHILIM

Though "Amy" did beg Jesus, along with the Nordics, to come to her
aid against her reptilian oppressor, the spiritual emphasis of UFOs is
not generally biblically oriented. In an article in *The Christian Research
Journal*, Bobby Brewer of the Christian Research Institute (CRI), a fundamentalist think tank, decries the lack of Christian references in UFO
encounters, stating that they smack of Eastern mysticism and New Age
ideology and make no reference to Jesus—Amy's cry for help not with-
standing. Thus, he declares, those bearing UFO messages of brotherhood and love are, essentially, false prophets. Though acknowledging
the reality of UFOs, Brewer doesn't acknowledge extraterrestrial intelligence (abbreviated as ETI) and says, "A dogmatic belief in ETI simply
cannot be justified from Scripture." (Of course, we really can't justify
the second law of thermodynamics from Scripture, either.) He views
the implication that a superior extraterrestrial civilization is interested
in our progress and could help us solve such pressing problems as the
threat of nuclear war or environmental degradation as antagonistic to
the Christian message:

> The presupposition that an advanced alien race could solve the world's problems
> is misleading and deceptive. Such a hope is a cloud without rain. Christ alone is
> able to provide a solution to the problem of sin, and we have been commissioned
> to spread this good news to the uttermost parts of our planet. We must focus our
> attention on the Gospel.[23]

Although Brewer's article in the *Christian Research Journal* is somewhat restrained, his acceptance that UFOs are real, combined with his

rejection of the idea of extraterrestrial intelligence, implies a satanic source for UFOs. Fundamentalists often quote a passage in 2 Corinthians, in which Paul warns of false apostles, as being applicable to the benign messages of UFO encounters (1 Cor. 11:13–15):

> For such men are false apostles, deceitful workers, disguising themselves as apostles of Christ. And no wonder, for even Satan disguises himself as an angel of light. Therefore it is not surprising if his servants also disguise themselves as servants of righteousness; whose end shall be according to their deeds.

Thus the benign Nordics and grays, masquerading as angels of light, are satanic. So rather than dismissing UFO encounters as New Age fluff, many fundamentalist Christians regard them as satanic manifestations.

Considering that fundamentalists, as Protestants, generally embrace the principle of *sola scriptura* ("scripture only")—meaning that all doctrine is to be based on the Bible—it is surprising that the sole biblical support for the notion that UFOs are demonic in origin is the rather tortured interpretation of a single, rather minor and obscure, biblical passage, Genesis 6:2–4, rendered in the King James Version (KJV) translation into English as follows:

> And it came to pass, when men began to multiply on the face of the earth, and daughters were born unto them that the sons of God saw the daughters of men, that they were fair and took them wives of all which they chose. And the LORD said, "My spirit shall not always strive with man, for that he also is flesh: yet his days shall be an hundred and twenty years. There were giants in the earth in those days and also after that, when the sons of God went in unto the daughters of men, and they bare children to them. The same became mighty men, men of renown, which were of old.

It's rather obvious, looking at this passage, that Genesis 6:3, regarding the 120-year lifespan of human beings is out of place and disrupts the flow of the other verses. It is, possibly, an interpolation. If we remove it, the passage flows more naturally:

> And it came to pass, when men began to multiply on the face of the earth, and daughters were born unto them that the sons of God saw the daughters of men, that they were fair and took them wives of all which they chose. . . . There were giants in the earth in those days and also after that, when the sons of God went in unto the daughters of men, and they bare children to them. The same became mighty men, men of renown, which were of old.

This passage was considered problematic by the Church in the fourth century, because it implies that angels have bodies and had acted much the way the pagan gods did. Accordingly, the Church fathers of that day invented what was called the "Sethite" doctrine—that "sons of God" meant men of the godly line of Seth, the third son of Adam and Eve, and that "daughters of men" meant women of the ungodly line of Cain. However, such an interpretation would not explain why their offspring were giants. To this, defenders of the Sethite doctrine would answer that the word translated as "giants" in the KJV is, in Hebrew, *nephilim*, which may mean "fallen ones."

Still, the passage refers to the Nephilim as mighty men, men of renown, which were of old. In his first century recap of Genesis in *Antiquities of the Jews*, Josephus says of the Nephilim that

> many angels of God accompanied with women and begat sons that proved unjust and despisers of all that is good on account of the confidence they had in their own strength; for the tradition is that these men did what resembled the acts of those whom the Grecians call giants.[24]

By the time of Josephus, the first century of our era, the angels who had sexual relations with mortal women had come to be regarded as specifically *fallen* angels, though the passage in Genesis 6 makes no reference to that. One of the main reasons for their fallen status was the proliferation of Jewish apocalyptic literature beginning ca. 200 BCE. Collectively called the *Pseudepigrapha* ("falsely inscribed"), most of these books had in common tales of an ancient failed angelic revolt against God, prophecies of a future final conflict between God and the fallen angels, and attribution of the text to a patriarch, usually from the book of Genesis. Among the books of the Jewish Pseudepigrapha are the Testament of Abraham, the Assumption of Moses, and the Testament of the Twelve Patriarchs. However, the most influential of these was the Book of Enoch, named for an obscure antediluvian patriarch. Other than relating that Enoch was the son of Jared and father of Methuselah, all that Genesis says of him is this (6:23, 24):

> And all the days of Enoch were three hundred sixty and five years. And Enoch walked with God and was not; for God took him.

The cryptic statement that Enoch walked with God and was not, for God took him, implies that Enoch did not die but that rather he was bodily assumed into heaven. Thus the author (or authors, since the book shows signs of addition and revision by more than one author) of the Book of Enoch was free to use him as the narrator and witness of great past events and the prophet of great future events. After Enoch is taken up to heaven, he witnesses an angelic revolt led, variously, by Shemihazi and Azazel. These are leaders of the "watcher" angels, assigned by God to watch over and guide the human race after the expulsion of Adam and Eve from the Garden of Eden. Instead, the watcher angels corrupt humankind, teaching them how to make weapons of war, teaching the women how to use cosmetics and perfumes to (deceptively) enhance their beauty and desirability, and, finally, committing the ultimate act of disobedience: having sexual relations with mortal women.

One would think that the Nephilim, "the mighty men, men of renown, which were of old," would all have been wiped out in the flood. However Genesis 6:4 says (emphasis added):

> *There were giants in the earth in those days and also after that, when the sons of God went in unto the daughters of men.*

So the giants remained after the biblical flood. In fact, in the opening chapters of Deuteronomy, they are variously referred to as the Nephilim, Rephaim, Zamzumim, Emim, and the Anakim. One of the last of the Rephaim is Og, king of Bashan, of whom Deuteronomy says the following (Deuteronomy 3:11):

> For only Og, king of Bashan remained of the Rephaim; behold, his bedstead was of iron: Is it not in Rabbath of the Ammonites? Nine cubits was the length thereof and four cubits the breadth of it after the common cubit.

A cubit is roughly a foot and a half. So Og's bedstead would have been 13½ feet long and 6 feet wide, making Og about 12 feet tall. The Philistine giant Goliath of Gath, along with his brothers, is supposed to have been one of the Anakim. According to 1 Samuel, Goliath was six cubits and a span in height (17:4). A span, half a cubit, is nine inches, so Goliath would have been nine feet nine inches tall—a giant indeed. However his height in the Septuagint (abbreviated LXX), the Hebrew scriptures translated into Greek *ca.* 100 BCE, is given as four cubits and a span, or six feet nine inches tall. Protestant Bibles, such as the KJV, use the Masoretic Text (abbreviated MT) as the source of their Old Testament. The MT is the official Jewish Bible, a purified canon edited by a rabbinical college between CE 600 and 900. Thus the LXX represents an earlier tradition than does the MT. Accordingly, Goliath could be seen as merely exceptionally tall, a giant in less than supernatural terms.

Regardless of how large we view the Nephilim as having been, however, they are part of an ancient tradition in which divine beings produced mixed offspring by marriages with mortal women. The apocalyptic overlay of the Nephilim being the offspring of mortal women and *rebellious* angels, coupled with a fundamentalist dogma that the beings of UFO sightings must be of satanic origin, has resulted in certain fundamentalists' seeing them as the Nephilim, living in another dimension and attempting to return to ours to subjugate the human race, in the process suborning the gullible with an evil New Age agenda. All this is seen as part of an end-times scenario. One of the chief proponents of this view is Chuck Missler, sometimes writing partner of Hal Lindsey. Investigative reporter Richard Abanes says the following of Missler:

> A final aspect of Missler's prophetic view of the end-times that must be mentioned is his preoccupation with UFOs. He believes that UFO occupants are actually demonic spirits invading our earthly dimension in hopes of creating a hybrid race via sexual intercourse with women (and possibly men). These inter-dimensional angels of darkness have supposedly committed such despicable acts in the past. Missler contends that Genesis 6:1–4 clearly describes fallen angels ("Sons of God") taking women ("daughters of men") as wives in order to produce a race of giants Nephilim (Hebrew for "mighty ones") Missler comes to a predictable conclusion: "I expect the Antichrist, when he shows up to be either an alien or connected with one."[25]

Missler's quote is from his article "The Return of the Nephilim," published in his newsletter, *Personal Update.*[26] We should stress here that although many fundamentalists do see UFOs as a manifestation of the satanic, not all of them see the UFO occupants as either the Nephilim or fallen angels attempting to re-create the Nephilim. For example, Richard Abanes, who finds Missler's views to be on the fringe, is himself a fundamentalist Christian.

The Roman Catholic church is less obsessed with millennial prophecies than are many fundamentalist Protestants and generally accepts the
idea that God would have populated planets in other solar systems with
intelligent life forms. Some of these, the church speculates, might not
share humanity's fallen nature. For all that, there are Roman Catholics
whose view of UFOs mirrors that of fundamentalist Protestants. Here
is what James Hough has to say in an article titled "What Is the Catholic
Church's View on Aliens?":

> On the other hand, most "encounters" with "aliens" that you read about are either
> one of two things. Either they are the result of an overactive imagination OR (and
> this is very real) they are interactions/viewing/or whatever of demons. When
> God created angels, He created many of them—look at the number of planets
> He created! He is a loving God who is not stingy with His gifts. A multitude—
> perhaps millions, perhaps trillions, perhaps more, we have no idea—rebelled
> against Him and wanted to be "gods" themselves, deciding what was good and
> evil. They were thrown out of heaven, and hell came into existence. Because of
> God's great love, and the fact that He can always bring a greater good out of evil,
> He has allowed them to tempt us. We have free will, and a choice. In allowing the
> demons (fallen angels) to tempt us, they have manifested themselves throughout
> history. As they are fallen, they are twisted, and imperfect in some way.[27]

So, although more restrained than certain fundamentalists, certain
Roman Catholics share an antagonistic view toward extraterrestrial
abductions, considering them the deceptive works of fallen angels
rather than merely dismissing them as the products of sleep paralysis,
compounded by low-grade science fiction via movies and television and
distorted by false memories via hypnosis. Perhaps such a view is partially justified by the fact that otherworldly abductions antedate UFOs
by centuries.

SUPERNATURAL ABDUCTIONS

In 1645, in Cornwall, a teenage girl named Anne Jeffries was found
unconscious. Once she was revived, she said that she had been assaulted
by little men. She said she was unable to move and that the little men
took her up to their castle in the air, molested her, then sent her home. As
Anthony Enns notes in his 1999 article "Alien Abductions: A Return to
the Medieval," this incident and others from before the 20th century are

remarkably similar to modern alien abduction stories. Although direct
sexual molestation isn't always experienced by abductees, the intrusive
nature of the abduction may involve indirect sexual violation. Specifically, Enns notes the following:

> But there are other, even more disturbing similarities between pagan myths and
> abduction narratives, such as the aliens' use of long needles inserted into the
> vagina, anus, navel or abdomen. Similarly, the penis of the incubi, a demon who
> visited women in their dreams and molested them, was said to be long, metallic,
> and "ice-cold" Russell Robbins cites the testimony of Antide Colas, a witch who
> claimed that her incubus entered through a hole below her navel, and she had a
> scar there as evidence.[28]

Remember the needle inserted into Betty Hill's navel. There are also
similarities between the reptilians attempting to produce a hybrid, the
Nephilim as hybrids between angels and humans, and the supernatural
visitations of the past:

> Other similarities include the collection of eggs and semen and the insemination
> of women (supposedly for the purposes of creating a hybrid species). Thomas
> Aquinas was the first to document that "devils do indeed collect human semen"
> (Robbins 255). Aquinas believed that the succubus collected the semen and
> then transformed into an incubus, thus transferring the semen to a woman.
> This concept was supported for hundreds of years until Heinrich Kramer, in his
> *Malleus Maleficarum* (1486), described the incubus' ability to mate with women.
> (This idea reappears in the 17th century in Fr. Sinistrari's *De Daemonialitate et
> Incubis et Succubis*.)[29]

In fact, along with the Nephilim, heroes of many cultures were divine—
human hybrids, such as Heracles and Achilles. The sexual act of divine
procreation with a mortal female was often in the form of a sexual assault,
as in the case of Leda and the swan (which was Zeus in disguise).

In 1989 folklorist Thomas Bullard compared abduction narratives
with older supernatural tales of being kidnapped and transported to the
otherworld. He noted that they held in common eight episodes:

· Capture
· Examination
· Conference
· Tour

- Otherworld journey
- Theophany (encounter with a divine being)
- Return
- Aftermath

Theophany is particularly significant when considering the prophetic warnings of the extraterrestrials and their injunctions to the abductees to carry what is, in essence, a divine message to the rest of the human race.

Another feature common to supernatural and alien abduction narratives is the feeling of being completely paralyzed while the event is taking place and of trying to scream or call for help but being unable to make a sound. Here we find a link between myth and human physiology.

SLEEP PARALYSIS AND THE HYPNAGOGIC STATE

Now let us consider sleep paralysis and the hypnagogic state, so eloquently evoked by "Judy." When we fall asleep and begin to dream, experiencing what is characterized as REM (rapid eye movement) sleep, our nervous systems institute a safety feature: the body is somewhat paralyzed at the level of the pons during REM sleep, apparently to keep the body from acting out our dreams and possibly injuring us. The pons is that portion of the brain stem, just above the medulla oblongata, connecting the brain and spinal cord. Hence a suppression of the pons function effectively, but only temporarily, disconnects the brain from the voluntary muscles. Typically, sleep paralysis is experienced when someone's sleep is disturbed and he or she wakes suddenly from REM sleep. In the hypnagogic state the brain is awake, but the pons is still suppressed, causing the newly awakened sleeper to experience paralysis and an inability to breathe deeply, which in turn leads to a sensation of suffocation. This, in turn, leads to panic. In the hypnagogic state, one may sense the presence of another person or being in the room. In ancient times, this was experienced as a demon. Today it is often experienced as alien abduction. The disassociation of brain and body experienced in this state can also give rise to out-of-body experiences.

During sleep paralysis, owing to the partial suppression of the voluntary control of breathing, the person in that physiological state can feel not only as if he or she is suffocating but also as if something or someone is pressing down on his or her body. This has given rise to a rich world-wide folk tradition of the incubus, the succubus, and the nightmare. Related to these motifs is that of the demon lover, particularly because both the incubus and succubus have sexual connotations. The demon lover's assault on the object of its desire is both sexual and oppressive. The word *incubus* derives from the Latin verb *cubare*, "to lie," coupled with the prepositional prefix *in*, which, in Latin, can mean either "in" or "on." Thus *incubare* means "to lie upon," whereas *succubare* (*sub* + *cubare*) means "to lie under"—the succubus being the feminine counterpart of the incubus.

Like the incubus, the nightmare was seen as lying upon or even trampling its victims. In chapter 16 of the *Ynglinga Saga*, the mythic material prefacing *Heimskringla*, the medieval Icelandic chronicle of the kings of Norway by Snorri Sturluson (1179–1241), a king named Vanlandi is killed by such a spirit. When he lies down to sleep, he suddenly cries out that Mara (the nightmare) is treading on him. His men try to save him by holding on to his upper body, but she treads on his legs. When they lay hold of his legs, she crushes his head. The Icelandic Mara may well be related to the Irish Morrígan, seen as a goddess of death and battle and sometimes depicted with a mare's head. The oppressive spirit that sits upon sleepers, threatening to crush the life out of them, is known in China as *pin yin gui ya chung*, meaning "ghost pressing on body." Among the Hmong people of Southeast Asia the spirit is called *dab tsog*, or "crushing demon." In Turkish folklore the demon is called *kara basan*, meaning "black presser." In Japan the experience is called *kanashibari*, "bound in metal," and in Korea it's called *gawee nulim*, "being pressed by scissors." The Shona people of Zimbabwe call the experience of being pressed down in sleep *madzikirira*, or "witch pressure." The Yoruba of Nigeria call it *ogun oru*, or "night war," and see it as a battle between one's spirit-world spouse and the earthly spouse. The Anglo–Swiss artist Henry Fuseli (1741–1825) captured this image in his famous 1781 painting *The Nightmare* (Fig. 7.3).

Thus the experience of "Judy," awakening in darkness and feeling paralyzed while experiencing the sensation of an intrusive presence in the room, is based on human physiology and has a rich mythic background.

FIGURE 7.3. An image of hypnagogic paralysis: *The Nightmare*, by Henry Fuseli, 1781, original painting in the Detroit Institute of Art. (Public domain; image by Hohum, via Wikimedia Commons.)

In our modern mythology, space aliens have replaced demons, incubi, succubi, and nightmares. Consider also that the paralysis experienced by Betty Hill was more likely a product of waking in the midst of a nightmare than anything that happened on the road near Indian Head. Likewise, Barney's experience of paralysis was likely part of a hypnagogic state induced by hypnosis.

SUMMARY

What has happened to many of those who believe they were abducted by alien UFO occupants is a profound experience, embedded in the physiology of their nervous systems, and is often a transforming experience, awakening them to a consciousness of the profound consequences of human actions, whether these lead to nuclear war or environmental destruction. It also may well open the consciousness of those otherwise entrained in the petty and the mundane to an appreciation of the cosmic and the numinous. Although the abduction experience seems initially to be a phenomenon beginning in the middle of the 20th century, the similarities of alien abductions to the supernatural abductions of earlier centuries tells us that such experiences are part of a universal mythic tradition embedded in the human psyche and neurophysiology. Such narratives embody deep emotional fears and longings common to all humanity. Thus those of us who doubt the scientific validity of alien abductions, characterized as they are by a singular lack of supporting evidence, should not do so with either contempt or condescension. Rather, lest these internal realities become fodder for the purveyors of cheap sensationalism, it must be our quest, and our duty, to subject the emotional reality of alien encounters to the rigors of objective reasoning and the demands of evidence—but to do so with compassion.

8

The Mythos of Ancient Aliens

In 1968 a sensational book came out that rocked the world. It showed what appeared to be compelling evidence of frequent visitations to Earth by extraterrestrial beings throughout human history, but particularly in ancient times. The book also strongly suggested that these visits were purposeful and that the extraterrestrials had shaped human civilization. According to this book, the "gods" of many mythologies, particularly those gods who gave us fire and taught ancient peoples the technology of metalworking, were in fact space aliens. That book was, of course, *Chariots of the Gods?*, by Erich von Daniken.

After the initial sensation provoked by the book had died down, another book, *The 12th Planet*, written by the late Zecharia Sitchin (1920–2010), appeared in 1976. Sitchin argued that an undiscovered planet, Nibiru, exists beyond the orbit of Neptune. Nibiru has a highly elongated, elliptical orbit that brings it close to the sun roughly every 3,600 years. He argued that Marduk, the Babylonian king of the gods, was actually an alternate name for Nibiru. He further argued that on one of these passes, one of the moons of Nibiru/Marduk collided with another planet, called Tiamat, originally orbiting between Mars and Jupiter. This collision split Tiamat in two, and half of it hit Nibiru/Marduk itself. That half was shattered by the impact and formed the asteroid belt; the other half again collided with one of Nibiru's moons, was knocked into a closer orbit around the sun, and became the Earth. Sitchin also said that Nibiru/Marduk was inhabited by a technologically advanced humanlike race called the

Anunaki (a class of deities from Mesopotamian mythology) and that, beginning about 450,000 years ago, some of these Anunaki were sent to Earth to mine minerals, particularly gold. These Anunaki, fed up with the primitive working conditions on our planet, mutinied. Enki, one of the leaders of the Anunaki, suggested that they replace these workers by creating slaves to work in the gold mines. They did this by genetic engineering, mixing their genes with those of the most advanced native hominid at the time, *Homo erectus*, thus creating *Homo sapiens*—us. Of course, things didn't go as planned. Sitchin attributed the evil wind that destroyed Ur (ca. 2000 BCE) in the *Lament for Ur* to fallout from a nuclear war between rival factions of the Anunaki. Not surprisingly, Sitchin also equated the Anunaki with the biblical Nephilim.

A TRADITION OF CATASTROPHE— AND PSEUDOSCIENCE

By mingling von Daniken's ancient astronauts, through the reinterpretation of Mesopotamian myths, with the motif of colliding planets, Sitchin reinvigorated a scenario earlier proposed by Immanuel Velikovsky (1895–1979) in *Worlds in Collision* (1950) and subsequent works. Velikovsky theorized that Earth experienced a number of near collisions with Venus, which he characterized as a comet that erupted out of Jupiter. This would have been an unlikely event considering that Venus, like Mercury, Earth, and Mars, is a rocky planet, and Jupiter is a gas giant. It is noteworthy that Velikovsky was not a scientist. He was a psychologist and psychoanalyst. He theorized that the myths of ancient people were the expression of the experience of these near collision events as repressed memories. That is, the events were so traumatic that memory of them was repressed but bubbled up from the subconscious in mythic form.

Among the disruptions caused by near passes of Venus, according to Velikovsky, were the plagues brought on Egypt and other miracles of Exodus. For example, he explained manna falling from the sky to nourish the Israelites in the desert, as carbohydrates formed in our atmosphere when it mingled with the hydrocarbon atmosphere of Venus. The hydrocarbons were partially oxidized into carbohydrates (sugars and starches), which subsequently condensed and precipitated out of the sky as manna. Velikovsky's scientific illiteracy reveals itself in this bizarre scenario. First of all, atmospheres aren't very thick. Earth's atmosphere extends only 100 kilometers (about 62 miles) from the surface of the planet, and 75% of its substance is within 11 kilometers (6.8 miles) of the surface. Considering that Earth's diameter at the equator is 12,756 kilometers (7,926 miles), our atmosphere is but a thin film covering its surface. Venus, having a diameter of 12,092 km, is nearly the same size as Earth. Although its atmosphere, at 250 km, is 2.5 times the depth of ours, it, too, is only a thin film around its planet. For the atmospheres of the two planets to mingle, the planets would have to be nearly touching. The gravitational pull from Venus at such a close passage would probably have torn the crust of the Earth apart badly enough to extinguish all life on its surface. Even if such a close passage were possible, there is another problem that makes the creation of manna by mingling the atmospheres impossible: the atmosphere of Venus *isn't* made of hydrocarbons. It's almost entirely carbon dioxide. Oxygen plus carbon dioxide doesn't add up to manna.

Velikovsky was one of the more recent in a long tradition of authors who have theorized that celestial catastrophes shaped human history in ancient times. Probably the earliest of these was German linguist Johann Gottlieb Radlof (1775–1846). He theorized that a large planet between Jupiter and Mars was destroyed by a collision with a comet, forming Venus and the asteroid belt. He saw this event as the source of myths about battles in the sky between the gods. The English translation of the title of his 1823 book on the subject is *The Destruction of the Great Planets Hesperus and Phaethon, and the Subsequent Destruction and Floods on the Earth; along with New Information about the Mythology of Ancient Peoples.*

Radlof was followed by the American writer, Ignatius Donnelly (1831–1901). Donnelly's most influential work was *Atlantis: The Antediluvian World*, published in 1882. It essentially created the new myth of Atlantis as the fountainhead of all ancient civilizations. (In the original myth, found in two of Plato's dialogues, *Critias* and *Timmaeus*, Atlantis was an evil empire oppressing the world.) In the following year, he published *Ragnarok: The Age of Fire and Gravel*, in which he claimed that the sinking of Atlantis was precipitated by a collision between Earth and a comet.

Hans Schindler Bellamy was a pseudonym used by an Austrian writer named Rudolf Elmaver von Vestenbrugg, who also wrote under the name Elmar Brugg (1901–82). He expounded a planetary collision theory that the moon was originally an independent planet captured by the Earth (which had just lost its own original moon) 50,000 years ago. This theory was based on the ideas of Hans Hoerbiger (1860–1931), a mining engineer and amateur astronomer. "Bellamy" published his theory in a 1936 book titled *Moons, Myth and Man*. He also wrote *The Book of Revelation Is History*, in which he said the Book of Revelation was a coded description not of future events but rather of the catastrophes that ensued when Earth captured its present moon. The idea that the Bible is a code was more fully developed by Michael Drosnin in his 1997 book *The Bible Code* and subsequent works. Bellamy's theory might well be part of a bizarre complex of Nazi pseudoscience. At his *Atlantipedia* website, Tony O'Connell says of Bellamy/von Vestanbrugg:

> During the Nazi reign he wrote a number of regime friendly works under the name of Elmar Vinibert von Rudolf with the SA rank of Obersturmführer including a volume of SA stories covering the war years 1939/40. After the war he was usually published in the German language as Elmar Brugg.[1]

In fact, Hoerbiger's theories, on which Bellamy elaborated, were part of the *Ahnenerbe*, a complex of pseudoscience promoted by Heinrich Himmler and the SS to find the origins of the Aryan race. The Nazis preferred Hoerbiger's acceptably "Aryan" theories to the "Jewish" science of Albert Einstein.

All these authors have many things in common. In particular, they were not scientists, and they based most of their theories on a special interpretation of ancient myths. And in many cases, their books were highly popular among a lay audience that was (unfortunately) as scientifically illiterate as they themselves were. For example, Zecharia Sitchin's books have sold millions of copies worldwide and have been translated into more than 25 languages. Although the works of Radlof, Donnelly, Bellamy and Velikovsky did not refer to ancient aliens' shaping our biology and history, they, along with the works of von Daniken, Sitchin, Drosnin, and "Rael," are part of a complex of pseudoscience involving lost continents, European origins of native American civilizations, New Age prophecies, and—of course—visitations by space aliens.

THE "PROOF" GOES "POOF!"

Let us now return to Erich von Daniken. It would be tiresome and some-what redundant to debunk *Chariots of the Gods?* in great detail. That job has already been done to death, beginning with Clifford A. Wilson's 1973 book *Crash Go the Chariots*. Most of the evidence in von Daniken's book amounts to nothing more than seeing flying chariots as spaceships and the like. However, von Daniken did present three compelling pieces of physical evidence that if not examined in detail would seem to reflect technology and knowledge beyond that of ancient civilizations and even of people living in the 1500s. These are the Baghdad battery, the Delhi iron pillar, and the Piri Reis map. The Baghdad battery seems to show that the Parthians, who were rivals of the Romans and a power in the Near East from 247 BCE to CE 224, had electrical capabilities. The Delhi iron pillar, erected by King Chandragupta II (ca. 380–ca. 415) has, mysteriously, not greatly rusted in more than 1,600 years. The Piri Reis map, made in the early 1500s, seems not only to show much of the Americas, particularly the Atlantic coast of South America, but also to depict accurately part of the coast of Antarctica. Because these regions were unknown at the time when this map was made, and because of its orientation, the implication is that the map depicts a view of part of the world from space. These would seem to be stunning demonstrations that someone on our planet had technology and knowledge far in advance of that held by civilizations of the day.

In addition to these three, highlighted in *Chariots of the Gods?*, we will examine five other stalwarts of ancient alien theorists. The first of these are the Nazca lines, huge inscriptions of figures whose forms can be made out only if viewed aerially. Second are ancient fortresses in Scotland made of stone that has been vitrified—that is, their surface has been turned to glass, a process that requires intense heat. Those arguing that extraterrestrials were actively involved with the human race claim that this vitrification is the result of atomic blasts from nuclear weapons either aimed at humans by extraterrestrials or aimed by warring extraterrestrials at each other. Along with the vitrified Scottish forts, ancient alien theorists also assert that the ruins of Mohenjo-Daro show evidence of ancient atomic war. We will examine those claims. We will also consider the mysterious crystal skulls allegedly from ancient Mayan sites in Central America, yet seemingly created by a technology far in advance of that of the Maya. Finally, we will examine the claims made by von Daniken, David Childress, and Georgio Tsoukolos about the ruins of Puma Punku at Tiwanaco, Bolivia.

Once examined, all these seemingly stunning pieces of evidence have explanations that, though interesting, are quite terrestrial. None of these involves technology beyond that ancient civilizations were known to possess, nor did any require knowledge that could only be gained by looking down at the Earth from an aircraft, let alone a spaceship.

The Baghdad "Battery"

Did the ancient Parthians have electricity? What is this mysterious battery found in the Baghdad Museum? Here's the actual history of the object in question (Fig. 8.1). In 1938 Wilhelm König, a professional painter, who was an assistant to the leader of the German Baghdad Antiquity Administration and who later became director of the National Museum of Iraq, found and examined a curious clay pot that was part of the museum collection. When he returned to Germany in 1940, he wrote a paper arguing that the pot, with its stopper and metal rods, might have been used as a galvanic cell for electroplating gold onto silver objects. So what exactly is the Baghdad battery? Well, it's a clay pot approximately five inches long with an asphalt stopper through which passes an iron bar surrounded by a copper cylinder—and that's it. How could it be a battery? If the clay pot were filled with an acid, such as vinegar, an electrochemical reaction could take place. The actual history of the artifact becomes inflated in retellings by those addicted to finding mysteries in ancient civilizations. At one website of this ilk, we are told that workers excavating a 2,000-year-old village near Baghdad found the object. The website refers to König as a German archaeologist although he was actually a painter. The website doesn't say that König theorized that the artifact was used for electroplating. Rather, it says that he "came to a surprising conclusion that the clay pot was nothing less than an ancient electric battery." The website goes on to say the following:

> In 1940, Willard F. M. Gray, an engineer at the General Electric High Volatage Laboratory in Pittsfield, Massachusetts, read of Konig's theory. Using drawings and details supplied by German rocket scientist Willy Ley, Gray made a replica of the battery. Using copper sulfate solution, it generated about half a volt of electricity.[2]

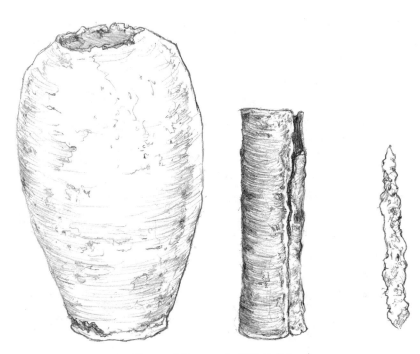

FIGURE 8.1. The "Baghdad battery." (Drawing by T. D. Callahan.)

Although a half a volt indicates that an electrochemical reaction took place, it hardly qualifies the artifact as an electric battery. By comparison, AAA batteries such as are used in television remotes, generate 1.5 volts, three times what Gray got out of his replica. However, it is possible that the artifact was used to electroplate jewelry. Concerning this, the website proclaims the following (emphasis added):

> In the 1970s, German Egyptologist, Arne Eggebrecht built a replica of the Baghdad battery and filled it with freshly pressed grape juice, as he speculated the ancients might have done. The replica generated 0.87V. He used current from the battery to electroplate a silver statuette with gold. *This experiment proved that electric batteries were used some 1,800 years before their modern invention by Alessandro Volta in 1799.*[3]

Notice that the article, after portraying König as an archaeologist and calling the artifact an electric battery, makes an unsupported leap of logic: Eggebrecht made a replica of the Baghdad battery that generated nearly a volt of electricity and used it to electroplate a silver statuette with gold. Therefore the experiment *"proved that electric batteries were used some 1800 years before their modern invention by Alessandro Volta in 1799."* Actually, the experiment proved no such thing. Instead, it showed that the Parthians *might* have used an electrochemical reaction to cover jewelry and other small objects with a thin film of gold. Thus, by a *craft* rather than a science, they might have figured out that if one dissolved gold dust into an acid solution, a chemical reaction involving iron, copper, and the acid would take place so that an object suspended in the solution would be coated with a thin film of gold. They could easily have used this gilding process without understanding anything about generating electricity. One way to test the proposition that the Parthians actually did use electric batteries is to look for a supporting technology: Did the Parthians have flashlights or other battery-operated devices? No, they did not.

Did the Parthians or other ancient peoples even use a chemical electroplating process for gilding? As it turns out, even that is doubtful. Ancient objects thought to have been electroplated probably weren't:

> König himself seems to have been mistaken on the nature of the objects he thought were electroplated. They were apparently fire-gilded (with mercury). Paul Craddock of the British Museum said "The examples we see from this region and era are conventional gold plating and mercury gilding. There's never been any irrefutable evidence to support the electroplating theory.[4]

Nor is there any indication from any ancient culture of an electric grid or any means of generating electricity. Another possibility is that the "batteries" might have been used to give worshippers a shock when they touched an idol. But all this is conjecture, as Dr. Aaron Sakulich notes:

> There are, I am sorry to say, a number of things that don't quite add up about the story, and that are just plain logically dishonest: No one can agree where the battery came from. Some say that Wilhelm Konig found it during an archaeological dig near Khujut Rabu, and others say that he found it in the basement of the Baghdad Museum when he took over as curator. As always, the more versions of a story, the more suspicious you should be about it.[5]

Not only is where the battery came from in dispute, but so also is when. Sakulich continues:

No one can agree on when the battery came from. Most authors cite it as being from the Parthian era, some time between 250 BC and 225 AD. Those knowledgeable about art point out that the clay pots are made in the style of a people called the Sassanians, who lived from 250–650 AD or so. If these artifacts are as amazingly important as paranormal enthusiasts would have you believe, why is there a range of 900 years in the date of their origin? What's causing the trouble?[6]

Finally, nobody can really be sure what its function was. Sakulich continues his devastating critique:

> It is a fact that the "Battery" is made of a clay pot with two pieces of metal and an asphalt stopper. The only evidence that it would have functioned as a battery is that chemical tests showed that at one time the pot "contained an acid substance." That proves one thing, and one thing only: that it contained an acid substance. It implies, but by no means proves, that it was capable of functioning as a battery.[7]

So at this point, the Baghdad "battery" is an artifact of indeterminate age, produced by either the Parthians or the Sassanid Persians for an unknown purpose.

However, this is not the last word on the subject. In an update on the Baghdad "Battery," Lenny Flank hammers more nails into the artifact's coffin:

> There are also problems with the construction of the presumed "battery." To function as a battery, the interior would have been filled with an acidic liquid, and this would need to be replaced often as it ran out. But the jar itself was sealed with asphalt, and the copper tube on the inside was also sealed at both ends—an impractical arrangement for a battery, which would make it enormously difficult to refill the liquid electrolyte. The presumed "battery" also has no terminals—while the iron rod did project outside of the asphalt plug, the copper tube did not, making it impossible to connect wires to make a circuit. . . . Copper tubes were often used as protective covers for papyrus scrolls. . . . So what was the Baghdad Battery, if it was not actually a battery? The key seems to be the papyrus fibers found inside some of the jars. Most archaeologists who have looked into the putative "battery" have concluded that it is nothing more than a now-decayed sacred papyrus scroll wrapped around an iron or wooden rod and placed inside a sealed copper tube (or sometimes a glass flask), which was then itself stored inside a clay jar and plugged with asphalt to protect it from water and weather. So, the consensus of archaeological science is that the "Baghdad Battery" is in fact not a battery at all, but simply a storage jar for a scroll.[8]

So, because the so-called battery was actually a storage jar for a scroll, with a fizzle—not even a mild shock—the first "proof" goes "poof!"

FIGURE 8.2. The Delhi iron pillar. (By Anupamg (Own work) [CC BY-SA 3.0 (http://creativecommons.org/licenses/by-sa/3.0)], via Wikimedia Commons.)

The Delhi Iron Pillar

The freestanding iron pillar in Delhi is something of greater substance (Fig. 8.2). It was erected about 1,600 years ago by King Chandragupta II (ca. 380–ca. 415). It is made of iron yet has not rusted significantly in all that time. Isn't this evidence of a technology superior to our own? As it turns out, the ability of the iron pillar to resist rusting derives from a technology *less* advanced than ours. A 2002 article in a technical journal, *Current Science,* by R. Balasubramaniam of the Department of Materials and Metallurgical Engineering of the Indian Institute of Technology points out that inferior means of removing slag in older methods of iron smelting left phosphorus in the iron, which forms a protective film when exposed to the air.

In modern blast furnaces, limestone (calcium carbonate) is used as a fluxing agent—that is, a material having a lower melting point than that of the material to be melted. Alternating layers of an iron ore (various iron oxides)/limestone mix and coal are hit with blasts of superheated air, which decomposes the limestone into quicklime (calcium oxide) and carbon dioxide, melting the lime in the process. The molten lime reacts with the iron oxide, communicating its heat to the ore, which allows the iron oxide to melt at a far lower temperature than its usual melting point. Oxygen in the superheated blast of air rapidly burns the coal to carbon dioxide. The carbon dioxide rising up through layers of coal reacts with unburned carbon, producing carbon monoxide. The carbon monoxide reduces the iron oxide to metallic iron and is converted to carbon dioxide in the process. The molten iron seeps down to the lower levels of the blast furnace, as does the molten lime. The lime is less dense than the iron and floats on top of it along with impurities such as silicon, phosphorus, and aluminum. This mix of lime and impurities is called slag. It is drawn off, leaving relatively pure metallic iron.

Limestone wasn't used in ancient and medieval iron smelting. The iron ore was heated to a malleable state but wasn't melted. The slag was often removed by hammering the malleable iron on a forge, as Balasub-ramaniam states:

> The ancient Indian smiths did not add lime to their furnaces. The use of limestone as in modern blast furnaces yields pig iron that is later converted into steel; in the process, most phosphorus is carried away by the slag. The absence of lime in the slag and the use of specific quantities of wood with high phosphorus content (for example, Cassia auriculata) during the smelting induces a higher phosphorus content . . . than in modern iron produced in blast furnaces. . . . This high phosphorus content and particular repartition are essential catalysts in the formation of a passive protective film of misawite (d-FeOOH), an amorphous iron oxyhydroxide that forms a barrier by adhering next to the interface between metal and rust.[9]

Ironically, if the Delhi pillar had been made using a modern blast furnace, the more effective removal of phosphorus would have left it vulnerable to atmospheric rusting. The phosphorus left by the less efficient removal of slag formed a protective rust-resistant film called misawite that has protected the pillar for 1,600 years.

The Piri Reis Map

In 1513 an admiral in the navy of the Ottoman Empire, known as Piri Reis, drew a map of the world depicting many parts of the Atlantic coast of South America with surprising accuracy. "Reis" is actually a title of military rank. So, Piri Reis means "Admiral Piri." His full name was Haci Ahmed Muhi Aldin Piri, and the word *Haci* is actually an honorific. It is the Turkish version of the Arabic *Hajji*, denoting one who has undertaken the *Hajj*, the pilgrimage to Mecca, one of the five duties enjoined on observant Muslims should they have the means to make the journey. For simplicity, we'll just refer to him as Piri. As well as being an admiral, he was a cartographer and is best remembered today for his map, from his *Kitab-ı Bahriye*, which translates as *Book of Navigation*. A portion of his 1513 map was discovered in 1929 at the Topkapi Palace in Istanbul.

The map (Fig. 8.3) seems to show a portion of the coast of Antarctica known as Queen Maud Land. Not only does this appear to be part of Antarctica, but it also depicts that part of the continent's coast as it would appear were it ice-free. It also shows it as connected to South America. The Piri Reis map was used by historian Charles Hapgood (1904–82) as a support for his theory that Earth's crust frequently shifted and that an ancient, worldwide civilization had once existed that had been destroyed by a radical pole shift. In 1958 Hapgood published his first book on the subject, *The Earth's Shifting Crust*, which featured a foreword by Albert Einstein. In the book, he opposed the theory of continental drift. His most famous work is *Maps of the Ancient Sea Kings*, published in 1966, just two years before von Daniken's *Chariots of the Gods?* came out. In this work Hapgood cited the Piri Reis map because Piri was reputed to have based some of his material on ancient maps. So, assuming that Piri's map showed Antarctica still connected to South America, as it had originally been, Hapgood argued that it had been copied from maps of the ancient civilization that he claimed had flourished before the shift of the poles, which he saw as having happened about 11,600 years ago. He argued that an ice-age civilization had mapped that part of Antarctica.

Actually, the Drake Passage between South America and Antarctica opened about 23 million years ago at the close of the Oligocene Epoch.

FIGURE 8.3. The Piri Reis world map, 1513, held at the Topkapı Palace Museum, Istanbul, Turkey. (© Heritage Image Partnership Ltd./Alamy stock photo.)

Prior to that time, Antarctica had been connected to the tip of South America. Thus the Brazil Current, which sweeps tropically heated waters south along the coast of Argentina, would have continued down across the Atlantic coast of Antarctica, warming it and carrying cold Antarctic waters into the Benguela Current, which sweeps north along the South Atlantic coast of Africa. Likewise, in the Pacific, the South Equatorial Current and East Australia Current would have brought warm ocean water to Antarctica and carried cold waters east from Antarctica to be swept north in the Peru Current. What directed cold Antarctic waters that were flowing east to flow north in the Pacific, and also directed the warm waters of the Brazil Current south along the Antarctic coast, was the land bridge between South America and Antarctica.

Fossil evidence shows that much of Antarctica was covered by temperate forests until the Drake Passage opened up. With no land barrier to direct the Pacific waters north, they continued to flow east, forming the

Antarctic Circumpolar Current. This current effectively blocks ocean currents such as the Brazil Current from reaching the coast of Antarctica. Nor do the Antarctic waters flow north to carry away the polar chill. Rather, the Antarctic Circumpolar Current carries the chilled water around Antarctica in an endless loop, locking the continent in a deep freeze. So Hapgood, despite being a historian rather than a scientist, did get it right: Antarctica was once much warmer than it is now, with an ice-free coast. However, the event that froze that continent occurred 33 million years ago, not a mere 11,600 years ago. Thus no ancient map made by any human being would ever have showed Antarctica joined by a land bridge to South America.

Possibly seizing on Hapgood's belief in a recent ice-free Antarctica, von Daniken and others have claimed that the coast of Antarctica shown on the map can represent only a view of the region from space and that the coast of Antarctica was mapped using ground-penetrating radar. But does it truly show the coast of Antarctica? If it does, why is that coast connected with South America, as it would have been more than 23 million years ago? Among many who have debunked the claims made by von Daniken, Steven Dutch of the University of Wisconsin at Green Bay notes the significance of a marginal note by Piri on the map:

> There's also a marginal note opposite South America that says "It is related by the Portuguese infidel that in this spot night and day are at their shortest of two hours, at their longest of twenty two hours. But the day is very warm and in the night there is much dew." That would indicate a far southern latitude, but note that the report explicitly comes from the Portuguese, not from arcane ancient sources. It's possible that some Portuguese expedition was blown very far south, not to Antarctica where the days are rarely "very warm," but perhaps to 50° south or so.[10]

Two points in this marginal note are damning to the claims of those promoting the map as being either of ancient origin or the product of extraterrestrial super science. First, consider the claim that the map was made before any Europeans could have mapped the Americas. This claim if falsified by Piri's noting things "related by the Portuguese infidel." In other words, Piri based his map on the explorations of South America already made by Portuguese explorers, as Dutch points out. In fact, Piri's 1513 map is *not* the oldest surviving map of the Americas. That honor goes to the map made in 1500 by Juan de la Cosa (1450–1510). This is hardly surprising when one considers that Juan sailed with Columbus in 1492: he was the owner and captain of the *Santa Maria*. The second damning blow is the Portuguese description of the region supposed to be part of Antarctica: "But the day is very warm and in the night there is much dew." This obviously can't be Antarctica. So the reason why the coastline claimed by von Daniken and others to be part of Antarctica is attached to South America is that it does *not* represent Antarctica at all.

Yet another blow to the supposed extraterrestrial origin of the map is that although it fairly accurately depicts some of the coastline of Brazil, it also depicts the mythical island of Antillia, which gave its name to the Antilles. Thus the map is not all that accurate, even showing a mythical island in the Caribbean. Much of its information is based on Spanish and Portuguese sources. It does *not* depict the coastline of Antarctica. Yet another "proof" goes "poof!"

The Nazca Lines

About 200 miles south of Lima, Peru, near the town of Nazca, lies a gravel-covered desert plain, 310 square miles of which are covered with a mix of lines, geometric shapes, and figures of animals and plants scratched into the underlying soil (Fig. 8.4). There are more than 800 straight lines, some of them 30 miles long. The animal and plant figures vary from 50 to 1,200 feet long. The figures can be seen only from the air. Accordingly, von Daniken asserts that they had to have been made by beings that could see them from high in the sky. He believes them to be demarcations for a landing field. As to the Nazca plain being used as a landing field, Maria Reiche, who spent 30 years researching the Nazca lines and protecting them from vandalism, points out that the underlying ground there is quite soft. She says, "I'm afraid the spacemen would have gotten stuck."[11]

Still, how are we to explain these amazingly straight lines—the accurate geometric figures drawn by the nonliterate people of the Nazca culture between CE 1 and 700? How could they have drawn 1,200-foot-long stylized animals, recognizable as figures only from the air, without being directed by those flying above? As to the straight lines, this can be done by tying a rope to a peg hammered into the ground and pulling it

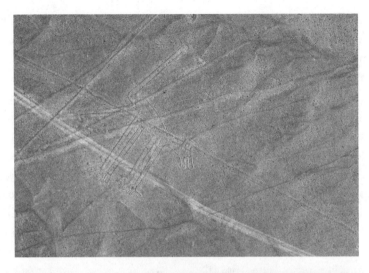

FIGURE 8.4. The Nazca lines. (A) The dog. (© iStock.com/AAR Studio.) (B) An "astronaut." (© iStock.com/harryjco.)

taut. This serves as a guide along which a straight line is drawn. Wooden stakes found marking the termination of some lines have been carbon-dated to CE 525 ± 80 years, indicating that this is indeed how they were made. As to the geometric shapes, one can draw a circle in a manner similar to drawing a straight line. Simply drive a peg into the ground, tie one end of a rope to it and tie the other end of the rope to another peg.

Pull the rope taut and walk it around the fixed end, dragging the free peg through the dirt as you go. Maria Reiche has found holes and stones in the center of circles at Nazca, indicating that this was the method used to draw them. Other geometric shapes found at Nazca consist of squares centered on circles. No math is required to draw a straight line through the center of the circle or to construct a line perpendicular to it. These two lines can be used to draw a square. Using just rope and boards, modern crop circle makers have created spectacular geometric shapes. Again, advanced technology isn't required.

But how could people on the ground have drawn the huge figures of animals, which they would have been unable to see except from the air? Maria Reiche found six-foot square plots in the centers of these figures where preliminary drawings were made, which were later enlarged. *Skeptic* author Joe Nickell, along with some members of his family, decided to try to reproduce one of the Nazca figures, a 440-foot-long condor, using simple tools, starting with a small drawing that they then enlarged:

> The method we chose was quite simple. We would establish a center line and locate points on the ground drawing by plotting their coordinates. That is, on a small drawing we would measure along a center line from one end (the bird's beak) to a point on the line directly opposite the point to be plotted (say a wingtip). Then we would measure the distance on the ground from the center line to a desired point. A given number of units on the small drawing would require the same number of units—larger ones—on the large drawing. We used a set of sticks, set up at right angles, for sighting, and two lengths of marked and knotted cord for measuring.[12]

Nickell and his relatives were able to reproduce the drawing of the condor fairly easily in less than two days.

Were the giant drawings made to communicate with beings from the heavens? Possibly, but not likely as landing coordinates for extraterrestrials. The landing strip theory espoused by von Daniken is untenable not only because of the softness of the soil on the Nazca plain but also because giant line drawings of spiders, monkeys, and condors aren't likely to have been the means by which space aliens would have communicated. However, monumental pictures of animals that might have been totemic would be a reasonable way to communicate with the gods. Peru isn't the only place in the world one finds giant figures that must be viewed from above. The ancient Britons carved such figures into chalky soil, such as the white horse of Uffington, dated somewhere between 800 BCE and CE 100. Another remarkable figure is the Cerne Giant, a 180-foot-long human figure carved into the chalk of a hillside above the village of Cerne Abbas, near Dorchester. The figure is that of a naked man brandishing a club. It is thought to be either Celtic in origin or a Roman depiction of Hercules. However, it is hard to date and may have been carved as a political satirical figure in the late 1600s. The last of these seems an unlikely basis on which to carve a monumental figure. Assuming that it is Celtic or Romano-Celtic, it seems unlikely that it was used to communicate with space aliens unless it was meant as interstellar graffiti—for the naked Cerne Giant sports a prominent erection.

Yet another assertion von Daniken makes relating to aerial views of the Earth has to do with the island of Elephantine on the Nile River, near the Aswan Dam:

> Every tourist knows the Island of Elephantine with the famous Nilometer at Aswan. The island is called Elephantine even in the oldest texts, because it was supposed to resemble an elephant. The texts are quite right—the island does look like an elephant. But how did the ancient Egyptians know that? This shape can be recognized only from an airplane at a great height, for there is no hill offering a view of the island that would prompt anyone to make that comparison.[13]

It's impossible to tell whether von Daniken failed to look at an atlas showing details of the Nile or whether he only hoped that none of his readers would. The island of Elephantine looks nothing like an elephant, or any other beast for that matter. In fact, it was called Elephantine or "Elephant Island" because, as it lay at the border of Egypt and Nubia, it was where the Nubians came to trade ivory to the Egyptians—hence the name "Elephant Island." Yet another "proof" goes "poof!"

ANCIENT ATOMIC WARFARE: VITRIFIED
SCOTTISH FORTRESSES AND MOHENJO-DARO

Those arguing that space aliens were actively involved in human history cite vitrified Scottish fortresses as evidence that these aliens waged nuclear war in ancient times. Vitrification is the process of turning silicate rocks into glass by intense heat. Vitrified fortresses are structures built of rocks piled without the use of mortar and that rather have been fused together, *in places*, by intense heat, melting some of the rock into glass. This requires prolonged heat at about 1,100 °C (2,012 °F). Because such high temperatures imply the presence of vast amounts of energy, advocates of the ancient aliens thesis often jump to the conclusion that they were vitrified by atomic bomb blasts. Here is David Childress on the subject (emphasis in the original):

> One of the best examples of a vitrified fort is Tap o'Noth, which is near the village of Rhynie in northeastern Scotland. This massive fort from prehistory is on the summit of a mountain of the same name which, being 1,859 feet (560 meters) high, commands an impressive view of the Aberdeenshire countryside. At first glance it seems that the walls are made of a rubble of stones, but on closer look it is apparent that they are made not of dry stones but of melted rocks! What were once individual stones are now black and cindery masses, fused together by heat that must have been so intense that molten rivers of rock once ran down the walls.[14]

The Royal Commission on the Ancient and Historical Monuments of Scotland (RCAHMS) has a far more sober description of Tap O' Noth (bracketed material added):

> The Iron Age hillfort on Tap O' Noth is one of the largest in Scotland, consisting of 21 ha [hectares] enclosed by a stone rampart. More than 100 house platforms have been recorded between the rampart and a massive wall that further protects the hill's summit. This stone and timber wall, more than 6m in width and 3m high, is vitrified in places—the stones have fused together through intense, prolonged heat. The extremely high temperatures generated by the burning timbers causes the surrounding stone to melt, and this phenomenon has been observed at many forts.[15]

Note that the fort is vitrified *in places* and that the heat was not only intense but also *prolonged*. Accordingly, it is unlikely that "molten rivers of rock once ran down the walls." In 1937 archaeologists Wallace Thorneycroft and V. Gordon Childe performed an experiment in with they piled up a six-foot-high wall of rocks interlaced with horizontal timbers. They then piled brushwood against the wall and set fire to it. The rubble might have reached a temperature of 1,200 °C. After three hours, the wall collapsed, and part of the rubble was found to be vitrified. Although this demonstrated that burning by forces assaulting the wall could account for some vitrification, it wouldn't explain its widespread incidence. There are about 60 vitrified forts in Scotland, as well as others in other parts of Europe and also Turkey. Other theories about the vitrification of these hill forts are that it was deliberately done by those who built them either for defense or for ritual purposes. None of these theories is entirely satisfactory. However, the 1937 experiment did demonstrate that advanced technology and nuclear weaponry were not required for vitrification.

Of course, the experiment also didn't prove that the forts were *not* vitrified by atomic blasts. So let's look at what we would expect to find if atomic blasts turned at least some of the stone of these forts into glass. One thing we would expect is that one side of the fort, the side facing the blast, would be heavily vitrified, whereas the other side would not. We would also expect to see a large, continuous area of the wall fused into glass. Finally, we should expect to find considerable amounts of shocked quartz in the vicinity of the forts. We find none of these. The vitrification of the forts is spotty, and it isn't on just one side.

Shocked quartz—which results when, as a result of intense pressures, the microscopic crystalline structure of quartz is deformed along planes inside the crystal. Shocked quartz was discovered after underground nuclear bomb testing, which caused the intense pressures required to form it. Meteor and asteroid strikes also produce shocked quartz. Its absence at the vitrified forts indicates, along with the pattern of vitrification, that whatever vitrified these stone hill forts wasn't ancient atomic blasts.

An even more striking claim for ancient atomic warfare on the part of the ancient alien theorists is that the ancient city of Mohenjo-Daro (Fig. 8.5), one of the main centers of the Indus Valley Civilization (flourished 2600–1500 BCE), was destroyed by an atomic bomb blast. Typical of the claims of these theorists are these: (1) at an epicenter within the city, 150 to 180 feet wide, everything is melted and fused—that is, vitrified; (2) masses of skeletons were found, lying in the street, some holding hands, as though struck down in a single terrifying instant; (3) areas of the city are still highly radioactive, as are many of the skeletons found there.[16] Certainly, were these claims true, they would show striking evidence that Mohenjo-Daro was destroyed by an atomic bomb blast ca. 1500 BCE—that is, 3,500 years ago.

The problem is that not one of these claims is true. There is no epicenter of vitrification in Mohenjo-Daro. There are piles of vitrified

FIGURE 8.5. Ruins of the Indus Valley city of Mohenjo-Daro, which is in remarkably good condition for a city supposedly destroyed by an atomic blast from aliens. (By Quratulain (Own work) [CC BY-SA 3.0 (http://creativecommons.org/licenses /by-sa/3.0)], via Wikimedia Commons.)

material—namely, pottery—that was vitrified in kilns. This vitrified material is called frit. Notably, the vitrified material found at Mohenjo-Daro is *not* trinitite, the type of vitrified sand produced by atomic bomb blasts. Not only is there no epicenter of vitrification, but there is also no blast epicenter—no ground zero. Furthermore, the buildings at Mohenjo-Daro were made of mud brick.[17] After 3,500 years, some of these mud-brick walls are still 15 feet high. Certainly, had these walls been hit by an atomic blast, they would have been flattened.

The masses of skeletons certainly attest, however, to some sort of catastrophe—a massacre if not a nuclear blast. Or do they? In fact, a grand total of 37 skeletons have been found in Mohenjo-Daro, in separate groups from different periods, some 1,000 years apart. The skeletons show definite signs of deliberate burial. But what of the claim by ancient alien theorists that masses of skeletons were found lying in a street? The street in question lies at one level of strata, and the skeletons are buried *above* that level—not in what was in their day a street.

What about the claim of heightened radiation at Mohenjo-Daro, particularly the radioactive skeletons?

> It has been claimed that the skeletons, after thousands of years, are still among the most radioactive that have ever been found, on a par with those of Hiroshima and Nagasaki.[18]

Note that the website gives no attribution for the claim. It doesn't say that a specifically named noted authority states that the skeletons were radioactive. Rather, using the passive voice, it just says "It has been claimed." This, however, is a minor flaw compared to the gross level of scientific illiteracy manifested in the claim. Consider that the radiation left over after an atomic blast is from fission products—the remains of the bomb's uranium or plutonium, the nuclei of which were split in two during the explosion. The most common of these fission products, radioactive isotopes of elements from the middle of the periodic table, are cesium-137, iodine-131, and strontium-90. Iodine-131 decays rapidly, having a half-life of a little longer than eight days. Cesium-137 has a half-life of 30 years and strontium-90 one of 28 years.

To understand the significance of this, we need to understand what a half-life is. We cannot predict when any given atom of a radioactive element will decay. However, we can predict that by a certain time, a given fraction of these atoms will have decayed. The simplest and largest single fraction is one-half. Thus a half-life is the period of time by which half of the atoms in a given radioactive isotope will have decayed. Thus in the first half-life of the fission products created in an atomic blast, 50% of them will have decayed; similarly, 50% of those remaining—that is, 25% of the original amount—will have decayed by the end of the second half-life. Thus after the third half-life, only 12.5% of the isotope will remain. This drops to 6.25% after the fourth half-life, 3.125% after the fifth, and 1.5625% after the sixth. Because 1.5625% is a bit cumbersome, let's round it up to 1.6%. Thus after seven half-lives, a bit less than 0.8% of the original material remains; after eight half-lives, less than 0.4% is left; and after nine half-lives, only slightly less than 0.2% is left. After 10 half-lives, only a bit less than 0.1—to be precise, 1/1024—of the original material is left. As a rule of thumb, after 10 half-lives, the radioactive isotope is considered to have all decayed away. Because cesium-137 has a half-life of 30 years and strontium-90 has a half-life of 28 years, they would have vanished from Mohenjo-Daro after 300 years. *Ergo*, no skeleton dating from 3,500 years ago —more than 100 half-lives of cesium-137—would be intensely radioactive.

Of course, with a half-life of 24,110 years, plutonium-239 would still be present from an ancient atomic bomb blast at Mohenjo-Daro. Although most of it would have fissioned in the explosion, there might be some lying about. Would it be in the skeletons? No. In fact, had the people of that ancient city been killed in an atomic explosion, their skeletons wouldn't have even been laden with strontium-90, that notorious "bone-seeker" of atomic fallout. The reason for this further highlights the scientific illiteracy of von Daniken and the others of the *Ancient Aliens* crew: they seem to have confused radiation *sickness* with radiation *poisoning*. Radiation sickness is caused by exposure to large amounts of radiation from outside the body, as in the case of those exposed to an atomic bomb blast. Gamma ray photons pass through the body damaging the cells in a single instant. The cells most vulnerable to being thus radiated are the fast-growing cells of the human body. These are the cells of the blood cell forming tissues of the bone marrow; the skin, nails, and hair; and the lining of the digestive tract. Those suffering from radiation sickness from an atomic blast experience their hair and nails falling out, their skin sloughing off, and, most important, their bodily fluids leaking out through the devastated walls of their digestive tract. Radiation sickness is characterized by nausea, vomiting, diarrhea, and dehydration. However, radiation sickness does not deposit radioactive material in the body.

Radiation *poisoning* results from the ingestion of radioactive materials, such as those left behind in nuclear fallout. Because strontium is in the same family of elements as calcium, the body deposits ingested strontium-90 in the bones and teeth. There it irradiates the blood cell–forming tissues of the bone marrow and the bone itself and causes, after exposure over a period of years, either leukemia or bone cancer. Radiation poisoning is long-term and chronic. Radiation sickness is acute. Had the skeletons at Mohenjo-Daro been laden with strontium-90 at the time their owners had died, they would have been victims of long-term fallout rather than an atomic blast, but even so, after 3,500 years, all the strontium-90 in their bones would have decayed away.

Thus neither the vitrified Scottish fortresses nor Mohenjo-Daro show any signs of having sustained damage from nuclear weapons. Furthermore, the arguments of the ancient alien theorists are rife with pseudoscience and scientific illiteracy. Under scrutiny, with a half-life measured in nanoseconds, yet another "proof" goes "poof!"

CRYSTAL SKULLS

The 2008 film *Indiana Jones and the Kingdom of the Crystal Skull* gave new life to an enduring New Age/Atlantis/ancient alien motif—that of the (allegedly) Mayan crystal skulls (Fig. 8.6). Supposedly first discovered in Central America in the 19th century, crystal skulls were exhibited at the British Museum in London, the Museum of Man in Paris, and the Smithsonian Institution in Washington, DC. They are certainly atypical of Mayan artifacts, their existence seemingly without reasonable explanation considering what we know of the Maya and other Central American civilizations. Not unlike the irritant about which the oyster builds a pearl, this element of mystery served as the nucleus about which grew a corpus of claims, legends, and assertions: the skulls were relics of lost Atlantis, they had healing powers, they were evidence of ancient alien visitation. Speaking on a Sci-Fi Channel documentary bearing the same title as his book, Chris Morton, coauthor of *The Mystery of the Crystal Skulls* (2002), said this of the seemingly Mayan artifacts:

> One has to remember that the crystal skulls are made of the same type of material that we now use in our microchips; and if you think of how much information that we now store inside a tiny quartz crystal so that it can be a chip inside a computer, just think how much information might be stored inside a crystal skull.[19]

Alas! scientific and technological illiteracy strikes again. No, our computer microchips are *not* tiny quartz crystals. They are made from metallic silicon into which minute quantities of impurities have been introduced to make them into semiconductors. Quartz, like sand and glass, is made of silica—silicon *dioxide*. Saying that the crystal skulls are made of the same material as microchips is even more inaccurate

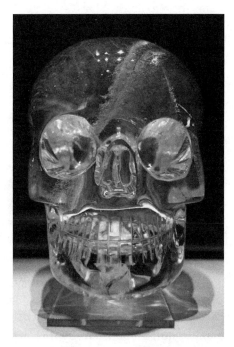

FIGURE 8.6. Crystal skull at the Musée du Quai Branly, Paris. Eugène Boban, a controversial antique dealer, sold this piece to Alphonse Pinart, a young explorer. Pinart donated it to the Museum of Ethnography at Trocadéro, Paris. (© David Gallimore/Alamy stock photo.)

than saying graphite is made of the same material as diamonds. At least graphite and diamonds are both made of elemental carbon. Metallic silicon—that is, *elemental* silicon—is a far different substance from the silicon *dioxide* that makes up sand, glass, and quartz crystals.

Leaving aside Morton's microchip nonsense, however, we are left with the mystery of the origin of these curious crystal skulls—or are we? In an online article for *National Geographic*, Richard Lovett and Scott Hoffman revealed why the skulls are now considered nothing more than fakes:

> In addition, recent electron microscope analyses of skulls by the British Museum and the Smithsonian Institution revealed markings that could only have been made with modern carving implements. Both museums estimate that their skulls date to sometime in the mid- to late 1800s, a time when public interest in ancient cultures was high and museums were eager for pieces to display.

Its examinations and the fact that no such skull has ever been uncovered at an official archaeological excavation led the British Museum to extrapolate that all of the famed crystal skulls are likely fakes.[20]

So yet another seemingly striking artifact proves to be utterly without substance. Another "proof" goes "poof!"

PUMA PUNKU

A tactic used by von Daniken and other ancient alien theorists is to argue that some form of ancient monumental architecture, be it the pyramids or Stonehenge, was beyond the capacity of the ancients not only to build without advanced technology but even to conceive. Puma Punku, a temple complex in the ruins of Tiwanaco in the highlands of Bolivia, is their *pièce de résistance* in this regard, and it is a truly remarkable ruin (Fig. 8.7). Before considering the theorists' arguments, let us consider the actual site. Puma Punku consists of a terraced earthen mound 167.36 meters (549.1 feet) on its north–south axis by 116.7 meters (383 feet) on its east–west axis. On the northeast and southeast corners of the Puma Punku are projections that are 20 meters (66 feet) wide, extending 27.6 meters (91 feet) north and south from the rectangular mound. The eastern edge of the mound is occupied by what is called the *Plataforma Lítica*. It consists of a stone terrace that is 6.75 by 38.72 meters (22.1 by 127.0 feet) in area. This terrace is paved with multiple, enormous stone blocks, the largest of which is 7.8 meters (25.6 feet) long, 5.17 meters (17 feet) wide, and 1.07 meters (3.5 feet) thick and is estimated by the specific gravity of the red sandstone from which it was carved to weigh 131 metric tons. In the Quecha language of the Inca Empire, *Puma Punku* means "Door of the Puma." By a variety of dating techniques, including carbon dating of organic material underlying the mound, we know Puma Punku was built and occupied between CE 300 and 600—that is, from late Roman to early medieval times. Thus it significantly antedates the Inca Empire, which rose in the early 1200s and lasted until the Spaniards, under Pizarro, destroyed it in 1572.

Now let us look at the claims of the ancient astronaut theorists and how they have been debunked. On the *Ancient Aliens* series, season 4, episode 6, "The Mystery of Puma Punku," Georgio Tsoukolos, frequent

FIGURE 8.7. Puma Punku. (By Janikorpi (Own work) [CC BY-SA 3.0 (http://creativecommons.org/licenses/by-sa/3.0)], via Wikimedia Commons.)

contributor to the show, said, "Puma Punku is the only site on planet Earth, in my opinion, that was built directly by extraterrestrials." That's quite a claim. What do the ancient astronaut theorists use as support for this? Here are their arguments: Tiwanaco and the Puma Punku complex were built 17,000 years ago, well before the civilizations of Sumer and Egypt began. Furthermore, Puma Punku was made of granite and diorite blocks that weighed 800 tons. Granite and diorite are two of the hardest rocks on Earth, so the 800-ton blocks would have been impossible for a primitive people without writing and with only stone tools to either move or carve. In fact, they claim, the granite blocks show signs of precision cutting that could only have been made by power tools with diamond drills. On the episode, David Childress, examining one of the blocks, says, "To me it's clear power tools have been used on the block of stone." Adding to the mystery, Childress notes, many of the blocks are strewn about as though blown apart in a gigantic cataclysm. If all this were true, it would seem irrefutable that Puma Punku was built by extraterrestrial visitors well before the most ancient of civilizations appeared on Earth.

In point of fact, however, none of this is true. First, let's consider the dating of Puma Punku. As already noted, carbon-14 dating of organic material *underlying* the mound dates it at about 1500 years ago. Where did the ancient astronaut theorists get the idea Puma Punku dates from 17,000 years ago? This dating comes from the work of amateur archaeologist Arthur Posnaski (1873–1946), who did extensive work to describe Puma Punku and protect it from looting. Unfortunately, he also grossly overestimated its age by assuming that when it was built, certain entrances lined up so that the sun rising on the equinoxes would shine directly into the temple complex. Because they did not line up in such a way in the 20th century, he reasoned that this was owing to periodic changes in the angle of the Earth's axis and extrapolated that they did line up properly 15,000 years ago. Posnasky's assumption that the openings originally lined up with the equinoxes long ago runs afoul of a rather interesting fact, as noted by Jason Colavito:

> The temple actually aligns perfectly with the spring and autumnal equinox right now, bisected by the sun's beams on those two days. There is therefore no reason to propose a 13,000 BCE construction date to solve a problem that does not exist. Many other stones do not line up with equinoxes, most probably because they were never intended to.[21]

Now let's consider the 800-ton granite and diorite blocks that were too big for the native peoples of Bolivia to have moved and that couldn't have been cut with anything but diamond-tipped power tools. First, as originally noted, the largest of the blocks weighs only 131 metric tons. Although that is appreciable, it is far less than the inflated 800 tons claimed by the ancient astronaut theorists. Claiming that the blocks couldn't have been transported on log rollers, Giorgio Tsoukolos says the following on the *Ancient Aliens* Puma Punku episode:

> What nobody talks about is the irrefutable fact that we are up in altitudes 12,800 feet, which means we are above the natural tree line. No trees ever grew in that area, meaning no trees were cut down in order to use wooden rollers.[22]

True enough, Puma Punku is in a treeless region. However, the Andean timberline is quite high—thus available timber isn't that far away from the site. As *Ancient Aliens Debunked* points out, the ancient Egyptians had to import their wood from nearby Lebanon. In the case of Puma Punku, all the builders had to do to get logs for rollers was to hike down to a lower elevation. *Ancient Aliens Debunked* also asks another question: if the stones of Puma Punku were either levitated or airlifted, why do they have holes through which ropes can be threaded, and why do the stones have drag marks? Also, the stones are not, as the *Ancient Aliens* episode claimed, made of granite and diorite. They are mainly of red sandstone, a comparatively soft stone to carve. Some of the stones are faced with andesite, a harder, igneous rock—but one that is not nearly as hard as either granite or diorite.

Nevertheless, the theorists say, the primitive tribes of the Andes had neither writing nor metallurgy. Thus they could not have planned or built Puma Punku. This assertion ignores the fact that the Inca built monumental architecture had no writing system yet. Obviously, writing wasn't necessary to build the pre-Columbian monuments of South America.

However, although these tribes lacked a writing system, they did have bronze metallurgy. In fact, tools made of an alloy of copper, arsenic, and nickel have been found at the site of Puma Punku. The Andes Mountains are rich in native copper—that is, copper already in its elemental metallic state. In addition, domekite and algodonite, copper arsenide ores, occur in the Andes, as does nickeline, a nickel arsenide ore. Bronze is an alloy that is about 90% copper and 10% either arsenic or tin. The Andean civilizations, culminating in the Inca Empire, first made copper arsenide bronzes, later shifting to bronzes alloyed of copper and tin. So, as demonstrated by the tools found at the site of Puma Punku, the tribes living there had the bronze tools required to carve red sandstone and andesite. Also, the builders of Puma Punku didn't need diamonds to polish even very hard materials. Scraping hard surfaces with sand worked equally well in ancient Egypt and at Tiwanaco. The blocks at Puma Punku are remarkably well finished. However, there are also unfinished blocks there that have been roughly shaped by pounding stones and that were never finished by bronze tools or polished by sand. So no mysterious high technology was required to build Puma Punku. Did a cataclysm destroy it as David Childress claimed? Also no. Despite not having fancy tools, ancient craftsmen did produce remarkably smooth-faced stones. Just as medieval builders carried off marble from Roman ruins as building material, so have the locals carted off and dislodged many of the stones of Puma Punku, leaving a chaotic ruin in their wake.

We did not have to look long or hard to debunk the wild claims that von Daniken, Childress, Tsoukolos, and others have made about Tiwanaco and Puma Punku. Not only has *Ancient Aliens Debunked* already done much of this work for us, but so also has Jason Colovito, a tireless investigator and debunker. More than anything, the Puma Punku claims highlight the shallowness and possible dishonesty of the ancient astronaut theorists. Only by selectively ignoring disconfirming evidence and wildly exaggerating other evidence, such as in their claim that the blocks at Puma Punku weighed 800 tons, can the ancient astronaut theorists argue their case in this and many other instances. Almost as egregious as this is their insult to the amazing ingenuity and inventiveness of ancient peoples whose remarkable skill, patience, and craftsmanship enabled them, without the aid of modern technology, to erect monumental architectural forms that endure to this day.

GIANT STONE SPHERES

One common tactic of the ancient astronaut theorists is to seize on any mysterious or seemingly mysterious archaeological find, particularly one involving massive structures, assert that their creation was beyond the capabilities of anyone without advanced technology, and then, having created a mystery, plug in ancient aliens or the human recipients of their advanced technology—often the denizens of Atlantis—as the creators of the artifacts. Although many of what they've asserted to be mysteries aren't so mysterious after all, the giant stone spheres of Costa Rica (Fig. 8.8), some of them more than two meters (six feet) in diameter, do remain an enigma, at least in terms of what their purpose might have been. Using this rather minor point as a thin entering wedge, David Hatcher Childress, affecting a quasi–Indiana Jones style of dress, in a feature he wrote and directed titled *Ancient Technology in Central America*, asserts that the native inhabitants of Costa Rica couldn't have made these perfect stone spheres, that whoever fashioned them used power tools, and that the spheres were probably deposited in Costa Rica by a giant tidal wave, no doubt part of the cataclysm that destroyed Atlantis (or Mu, since the stones are found closer to the Pacific coast of Costa Rica than to the Atlantic), the original home of the spheres.

Although Childress doesn't directly assert that the stone spheres were made by space aliens, he does assert that they were made by the same people who built Tiwanaco and who thus were either extraterrestrial

FIGURE 8.8. Giant stone spheres, held at National Museum of Costa Rica.
(By Rodtico21 (Own work) [CC BY-SA 3.0 (http://creativecommons.org/licenses
/by-sa/3.0)], via Wikimedia Commons.)

colonists or their human protégés. As he did when speaking of that city
in the Andes in an *Ancient Aliens* episode, Childress asserts that the stone
spheres could only have been the product of high technology. He states
unequivocally, "In fact, the machining of these perfectly spherical stone
balls is something that is so difficult, it could only really be done with
power tools."

Were these spheres that perfect, he would have a reasonable point.
However, they are not. Although some of the spheres appear remarkably regular to the eye, they are not perfect—as can be seen from the
photograph of some of the stones at the National Museum of Costa
Rica, some of which are even visibly lopsided. Although considerable
skill is required to make a large stone that is spherical to the eye, such a
stone can be far from a perfect sphere and need not require machining.
John Hoopes, associate professor of anthropology at the University of
Kansas, believes that the people of Diquis culture of Costa Rica used a
well-known procedure called pecking and grinding to carve the stone
spheres, which are made of gabbro, an intrusive igneous rock.[23] For

all that, Childress sees in a stylized gold zoomorphic figurine associated with the spheres a representation of a backhoe. Suffice it to say such a conclusion owes more to imagination than to induction.

As further evidence that people native to the area couldn't have created the spheres, Childress states the following:

> One of the great mysteries of Central America, especially the area around Nicaragua and Costa Rica, is that we have very little evidence of organized city-states, of planned communities, and yet, here in the central jungles of Costa Rica, we find exactly that.[24]

Really? The Mayan city-states weren't located in Central America? Their cities weren't planned? In fact, at least one of their cities, Nixtun-Ch'ich',' in Guatemala, was laid out on of grid, a fact that directly refutes Childress's assertion. Nixtun-Ch'ich' was occupied between 600 and 300 BCE, whereas the Diquis culture of Costa Rica, the culture with which the stone spheres are associated, began much later, ca. CE 300, and lasted until the Spaniards destroyed the native cultures of Mexico and Central America ca. 1550. The stone spheres are found in conjunction with remains that have been radiocarbon dated to ca. CE 600.

Childress then asserts that mysterious stone spheres are found in many places across the globe, indicating the existence of a worldwide culture—further evidence of high technology and organization. He says specifically that the giant spheres are found on Easter Island, New Zealand, and the island of Malta in the Mediterranean. There are indeed giant stone spheres in New Zealand. However, they are not manmade. The Moeraki Boulders are concretions of what was originally mud and silt cemented by calcite, a form of calcium carbonate. They were formed in the Paleocene Period, 65.5 million to 56 million years ago (mya). Other spherical concretions of this type found in New Zealand are the Kouto Boulders, which are as large as 10 feet in diameter, and the Katki Boulders, some of which contain the bones of mosasaurs and plesiosaurs and thus date from the Cretaceous Period, 145 to 65.5 mya. Spherical concretions of similar size have been found in North Dakota, Utah, Wyoming, and Kansas in the United States and in Ontario, Canada. Some of the stones in Canada reach 20 feet in diameter.

There are ancient manmade spheres in Malta, possibly dating from ca. 3000 BCE, but their diameters measure inches rather than feet. There is on Easter Island, at one site only, one rounded stone, more egg-shaped than spherical, shaped by humans; it is surrounded by four smaller ones dating from CE 300 to 400. However, the largest stone only measures 2.5 feet in diameter. Thus the distribution of giant stone spheres is hardly global as Childress asserts. In fact, the only other large manmade stone spheres are found in Bosnia, formerly part of Yugoslavia. They might have been made 1,500 years ago. One of these is larger even than the Costa Rica spheres, having a 10-foot diameter.

Thus giant carved stone spheres are found in only two locations, Costa Rica and Bosnia—on opposite sides of the globe and dating from different periods. They are not perfect spheres made by power tools but were made by hand, and they were not made by either space aliens or their human protégés. Yet another "proof" goes "poof!"

ALLEGED DEPICTIONS OF UFOS

Vimanas

Another argument made by von Daniken and other ancient astronaut theorists is that there are numerous depictions of UFOs thinly disguised as chariots (hence the title of von Daniken's book) in ancient literature, as well as depictions of UFOs in art. Among the literary depictions, von Daniken asserts, are the *vimanas* of ancient Hindu epics:

> We shall not be very surprised when we learn in the Ramayana that Vimanas, i.e., flying machines, navigated at great heights with the aid of quicksilver and a great propulsive wind. The Vimanas could cover great distances and travel forward, upward and downward. Enviably maneuverable space vehicles![25]

Ancient depictions of *vimanas*, however, do not show anything remotely like a spacecraft. Rather, they depict either wheeled vehicles or castles floating in the air. There is a text that does describe in detail an engine using quicksilver (mercury) as a working fluid and describes some sort of engine. It is the *Vimanika Shasatra*. There's just one problem with this document as evidence of ancient extraterrestrial involvement with the civilization of India: the document is a fraud. The *Vimanika Shasatra* does not exist as an ancient text. Instead, it was channeled—dictated to the author from the spirit world—in 1918, though even that date may be too early. Chris White cites other problems with the *Vimanika Shasatra*:[26] The spirit who supposedly dictated the text claimed to be an ancient seer named Bharadvada, who is prominent in some ancient writings, so perhaps that is what is supposed to give this text credibility— that is, the idea that the ghost of someone ancient supposedly dictated it. But they're not even sure whether that version of the story is true, because the first mention of any of this came in 1952 and was made by the person who supposedly found and translated this text from 1918—so as far as anyone can tell, he could have made up the whole channeled-by-a-famous-ghost story in 1952.

So the *vimanas* of ancient Hindu texts turn out to be either flying, wheeled chariots or flying castles, and the notion that they represent spaceships powered by engines using liquid mercury as a working fluid is based on a 20th-century forgery.

The "Spaceship" of Ezekiel

A perennial example of a supposed ancient record of extraterrestrial contact, used by authors asserting that alien astronauts visited the Earth in antiquity, is to be found in the first chapter of Ezekiel, in which that prophet, in exile near Babylon, recounts a vision in this rather long passage (1:4–28):

> As I looked, behold, a stormy wind came out of the north, and a great cloud, with brightness round about it, and fire flashing forth continually, and in the midst of the fire, as it were gleaming bronze. And from the midst of it came the likeness of four living creatures. And this was their appearance: they had the form of men, but each had four faces, and each had four wings. Their legs were straight, and the soles of their feet were like the sole of a calf's foot; and they sparkled like burnished bronze. Under the wings on their four sides they had human hands. And the four had their faces and their wings thus: their wings touched one another; they went straightforward, without turning as they went. As for the likeness of their faces, each had the face of a man in front; the four had the face of a lion on the right side, the four had the face of an ox on the left side, and the four had the face of an eagle at the back. Such were their faces. And their wings were spread out above; each creature had two wings, each of which touched the wing of another, while two covered their bodies. And each went straightforward; wherever the spirit would go, they went, without turning as they went. In the midst of the living creatures was something that looked like burning coals of fire, like torches

moving to and fro, like a flash of lightning. And the living creatures darted to and fro, like a flash of lightning. Now as I looked at the living creatures, I saw a wheel upon the earth beside the living creatures, one for each of the four of them. As for the appearance of the wheels and their construction; their appearance was like the gleaming of chrysolite; and the four had the same likeness, their construction being as it were a wheel within a wheel. When they went, they went in any of their four directions without turning as they went. The four wheels rims were so high that they were dreadful and their rims were full of eyes round about. And when the living creatures went, the wheels went beside them; and when the living creatures rose from the earth, the wheels rose. Wherever the spirit would go, they went, and the wheels rose along with them; for the spirit of the living creatures was in the wheels. When those went, these went; and when those stood, these stood; and when those rose from the earth, the wheels rose along with them; for the spirit of the living creatures was in the wheels. Over the heads of the living creatures there was the likeness of a firmament, shining like crystal, spread out above their heads. And under the firmament their wings were stretched out straight, one toward another; and each creature had two wings covering its body. And when they went, I heard the sound of their wings like the sound of many waters, like the thunder of the Almighty, a sound of tumult like the sound of a host; when they stood still, they let down their wings. And there came a voice from above the firmament over their heads; when they stood still they let down their wings. And above the firmament over their heads there was the likeness of a throne, in appearance like sapphire; and seated above the likeness of a throne was a likeness as it were of a human form. And upward from what had the appearance of his loins I saw as were gleaming bronze, like the appearance of fire enclosed round about; and downward from what had the appearance of his loins I saw as it were the appearance of fire, and there was a brightness round about him. Like the appearance of the bow that is in the cloud on the day of rain, so was the appearance of the brightness round about. Such was the appearance of the glory of the LORD. And when I saw it, I fell upon my face and I heard the voice of one speaking.

What most often strikes UFO enthusiasts as reminiscent of a spaceship, other than the fact that everything that's in it is airborne, are the wheels. Before dealing with those specifically, however, let us consider the vision as a whole and see whether the imagery of the ancient religions of the Near East isn't a better fit as the source of Ezekiel's vision than the imagery of a spaceship.

First of all, Ezekiel says he saw the apparition as a cloud coming on a wind from the north. The word translated as "north" is, in the original Hebrew, *zaphon*, a word that means "hidden" or "dark" and that is associated with the north only as being a quarter of the compass that was gloomy and unknown. To the north of the civilized world, from the perspective of the Ancient Near East, lay wild barbaric lands out of which such nomadic peoples as the Cimmerians and the Scythians erupted to wreak havoc on the settled world. Thus the portent of the cloud is ominous or mysterious. However, *zaphon* is also the name of a locality—Mt. Zaphon in Lebanon, the home of the Canaanite gods. The patriarch of the Canaanite pantheon, El, was merged with Yahweh and is referred to in the Bible in the plural / intensive form, *Elohim*. In fact the two names were often coupled as *Yahweh Elohim*, translated in the Bible as "the LORD GOD." In co-opting El's name, Yahweh also co-opted his wife, Asherah, whose image is constantly being removed from the Temple by strictly Yahwist kings, such as Hezekiah and Josiah (2 Kings 18:4, 23:6). Yahweh also would have been seen by his worshipers as taking over the sacred mountain of the Canaanite gods, Mt. Zaphon. Thus the cloud that approaches from the north (*zaphon*) is both mysterious and holy.

Next let us consider the living creatures within the flaming cloud. Though having the form of men, they have two sets of wings, four faces, and feet like calves' hooves. This chimera-like anatomy has been glossed over by those seeing Ezekiel's vision as a spaceship. Their explanation is that the four faces are the faces of different alien beings looking out of different windows of the spacecraft. Considering Ezekiel's primitive mindset, the argument goes, he certainly couldn't have understood what he was seeing. However, for someone who was overawed by indescribable space aliens, Ezekiel was very precise and particular in describing what he saw. Furthermore, his imagery fits that of the myths of the Ancient Near East. In an ancient relief from the Assyrian city of Nineveh, the god Ninurta is shown as having four wings, just as the beings in Ezekiel's vision have. The wings of the four beings in the vision touched each other, which calls to mind the description of the cherubim in the Holy of Holies of Solomon's temple in 1 Kings 6:27, 28:

> He put the cherubim in the innermost part of the house; and the wings of the cherubim were spread out so that a wing of one touched the one wall and the wing of the other cherub touched the other wall; their other wings touched each other in the middle of the house. And he overlaid the cherubim with gold.

As with the cherubim flanking the Holy of Holies, the wings of the living creatures in Ezekiel's vision touch each other. Solomon overlays the cherubim with gold, whereas the living creatures sparkle like burnished bronze. As to the four faces of the creatures in Ezekiel's vision, these, too, hearken back to the cherubim and to Solomon's temple. Lions and bulls are frequently mentioned as part of the imagery of the temple interior (1 Kings 7:25, 29, 36). Cherubim were often depicted as having men's heads and the bodies of either lions or bulls—or sometimes a lion's fore parts and a bull's hindquarters (hence the calf's feet of living creatures of Ezekiel 1:7)—as well as the wings of eagles sprouting from their lion or bull shoulders.

The symbolism expressed in this depiction of the cherubim is one of supremacy: the eagle is the foremost bird, the lion the foremost wild beast, and the bull the foremost domestic beast. All these, coupled with human intelligence (symbolized by the man's head or, in Ezekiel's vision, a human face) were indications of the godlike power of the cherub. This word in the original Hebrew form is *krubh*, which is related to the Akkadian *karibu*, meaning an intermediary between gods and men—a divine messenger. The word for messenger in Greek, hence the word used to designate divine messengers in the Christian scriptures—which were written in Greek—is *angelos*, the source of the English word *angel*. So the living creatures of Ezekiel's vision were not bizarre four-faced space aliens. Nor were they a depiction of different alien species looking out of the portholes of a spaceship. They were angels. The thrones of Phoenician kings often had armrests in the shape of cherubim, signifying divine sponsorship of the king, and God is often depicted in the Bible as enthroned on the cherubim. That this is precisely what Ezekiel relates in his vision is made clear by the final description in this passage (1:26–28, emphasis added):

> And above the firmament over their heads there was the likeness of a throne, in appearance like sapphire; and seated above the likeness of a throne was a likeness as it were of a human form. And upward from what had the appearance of his loins I saw as were gleaming bronze, like the appearance of fire enclosed round about; and downward from what had the appearance of his loins I saw as it were the appearance of fire, and there was a brightness round about him. Like the appearance of the bow that is in the cloud on the day of rain, so was the appearance of the brightness round about. *Such was the appearance of the glory of the LORD.* And when I saw it, I fell upon my face and I heard the voice of one speaking.

God sits enthroned on a firmament resting on the cherubim. All the symbolism fits that of the Ancient Near East. The firmament over the living creatures, upon which God sits enthroned, represents the firmament of heaven, the dome of the sky to which the stars were fixed and over the surface of which the sun and moon traveled according to the Ancient Near Eastern model of the cosmos. If anyone has any doubts that Ezekiel's vision is of God enthroned upon the cherubim, consider what Ezekiel says in the final description in this passage (verse 28): "Such was the appearance of the glory of the LORD." In the very next verse, 2:1, the voice of the one enthroned says to Ezekiel: "Son of man, stand upon your feet, and I will speak to you." Always, in the Book of Ezekiel, when God speaks to the prophet, he opens by addressing him as "son of man" (*ben adam* in the original Hebrew).

Yet what is the significance of the wheels? As noted in the *Anchor Bible*'s commentary on Ezekiel, wheels before the invention of spokes were small solid disks with large axles. The large axle would have the appearance of a wheel within a wheel. Although the wheels of chariots and most other vehicles of Ezekiel's time had spokes, wheeled stands used within the temple often used the small old-fashioned wheels of earlier days. That the rims of the wheels were full of eyes again evokes the supernatural. The wheels are more metaphor than mechanical detail. So the wheels were yet another reference to temple imagery and holiness and were not flying saucers.

However, to those who see Ezekiel's vision as a UFO sighting, the phrase "a wheel within a wheel" suggests machinery. The wheel within a wheel is rendered as a "wheel in the middle of a wheel" in the King James Version (KJV) of the Bible, the word in Hebrew being *tavek*, meaning "to sever" or a bisection, in other words "in the center of." In *Chariots of the Gods?* Erich von Daniken wrote the following:

> Ezekiel says that each wheel was in the middle of another. An optical illusion!
> To our present way of thinking what he saw was one of those special vehicles the Americans use in the desert and swampy terrain.[27]

Von Daniken went on to say that the vehicle was possibly a version of an amphibious helicopter. But are these wheels made from some form of rubber, as we would expect in such a vehicle? No. According to Ezekiel's vision, they have the appearance of chrysolite (Revised Standard Version or RSV) or beryl (KJV). Chrysolite is a yellow-green form of the mineral olivine and is sometimes used as a gemstone. Beryl is also a gemstone. It is noted for its hardness and comes in green, yellow, pink, or white six-sided prisms. That means that light shines through it. The actual Hebrew word translated as chrysolite or beryl is *tarshiysh*, which can as well mean topaz (another yellow, transparent jewel) or merely a gem. The word is identical to the name of the city of Tarshish in Spain that symbolically stood for the uttermost west, or the ends of the Earth. So the substance called *tarshiysh* meant any exotic gemstone from far away. It seems odd that a machine that might be a form of amphibious helicopter would have wheels made of a yellowish translucent crystal.

Once we take away the "wheels within wheels" reference, is there anything in Ezekiel's vision that sounds like alien machinery? Possibly, depending on how certain words are translated. For example, the creatures' feet are not so much the feet of calves as they are those of bull-calves or young bulls. In Hebrew the word is *egal*. But Hebrew has few words, and each of those words has many meanings. The ancient Israelites understood which meanings to use from how the word was used in a sentence, but the meanings aren't always clear to us. *Egal* can also mean "round." Because the beings' legs were straight, could it be that the feet were round because the straight legs were actually metal columns with round bases? How can we know whether this is the case?

To answer that question, we must remember that parts of the description of what Ezekiel saw can't be viewed as separate from the rest of it, and we have to see whether the images in the vision relate to anything else in the Bible. If the legs were metal columns, wouldn't we expect the creature's wings to also be like aircraft wings? In fact, they don't seem to be either fixed wings or a picture of a whirling helicopter propeller. The Hebrew word translated as "wings" is *kanaph*, meaning, either a bird wing, an edge, or a flap of clothing.

If those who want to see the living creatures as space aliens would read more of Ezekiel, they would find an almost identical vision in Ezekiel 10. There the prophet identifies the beings as cherubim and says, in verse 15, that they were the living beings he saw in his first vision. Finally, let us consider what Moshe Greenberg said in his commentary on Ezekiel 1–20 in volume 22 of the *Anchor Bible*:

> Virtually every component of Ezekiel's vision can thus be derived from Israelite tradition supplemented by neighboring iconography—none of the above cited elements of which need have been outside the range of the ordinary Israelite.[28]

Ezekiel's vision, so exotic and otherworldly to us, was something that any of his compatriots in exile in Babylon would have understood easily. In fact, it is only when the vision is rendered in translation, rather than in the original Hebrew, that we today, owing to modern readers' unfamiliarity with the iconography of the Ancient Near East, mistake angels for spaceships.

The Abydos Helicopter and the Dendera Lightbulb

Two other records of seemingly modern technology in ancient times are iconic in nature. The first of these is a sandstone lintel (Fig. 8.9A) in the mortuary temple of Seti I in Abydos, Egypt. The second pharaoh of the 19th dynasty and the father of Ramesses the Great, Seti I, reigned from 1291 to 1278 BCE. Yet a photo of a lintel from his mortuary temple, widely circulated on UFO websites, clearly depicts a helicopter and what may be a motorboat and an airplane. That is, the *photo* plainly does depict this. The actual lintel, however, does not. Without the clean-up and doctoring to produce the clear image of a helicopter, the original image shows a large bird perched next to and slightly overlapping the supposed helicopter. In fact, the "helicopter" seems to have been the product of an original hieroglyph, partially eliminated, overlaid by a second image. This is called a palimpsest, a term used for an ancient document that has been erased and written over, in which portions of the earlier writing, drawing, or engraving still show through. Bruce Rawles shows the cleaned-up image next to the original, along with a diagram showing the likely stages of initial carving and later carving. One wonders, in any case, why—were we to accept the engraving as depicting a helicopter—there would not be an accompanying text describing the gods descending in such a craft. Because there is no such text, and because there is nothing in any of the other ancient Egyptian texts referring to such flying machines, it is highly unlikely that the engraving represents a helicopter.[29]

FIGURE 8.9. (A) The Abydos lintel, with the hieroglyphs that have been interpreted as "spacecraft" or "helicopters." (By Olek95 (Own work) [Public domain], via Wikimedia Commons.) (B) The Dendera "lightbulb." (Original uploader Twthmoses at English Wikipedia, transferred from Wikipedia to Wikimedia Commons [CC BY 2.5 (http://creativecommons.org/licenses/by/2.5)], via Wikimedia Commons.)

The second iconic record widely touted by ancient alien theorists is a relief inscription, seemingly depicting a lightbulb, from the temple of Hathor at Dendera, Egypt (Fig. 8.9B). Although the complex itself dates from the early days of Egyptian civilization, the existing temple of the goddess Hathor dates from late in the Ptolemaic Period, possibly from the time of famous Cleopatra VII (69–30 BCE). The relief in question shows a male human standing behind a lotus blossom from which extends a serpent enclosed in an elongated envelope shaped somewhat like a lightbulb. This envelope is supported by a pillar, from which extend two arms that support the serpent. On the History Channel's *Ancient Aliens*, season 4, episode 13, "Alien Power Plants," Giorgio Tsoukolos says this of the Dendera temple:

> In Egypt there was this underground crypt that was always secret, and only the high priests had access to the crypt. It's very hot in there and very narrow, low ceilings. And on the walls you have these reliefs of what look like ancient light bulbs. Because we have to question one thing: How did the ancient Egyptians light the insides of their tombs?[30]

Tsoukolos goes on to relate a personal experience that he says he had in just such a tomb:

> Also, there isn't enough oxygen in there to support or feed a flame. I was once in the king's chamber in the pyramid of Giza and somebody turned out the lights, and immediately we were in pitch darkness. And I said, "No problem. I'll just take out my lighter from my satchel." And I turned on my lighter, and it didn't work![31]

One wonders: if there were not enough oxygen in the chamber to support a flame from a lighter, how did Tsoukolos and his companions have enough oxygen to keep from suffocating? That archaeologists exploring these tombs in the early 20th century used torches to light their way gives the lie to Tsoukolos's claim. Further emphasizing that these spaces couldn't have been lit by torches or oil lamps, the narrator of the *Ancient Aliens* episode says that "nowhere on the ceilings is there the slightest evidence of soot or smoke residue." However, an Egyptian travel website shows a dramatic full-color picture of the ceiling of the Dendera Temple, half of which has been recently cleaned—and the other half of which is covered with black soot.[32] Thus the claim made by *Ancient Aliens* is false and must be viewed as an example either of profound and irresponsible ignorance or as a deliberate lie. In any case, Tsoukolos has conflated the relatively open and aboveground temple at Dendera, dating from the first century BCE, with the underground crypts of the Giza Pyramid, built thousands of years earlier, during the reign of Khufu or Cheops (reigned 2589–66 BCE). The temple at Dendera was not an "underground crypt," as Tsoukolos characterizes it. It was open to the outside, had plenty of oxygen, and was lit by torches.

For all that, we must still deal with a relief that looks somewhat like a lightbulb. To make their case stronger, the theorists on the *Ancient Aliens* episode had to mischaracterize the explanation of the image by Egyptologists. Here is their version of the conventional interpretation of the relief, given by David Childress:

> Egyptologists' explanation of this, and they have to have some explanation. There's got to be one and it has to be pretty mundane. It can't be an electrical device. Their explanation is that it is a lotus flower and what appears to be a bulb around it is the aroma of the lotus flower.[33]

This simply isn't true. The conventional explanation of the image by Egyptologists is that it represents one of the many creation stories of Egyptian myth. Each locality in Egypt had its own creation myth. The relief represents the one from Heliopolis (Gr. "city of the Sun"; the Egyptian name of the city was Iunu). According to this creation myth, a lotus flower emerged from the primordial watery abyss, just as lotus blossoms rise from under the surface of the water at dawn. From it emerged the god Atum, often represented as a serpent, surrounded by a bubble of air. Thus what the ancient alien theorists claim is a lightbulb surrounding a filament is, in fact, a stylized bubble of air surrounding a serpent, representing the sun god. That the image inside the alleged lightbulb is a serpent, and not the filament of a lightbulb, is quite obvious. It has a head replete with eyes.

It is further damaging to the lightbulb theory that in none of the Egyptian tombs, spanning many dynasties and thousands of years (from ca. 3000 to the first century BCE) do we find other representations of what could be electrical technology, writings describing electricity or actual electrical devices, preserved in the tombs' dry, sealed compartments. Yet another "proof" goes "poof!"

The Tolima Fighter Jets and King Pacal's Spaceship

Images of what seem to be advanced technology in a primitive culture also occur in South America, specifically in the jungles of Columbia. These are the so-called Tolima fighter jets, gold figurines that at first glance do indeed seem to resemble modern-day jet fighters. Among the

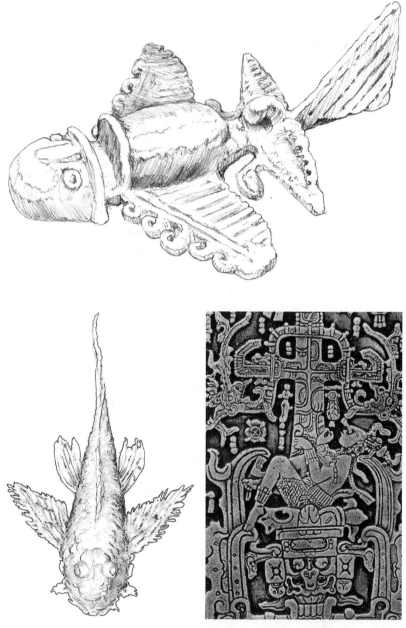

FIGURE 8.10. (A) The Tolima "fighter jet." (B) *Lithoxus lithoides*, one of the species of sucker-mouth catfish (A–B drawings by T. D. Callahan). (C) King Pacal's "spaceship." (©nickolae/Fotolia.) (D) 3D visualization of King Pacal's spaceship. (Sculpture by Paul Francis and Kristina Lucas Francis.)

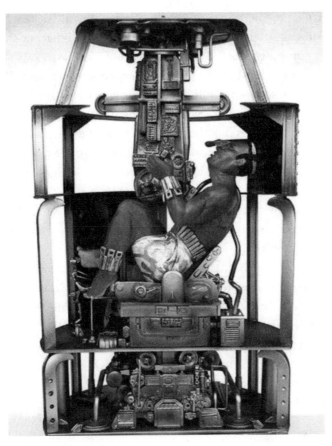

FIGURE 8.10. *Continued*

many gold figurines of the pre-Columbian Tolima culture, most are stylized representations of frogs, fish, insects, or other animals

At first glance, a few of these seem to have the vertical tail and horizontal wings of jet aircraft. However, it does seem odd that if these were
actual depictions of ancient alien jet fighters, they alone among the other
figurines of this culture demonstrate contact with ancient aliens. Surely
such a contact would have had a striking effect on the primitive native
culture, and the people of the Tolima region would have also made
figures of spaceships and other advanced machinery.

However, if these aren't representations of jet fighters, then what are
they? Is there anything in nature in that region resembling these figures?

As it turns out, there is. It is the sucker-mouth catfish (Fig. 8.10B). Because it is a bottom dweller, the underside of its tail is flat and the upper side sticks up, much like the tail of an airplane. This catfish also has two broad side fins, much like the seeming wings of the figurines. The figurines also sport somewhat bulbous eyes, fitting for a fish but rather out of place on a jet aircraft. One objection to the sucker-mouth catfish as the source of the images is that the fish has a dorsal fin whereas the images do not. However, the fish's dorsal fin is often relaxed flat against its back. In any case, the images, both those of the supposed jet fighters and those clearly of animals, are highly stylized rather than realistic. Considering the context of the culture and its imagery, the likelihood that the images represent sucker-mouth catfish is far greater than the likelihood that they represent fighter jets.

Yet another image from the ancient Americas supposedly representing knowledge of advanced technology by the ancients is the lid of the stone sarcophagus of King Pacal of the Mayan city of Palenque. Pacal reigned a remarkable 62 years, from 615 to 677. According to Erich von Daniken, the lid of his sarcophagus represents not his journey into the underworld, as most archaeologists interpret the carved relief (Fig. 8.10C), but rather Pacal as a pilot at the controls of a spaceship. Here is what von Daniken actually says of the image:

> In the center of that frame is a man sitting, bending forward. He has a mask on his nose, he uses his two hands to manipulate some controls, and the heel of his left foot is on a kind of pedal with different adjustments. The rear portion is separated from him; he is sitting on a complicated chair, and outside of this whole frame, you see a little flame like an exhaust.[34]

To make the sarcophagus lid seem more like a representation of an astronaut, von Daniken juxtaposed a photo of astronaut John Glenn in the same position in the Mercury space capsule. One striking difference between John Glenn and King Pacal is that Glenn is wearing a space suit whereas Pacal is dressed only in a kilt, wrist and ankle bracelets, jewelry, and an elaborate headdress that is nothing like a space helmet. Nor does King Pacal wear any shoes. Perhaps we should call him "the barefoot astronaut." Does Pacal wear a mask (presumably an oxygen mask) on his nose, as von Daniken asserts? No. There is an object in contact with his nose, but it is plug-shaped and not connected to any hose or other apparatus. It certainly bears no resemblance to an astronaut's facemask. Does the heel of his foot push a pedal, and do his hands manipulate controls? No, his heel rests against a stylized platform, and his hands, making stylized gestures, really aren't in contact with much of anything.

The Mayan iconography is so stylized that laypeople may readily dispute the archaeologists' interpretation the image as Pacal's descent into the underworld against the background of a cruciform world tree. However, considering the ornate stylization of the images, it is an even greater stretch to interpret them as aspects of a space ship. We would be particularly hard pressed to interpret the two grimacing masks floating at the top of the picture (assuming one views the lid as vertical) as anything to do with a space capsule. Even if we were to grant the ancient astronaut theorists extreme leeway in the interpretation of the stylized iconography, we must wonder that there is no other evidence that the Maya were acquainted with space-age technology. Unlike the Andean civilizations, those of Central America and Mexico had systems of writing and yet did not record anything that could be construed as extraterrestrial spaceships. Again, unlike the Andean civilizations, the Maya did not have bronze metallurgy. Suffice it to say that they also lacked spacecraft.

UFOs in Renaissance Paintings?

We are now left with late Medieval and early Renaissance paintings as the last hope of depictions of spacecraft in pre–20th century art. In a 2004 article in *Skeptic Magazine*, Diego Cuoghi gives examples of paintings that seem to show UFOs. In *The Annunciation* by Carlo Crivelli (1486), a thin beam of light shooting from a heavenly vortex goes through a dove into the head of the Virgin Mary. In the background of the *Madonna and Child with the Infant St. John,* attributed to Sabastiano Mainardi or Jacopo del Sallaio (late 1400s) a shepherd looks up at a curious disc-shaped cloud emitting beams of light. Large, disc-shaped objects that could be clouds but that seem too regular in shape dominate the skies in *The Miracle of the Snow* by Masolino da Panicale (ca. 1428). A small bright red UFO seems to be lying on the ground near a penitent St. Jerome as he kneels before a crucifix in a detail from *Scenes of Monastic Life* by Paolo Uccello painted between 1460 and 1465. Finally, two images from a 16th-century fresco of the Crucifixion, on the wall of the Vsoiki Decani Monastery in Kosovo, Yugoslavia, look very much like stylized spacecraft, each bearing a single occupant.

At first, all these seem to be representations of UFOs. However, Cuoghi points out that on closer examination, they clearly depict angels. Of the vortex seeming to send a beam of light into the Virgin Mary's head, Cuoghi writes the following:

> If the beam emanating from the radiant cloud in Crivelli's panting were actually intended to represent the action of a technological device—a Star Trek transporter, perhaps, or some sort of laser communications system—it would seem extremely odd that this advanced technology was employed, apparently, for the purpose of beaming a bird into the head of the Virgin. In the context of Christian religious iconography, the little white bird is far less bizarre. In fact, painters routinely placed doves in such beams of light, with the doves representing the Holy Spirit and divine guidance descending to the blessed contactee through a dove.[35]

Context is the operative word here. Without understanding Christian iconography, particularly that derived from the Gospel of Luke, one is likely to make all sorts of erroneous assumptions and assertions. The dove as a Christian symbol of the Holy Spirit goes all the way back to the gospels. In Luke's version of the baptism of Jesus by John the Baptist, Jesus is praying after being baptized when there is a sign from heaven (3:21, 22, emphasis added):

> Now when all the people were baptized, it came to pass that Jesus also being baptized, and praying, the heaven was opened. *And the Holy Ghost descended in bodily shape like a dove upon him*, and a voice came from heaven, which said, "Thou art my beloved Son, in thee I am well pleased."

Cuoghi also points out that a close examination of the cloud vortex shows the heads of angels looking down out of it. Closer examination of the disc-shaped cloud that the shepherd views in the background of the *Madonna and Child with the Infant St. John* also shows it to be an angelic apparition, particularly when we consider another background detail from that painting: the Nativity star, plus three smaller stars representing the threefold virginity of Mary (before, during, and after the virgin birth, the perpetual virginity of Mary having become part of Church dogma by that time). Likewise, the flotilla of disc-shaped clouds in *The Miracle of the Snow* is typical of angelic platforms frequently depicted in other paintings of the period. Closer examination also shows the red "UFO" lying on the ground next to St. Jerome is a cardinal's hat. In other paintings St. Jerome is often shown kneeling in prayer with the same red cardinal's hat near him on the ground.

Surely, however, the two rocketlike images from Vsoki Decani, each bearing a single occupant, *must* be UFOs, right? No. Again, a closer look at them reveals what they really are. First, they are on either side of the crucified Christ. Second, though we seem to be shown a cut-away revealing a man within each object, the man isn't dressed in a spacesuit, nor is there any representation of controls. The men appear to be dressed like classical Greeks. One of the objects is colored gold, the other silvery gray. The pointy projections are not the fins of an airship. Rather, they are rays, for one of these figures depicts the sun and the other depicts the moon. Yet another ancient astronaut "proof" goes "poof!"

The "Mysterious" Sumerians

The late Zecharia Sitchin did not claim much in the way of testable artifacts. Instead, he made various startling claims about the origins of civilization. According to Sitchin, civilization appeared abruptly and inexplicitly on Earth. To support this theory, Sitchin specifically claimed that Sumerian civilization appeared "suddenly" and thus can't be explained by any other means than intervention from above. In fact, he asserts that even our mere existence can't be explained by natural causes. In *The 12th Planet* he claims the following:

> Of the evidence that we have amassed to support our conclusions, exhibit number one is Man himself. In many ways, modern man—Homo sapiens—is a stranger to Earth. . . . And why does all living matter on Earth contain too little of the chemical elements that abound on Earth, too much of those that are rare on our planet?[36]

Did he mean that he couldn't understand why life isn't silicon-based? Carbon, oxygen, nitrogen, and hydrogen, the basic elements from which living things are made, are all common on Earth. Furthermore, humans and chimpanzees share more than 98% of their DNA, and primates share most of their genetic makeup with other mammals. Even a smattering of knowledge about biochemistry suffices to debunk Sitchin's assertion. Thus, starting with three unfounded, and false, assumptions—Sumerian civilization began suddenly, modern humans are alien to Earth, and living things are made of elements rare on Earth—Sitchin then proceeded to interpret ancient epics as supporting the bad science fiction that is the essence of his theory.

In his 2003 book *The Genesis Race*, Will Hart also asserted that the earliest civilizations started up suddenly, without any indication of a transition, from primitive Neolithic villages to full-blown civilization in possession of the wheel, metallurgy, writing, and irrigation systems. He says that before the Sumerians

> there simply were no previous human cultures that were experienced and
> knowledgeable enough to tame rivers for irrigation, nor were any of the
> Sumerians' contemporaries in other parts of the world attempting such a project.[37]

This echoes Sitchin's view of Sumer as "The Sudden Civilization":

> Having begun to use stones as tools some 2,000,000 years earlier, Man achieved
> this unprecedented civilization in Sumer circa 3800 B. C. And the perplexing
> fact about this is that to this day the scholars have no inkling who the Sumerians
> were, where they came from and how and why their civilization appeared. For its
> appearance was sudden, unexpected and out of nowhere.[38]

Let's examine what we actually know about the Sumerians. In the third edition of his book *Ancient Iraq*, the English translation of which was published in 1992, historian Georges Roux notes the following:

> The Sumerian civilization was never imported ready-made into Mesopotamia
> from some unknown country at some ill-defined date. Like all civilizations—
> including ours—it was a mixed product shaped by the mould into which its
> components were poured over many years. Each of these components can now
> be traced back to one stage or another of Iraqi prehistory, and while some were
> undoubtedly brought in by foreign invasion or influence, others have roots so deep
> in the past that we may call them indigenous.[39]

Roux goes on from this starting point to detail the gradual increase in technological sophistication from the Neolithic to the Bronze Age, which includes four stages of culture in the Chalcolithic or Copper and Stone Age. These earlier periods include the Neolithic Jarmo culture, appearing ca. 7000 BCE, during which there was progressive domestication of animals and plants, the growth of villages, and the invention of pottery and brick-making; the Hassuna Period, from 5800 to 5500 BCE, the first of the Chalcolithic stages; and the Samarra culture, from 5600 to 5000 BCE, during which irrigation agriculture first appears—thus putting the lie to Hart's assertion that nobody before the Sumerians had attempted such projects. The Halaf culture, from 5500 to 4500 BCE, featured villages with cobbled streets on which were used ox-drawn sledges. The people of this culture grew emmer, wheat, barley, einkorn, lentils, and flax and kept cattle, sheep, goats, pigs, and dogs. Finally, the Ubaid Period, from 5000 to 3750 BCE, featured houses and temples of increasing size and complexity. In addition, their large mud-brick platforms foreshadowed the earliest ziggurats. In the earliest Bronze Age civilization, the Uruk Period, from 3750 to 3150 BCE, the wheel and cylinder seals appear. By about 3500 BCE these people were using a form of pictographic writing that would eventually develop into cuneiform. The Uruk Period was followed by the Jemdat Nasr Period, from 3150 to 2900 BCE, and the Early Dynastic (ED) periods (ED I, 2900–2750; ED II, 2750–2600; ED III, 2600–2334) of Sumerian civilization proper.

Thus the Sumerians did *not* suddenly appear with a full-blown civilization. Rather, the archaeological strata show a gradual stepwise increase in the sophistication and complexity of cultures in southern Mesopotamia, starting with a Neolithic culture that was gradually domesticating plants and animals (Jarmo) and ending with the Bronze Age dynasties of the fully civilized Sumerians with writing, the wheel, bronze metallurgy, monumental architecture, and large-scale irrigation systems. It is particularly notable that irrigation agriculture first appears in one of the earliest Chalcolithic periods (Samarra) between 5600 and 5000 BCE.

Much of the evidence of earlier, Neolithic and Chalcolithic, cultures in the region of southern Iraq came to light comparatively recently. Sitchin may thus be excused for his erroneous assertions to some degree, they having been based on a lack of data. However, *The 12th Planet* has recently been republished without any acknowledgement of the more recent archeological discoveries in southern Iraq. Similarly,

Hart, writing in 2003 in *The Genesis Race,* had no excuse whatsoever for making his assertions.

Another point sometimes seized on by ancient alien theorists is that the Sumerian language is unique, not related to any other language on Earth. Consider this quote from the *Mysterious Universe* website:

> But here's the funny thing: we have absolutely no idea where the Sumerians acquired their language, or what they might have looked like. Their language, which we call Sumerian, was a linguistic isolate—it's the oldest known written language on Earth, and any languages it might have derived from or developed alongside have been lost to time. The Sumerian people were also, it can be reasoned, ethnically isolated; referring to themselves as the sag gigga ("black-headed people"), they appear to have had no concept of race. And figuring out what their ethnic identity might have been based on their art is a doomed effort, because their art was so stylized that a good case could be made that it portrays people of any ethnicity.[40]

Regardless of how stylized Sumerian art was, several skeletons from Sumerian gravesites have been found. For example, more than 550 nearly complete skeletons from the Sumerian city of Kish were sent to the Field Museum in Chicago, and other Sumerian human remains are at the Natural History Museum in London. Still others have been sent to Japan. Although it is true that the ethnicity of the original Sumerians is still in doubt, their remains are fully human.[41]

It's true that Sumerian is a linguistic isolate, a language not related to any other. But does this mean that it was the language of ancient aliens? Hardly. Although Sumerian is an isolate, so also are Basque and Etruscan. Yet the ancient aliens theorists haven't claimed extraterrestrial ancestry for either of these peoples. Zecharia Sitchin claimed that ancient Sumerian was the source language for all other ancient civilizations, an assertion falsified by Sumerian's status as an isolate.

Sitchin's Massacre of Mythology and Sumerian Texts

In the process of creating his ancient aliens mythos, Zecharia Sitchin conflated Sumerian, Akkadian, and Babylonian myths, freely mixing the older Akkadian creation and flood myth *Atrahasis* with the much later Babylonian creation myth *Enuma elish* and calling them both Sumerian.

To understand what a mishmash he made of these various Mesopotamian mythologies, we have to understand the players and the timeline of Mesopotamian history. The culture of the non-Semitic Sumerians, as already noted, gradually evolved into a full-fledged civilization about 3500 BCE. The Sumerians gradually fused with the Semitic Akkadians between 3000 and 2000 BCE. Under Sargon of Akkad (ca. 2279–ca. 2234 BCE), the Akkadian Empire controlled most of Mesopotamia. By this time, Sumerian was chiefly a liturgical language. For all that, there was a resurgence in Sumerian power between 2119 and 2004 in what is often called the Neo-Sumerian Empire. After the collapse of this state, the Amorites, a Semitic people from just west of the Euphrates invaded and set up kingdoms in Mesopotamia. One of these was the city-state of Babylon, which had risen to supremacy over its rivals by 1894. Under the Amorite king Hammurabi (ca. 1792–ca. 1750 BCE), Babylon ruled a small empire. Thus between the emergence of Sumerian civilization ca. 3500 and the beginning of the reign of Sargon of Akkad, ca. 2334, more than 1,000 years had passed. Nearly another 500 years passed between the time of Sargon of Akkad and the reign of Hammurabi. This is important because Sitchin cobbled together two epics, one Sumero–Akadian, the other Babylonian, that were likely about 1,000 years apart, to make up his ancient alien mythos.

Sitchin merged the story of *Atrahasis*, an older Sumero–Akkadian myth, with that of *Enuma elish*, a Babylonian creation epic in which Marduk, the patron deity of Babylon, was raised from a minor status to be the king of the gods. The earliest complete version of *Atrahasis* can be dated by its scribal inscription to the reign of Hammurabi's great-grandson, Ammi-Saduqa (1646–26 BCE). However, older fragments of the epic exist. In the story of Atrahasis, the Anunaki, a lesser class of gods, are either the laborers or, in other versions of the story, the masters of a still more inferior class of gods, the Igigi. When either the Igigi or the Anunaki rebel, the god Enki (notably the Sumerian version of the god the Akkadians called Ea) decides to create mortal workers out of clay. To animate the clay, the gods kill a god called Gishtu-e and mix his blood with the clay, just as Yahweh animated Adam, made from clay, with his breath. The new workers, the first humans, prove to be too difficult to handle, and the gods decide to wipe them out. After trying plagues and famine as a means to destroy them, the gods settle on a flood. However, Enki warns a pious king named Atrahasis about the coming flood and tells him to build a vessel filled with livestock and food. Thus Atrahasis and his household survive and repopulate the world. It is notable that Atrahasis means "Exceedingly Wise" and is an Akkadian name, whereas the deity who warns him of the flood bears the Sumerian name Enki. Thus the story is a Sumerian–Akkadian blend.

Enuma elish, which means roughly "When on high"—these being the opening words of the epic—has been dated anywhere from ca. 1895 to ca. 1550 BCE. In this creation epic, Ti'amat, a female sea dragon representing the saltwater depths, mates with Apsu, who represents the fresh waters. Their children, the gods, led by Ea (the Akkadian version of the Sumerian Enki), begin to take charge of the cosmos, angering Apsu, who wants to destroy them. However, Ea kills Apsu. Ti'amat begins to find the gods uncontrollable and generates other deities to destroy them. However, Ea's son, Marduk, who is even more powerful than Ea, kills Ti'amat and forms the heavens and the earth from the top and bottom parts of her body. At first, the gods who had supported Ti'amat are forced to labor in service to Marduk and his allies. However, Marduk frees them from this labor by having a mother goddess, Mami, mold mortals out of clay. Marduk then kills Kingu, a god whom Ti'amat had chosen for her consort after the death of Apsu, and uses his blood to animate the clay. Marduk's elevation to creator of the world and destroyer of the chaos dragon Ti'amat is part of a political myth, reflecting Babylon's rise to supremacy in Mesopotamia. In the Assyrian version of *Enuma elish*, found in the library of King Sennacherib (705–681 BCE) in Nineveh, Marduk is replaced by Asshur, patron deity of the Assyrians, reflecting yet another political shift.

Sitchin used *Enuma elish* as the source for his colliding planets Nibiru/Marduk and Tiamat and took the Anunaki and their rebellion, with the subsequent creation of humans, from *Atrahasis*, a far older epic. Then, with no basis either in myth or science, he converted the creation of humans from clay—animated by the blood of a killed god—in both epics into genetic manipulation of primitive hominids and the creation of humans as a hybrid race between the hominids and the Anunaki. This bastardized pastiche is neither ancient myth nor science. Rather, it is a pop culture creation of the 20th century.

Sitchin also cherry-picked selections from Sumerian texts that he could interpret as modern technology while ignoring those that couldn't be so interpreted. Consider his attribution of the evil wind that destroyed Ur, in the *Lament for Ur*, to fallout from a nuclear war. Here is a translation of lines 179 to 187 from the *Lament for Ur*:

> Enlil brought Gibil as his aid. He called the great storm of heaven—the people groan. The great storm howls above—the people groan. The storm that annihilates the Land roars below—the people groan. The evil wind, like a rushing torrent, cannot be restrained. The weapons in the city smash heads and consume indiscriminately. The storm whirled gloom around the base of the horizon—the people groan. In front of the storm, heat blazes—the people groan. A fiery glow burns with the raging storm.[42]

The storm and its evil wind, blazing heat and fiery glow could certainly be likened to a nuclear blast. However, this could also be a description of a sandstorm combined with fires set by those attacking the city. In any case, in lines 218 to 229, the description of the destruction of Ur seems much more in keeping with the technology of an ancient civilization:

> The heads of its men slain by the axe were not covered with a cloth. Like a gazelle caught in a trap, their mouths bit the dust. Men struck down by the spear were not bound with bandages. As if in the place where their mothers had laboured, they lay in their own blood. Its men who were finished off by the battle-mace were not bandaged with new (?) cloth. Although they were not drunk with strong drink, their necks drooped on their shoulders. He who stood up to the weapon was crushed by the weapon—the people groan. He who ran away from it was overwhelmed (?) by the storm—the people groan. The weak and the strong of Urim perished from hunger. Mothers and fathers who did not leave their houses were consumed by fire. The little ones lying in their mothers' arms were carried off like fish by the waters. Among the nursemaids with their strong embrace, the embrace was pried open.[43]

Not only are the people killed by axes, spears, and battle maces, but also the buildings, too, were destroyed by axes, which are wielded not by the Anunaki but rather by the Elamites and the people of Cimacki (lines 241–49):

> The good house of the lofty untouchable mountain, E-kic-nu-jal, was entirely devoured by large axes. The people of Cimacki and Elam, the destroyers, counted its worth as only thirty shekels. They broke up the good house with pickaxes. They reduced the city to ruin mounds. Its queen cried, "Alas, my city,"

cried, "Alas, my house." Ningal cried, "Alas, my city," cried, "Alas, my house. As for me, the woman, both my city has been destroyed and my house has been destroyed. O Nanna, the shrine Urim has been destroyed and its people have been killed."[44]

The Elamites were an ancient people situated just east of the southern end of the Tigris–Euphrates river valley. They were involved in a series of wars with the Sumerians and Akkadians. The Cimacki are described in another ancient Sumerian text, *Letter from Sin-iddinam to the god Utu about the distress of Larsa*, as a barbaric people living in tents—hardly a description of Sitchin's space-faring Anunaki.

Now let us consider Sitchin's identification of Nibiru as the 12th planet, with its highly elliptical orbit taking it out beyond the orbit of Pluto—an orbit that for some strange reason doesn't kill all life on the planet. Do Sumerian, Akkadian, and Babylonian texts identify Marduk with Nibiru and say that Nibiru is a planet beyond Jupiter? Here is what Dr. Michael S. Heiser, an expert on Mesopotamian languages, has to say about this identification. Referring to Tablet VII of *Enuma elish*, Heiser says the following:

> Here we have a specific Sumero-Akkadian text that says Nibiru is the name of Marduk. Below we'll see that Marduk was the name of the planet Jupiter. Nibiru can't be a planet beyond Pluto if it's Jupiter. Nibiru-Marduk-Jupiter has something to do with a crossing place—that is, this astronomical body itself isn't doing the crossing, but marks or is positioned at a crossing point. This is another point of contradiction with Sitchin's teachings, as he argues it is Nibiru that is mobile and "crosses" into the orbital paths of our solar system's planets. The texts do not say this.[45]

So, yes, Nibiru is another name for Marduk, and Marduk is indeed identified with a planet. However, that planet is *Jupiter*, not some planet with a weird elliptical orbit that takes it beyond that of Pluto. Furthermore, although it is associated with a *fixed* place of crossing, the planet Nibiru/Marduk/Jupiter is not crossing the paths of other planets. There is just enough accuracy in Sitchin's interpretation—Nibiru does refer to a crossing place; it is the Sumerian name for Marduk; it is a planet—to lend a spurious resemblance to what the ancient texts actually say to lead those wanting to believe in ancient astronaut theories to believe that

Sitchin had actual expertise in interpreting Sumerian and Akkadian writings.

Finally, Sitchin made a hash of interpreting Sumerian cylinder seals and deriving from his incorrect interpretations the erroneous notion that the Sumerians knew that the Earth revolved around the sun. Leaving aside the seeming failure of experts such as the late Samuel Noah Kramer, considered the foremost expert on Sumerian texts, to discover that the Sumerians had a heliocentric view of the solar system, consider that people knowing that the sun was still and that the Earth moved wouldn't likely depict the sun as moving in their sacred texts. Yet in the "Hymn to Utu," the Sumerian sun god, one line says, "My king Utu, you cross all the shining mountains like an eagle."[46] Furthermore, as Michael Heiser points out, what Sitchin took for the sun and the planets on a cylinder seal is actually a large star surrounded by smaller stars. Heiser points out that the Sumerians' symbol of the sun god was a central circle with four extended arms with wavy lines radiating out between them, or just the wavy lines. This entire symbol was itself enclosed in another circle. By contrast, what Sitchin took for the sun is a six-pointed star enclosing a circle. Thus it is likely to depict a constellation, not the solar system.[47]

Heiser also notes that the inscription on the cylinder seal translates as "Dubsiga, Ili-illat [is] your servant." That a man named Illi-illat is the servant of another man named Dubsiga has nothing to do with astronomy. Although cylinder seals sometimes depicted scenes from mythology, they were used as administrative tools, as a form of signature. In later periods, they were used to notarize documents inscribed in clay.

The Sirius Mystery

Yet another author to make the claim that Sumerian civilization appeared suddenly is Robert K. G. Temple, in his 1976 book *The Sirius Mystery*. The "mystery" is that the nonliterate Dogon people of Mali had knowledge of Sirius B, a white dwarf companion of Sirius A that is not visible to the naked eye. From this assertion and some rather bizarre interpretations of Dogon myth, Temple asserted that the Dogon had been given this knowledge by extraterrestrials from the Sirius system. These aliens from Sirius also caused the "sudden" emergence of the Sumerian and Egyptian civilizations. Oddly enough, the Sirius space aliens, called the Nommo, were not mammalian: they were aquatic beings with human torsos ending not in legs but rather in fishlike tails. They lived in water and were, apparently, amphibians. It is true that the Nommo of the Dogon, either ancestral spirits or deities, are, in their mythology, described as amphibious, hermaphroditic, fishlike creatures and are depicted as fishlike. The Dogon also refer to the Nommos both as "Masters of the Water and as "the Teachers." It is also true that many Mesopotamian gods are represented as having human torsos ending in fish tails and that the initial creatrix, Ti'amat, was a sea dragon. The idea of the amorphous sea, representing the unformed world of total potential, as the source of the gods and creation is widespread in world mythologies. Rivers, particularly in Greek myth, were often personified as serpent deities, similar to Ti'amat, and water in general as a source of life and creation was a common concept. However, to transpose mythic amphibious or part-human, part-fish gods into extraterrestrial beings involves considerable scientific illiteracy: it is highly unlikely that amphibians would develop the high metabolism required to support a large brain.

Still, how did the Dogon know of Sirius B, a star invisible to the naked eye? One possibility is cultural contamination. In 1893 French astronomers visited the Dogon. They were in Mali observing a total solar eclipse visible from equatorial regions on April 16, 1893. These astronomers, led by Henri Deslandres, stayed in the field five weeks and had many opportunities for contact with the Dogon, during which they could have imparted knowledge of Sirius B to the local people.

However, we must ask: did the Dogon even know of Sirius B? Temple relies heavily on the work of French anthropologists Marcel Griaule and Germaine Dieterlen, done in the 1940s. Griaule and Dieterlen described an important Dogon religious event called the *Sigu* ceremony, which was associated with Sirius, the brightest star in the nighttime sky. They call Sirius A *Sigu tolo*, *tolo* meaning "star." They also describe a companion star called *Po Tolo*, presumably Sirius B, and even tell of a third companion star (as yet unknown) called *Eme Ya Tolo*—or so Griaule reported.

There are a number of problems with this story, however. First, Griaule based most of this on the testimony of one informant. Second, there are considerable examples of cultural contamination in Dogon myth. For example, Walter E. A. Van Beek (1991) relates a Dogon story of one of their ancestors who drank too much beer and fell asleep with his genitals exposed. One of his sons laughed at him, but the other two showed respect. Walking backward so as not to look on their father's nakedness, they draped a cloth over him to cover his genitals. This is simply a retelling of the curious story of the drunkenness of Noah from Genesis (9:20–23). Another problem is that peoples being questioned by anthropologists often divine what answers their questioners are looking for and oblige them by saying what they think the anthropologists want to hear. When anthropologist Walter E. A. Van Beek did field work among the Dogon in the 1980s, he could not find any evidence that they knew anything of Sirius's being a double star:

> No Dogon outside the circle of Griaule's informants had ever heard of sigu tolo or po tolo, nor had any heard of eme ya tolo (according to Griaule in RP, Dogon names for Sirius and its star companions). Most important, no one, even within the circle of Griaule's informants, had ever heard or understood that Sirius was a double star (or according to RP, even a triple one, with B and C orbiting A). Consequently, the purported knowledge of the mass of Sirius B or the orbiting time was absent. The scheduling of the sigu ritual was done in several ways in Yugo Doguru, none of which has to do with the stars.[48]

So, once again, what initially seemed compelling evidence evaporates when scrutinized.

Robert Temple has claimed that astronomers have recently discovered perturbations in the orbits of Sirius A and B, indicating the existence of a third star in the Sirius system. This would validate Griaule's report that the Dogon spoke of a third companion star called *Eme Ya Tolo*. However, in a 2008 article in *Astronomy and Astrophysics*, astronomers J. M. Bonnet-Bidaud and E. Pantin, writing about the infrared imaging of the Sirius system, state the following:

> These deep images allow us to search for small but yet undetected companions to Sirius. Apart from Sirius-B, no other source is detected within the total 25" field.[49]

So there is no third star in the Sirius system, and Temple's thesis still fails.

DIRECTED PANSPERMIA AND THE ATTACK
ON EVOLUTIONARY BIOLOGY

The theme of space aliens as our creators surfaces again in the form of directed panspermia in Michael Drosnin's 2002 bestseller *The Bible Code II: The Countdown*. In 1997, Drosnin shocked the world by popularizing, in his book *The Bible Code*, the equidistant letter spacing (ELS) theory of Israeli mathematician Eliyahu Rips. Rips had, along with Doron Wiztum and Yoav Rosenberg, written a paper titled "Equidistant Letter Sequences in the Book of Genesis," which appeared in 1994 in the journal *Statistical Science*. Without going into excessive detail, the theory of ELS is that if the Hebrew text is squared off into a grid, letters put together vertically, horizontally, or diagonally will spell out encoded messages, which are often prophecies. The letters are chosen by equidistant spacing—for example, by choosing every fourth letter. Supposedly, one can find, embedded in code within the biblical text, prophecies of historical events written centuries or millennia before they happened. In *The Bible Code* Drosnin appeared to support the idea that this showed divine authorship of the Jewish scriptures. However, in his 2002 book, Drosnin claimed that the code was implanted by aliens who traveled through not only space but also time, enabling them to encode future events in the biblical text. Not only did these space aliens embed this code in the Bible, but Drosnin also asserted that they were our creators. They did this, of course, by genetic manipulation.

Essentially, this is the same New Age theory expressed by Jean Sendy, Rael, and Zecharia Sitchin. It is also a variant of the creationist theory of intelligent design, substituting space aliens for God. It amounts to an attack on evolution. Much of this attack on evolution, though dressed in New Age regalia, is depressingly similar to old-style creationist arguments that have been refuted time and again. For example, in *The Genesis Race*, Hart asserts that flowering plants suddenly appeared 100 million years ago. This is identical to the assertion, made by the late Duane Gish of the Institute for Creation Research (ICR) in the 1978 edition of his book *Evolution? The Fossils Say NO!*, that there were no primitive angiosperms (flowering plants) in the fossil record—that flowering plants appear suddenly as a well developed group in the middle and Upper

Cretaceous. For quite some time, no evidence could be found of the first angiosperms in the Jurassic or earlier. Then, in 1976, two paleobotanists, James A. Doyle and Leo J. Hickey, discovered a sequence of fossil leaves and pollens in the Lower Cretaceous documenting the origin, rapid evolution, and diversification of the angiosperms. Paleobotanists had wrongly assumed a longer time requirement for the early evolution of flowering plants and had been looking for the ancestors of angiosperms in the wrong strata.[50] Note that this extremely significant discovery was made a full two years before the 1978 edition of Gish's book was published. The discovery was big news in the scientific journals at the time. Had Gish bothered to do a little library research, he could have avoided being two years out of date with respect to the fossil record. Hart, writing in 2003, was 27 years out of date.

Another creationist shibboleth embraced in the New Age paradigm is a cavalier argument against the probability of evolution. Essentially, this amounts to disbelief either in the possibility of life's having arisen from prebiotic molecules by natural processes or the possibility of random mutation and natural selection's having produced great changes in living things. The source of this line of reasoning was none other than the late Francis Crick (1916–2004), codiscoverer, with James Watson, of the structure of DNA. In his 1981 book *Life Itself*, Crick wrote the following

> A few years ago [Leslie] Orgel and I reintroduced the idea—originally propounded by the twentieth century Swedish scientist Svante Arrhenius—that life evolved elsewhere and was wafted to earth in the form of spores. Arrhenius called this hypothesis panspermia ("seeds everywhere"). But Orgel and I suggested that the microorganisms more probably had arrived in a space vehicle "homed in" on Earth by a distant civilization. We called this concept "directed panspermia."[51]

The concept of panspermia was originally called the "cosmozoic" theory and was proposed in 1865 by German physician Hermann Richter (1808–76). The primary objection to panspermia (directed or otherwise) is that it only puts off the problem of the origin of life to another time and place.

Crick's hypothesis was subsequently picked up by the physicist Fred Hoyle (1915–2001), who, with mathematician Chandra Wickramasinghe, wrote a number of books claiming that life was seeded from outer space. In their 1990 book, *Cosmic Life Force*, for example, Hoyle and Wickrama-singhe wrote the following:

> A primary thesis of this book is that cells, among which is a viable faction, cells particularly in the form of freeze-dried bacteria, are prevalent everywhere in the galaxy. Any planet or planets that arise anywhere in the galaxy will therefore be in receipt of this cosmic bounty of life, and wherever the planetary conditions are favourable for survival such life could take root. With the billions of candidate stars that are available with conditions similar to the Sun, it does not stretch credulity too far to suppose that the survival and development of cosmic life on planets must be rather commonplace.[52]

Prebiotic compounds certainly exist in space, and it is widely accepted that some of those that eventually gave rise to life were delivered to our planet by way of comets and asteroids. However, the survival of bacteria in interstellar space, though quite possible, is purely theoretical. From this tenuous possibility the authors made what can only be characterized as a leap of faith:

> Evolution of life on the Earth can be viewed as a direct consequence of our continued exposure to cosmic genes, and had it not been for such exposure terrestrial life would not, as we see it, have proceeded beyond the stage of single-celled structures.[53]

In this view, not only is the initial delivery of primitive life from space, but extraterrestrial forces also continually fuel evolution. So rather than evolution's proceeding due to random mutations culled by natural selection, the major genetic changes are delivered to Earth as viruses and passed to living things as infections.

A number of objections to this theory naturally occur to the less credulous. How is it that a virus in space happens to be carrying the genetic material needed for a given species to evolve? Also, how does the virus find the right species? This wouldn't work that well in the random mixing we would expect to find in nature. The authors agreed that it can't be random:

> The alternative to assembly of life by random, mindless processes is assembly through the intervention of some type of cosmic intelligence. Such a concept would be rejected out of hand by most scientists, although there is no rational argument for such a rejection.[54]

Having leaped from science to metaphysics, the authors eventually revealed what sort of cosmic creator they had in mind in the last paragraph in the book:

> The creator has been given many shapes and names in the diverse cultures throughout the world. He has been called Jehovah, Brahma, Allah, Father in Heaven, God, in different religions, but the underlying concept has been the same. The general belief that is common to all religions is that the Universe, particularly the world of life, was created by a being of incomprehensibly magnified human-type intelligence. It would be fair to say that the overwhelming majority of humans who have ever lived on this planet, would have instinctly [sic] accepted this point of view in some form, totally and without reservation. In view of the thesis of this book, it would seem to be almost in the nature of our genes to be able to evolve a consciousness of precisely this kind, almost as if we are creatures destined to perceive the truth relating to our origins in an instinctive way.[55.]

So the fact that most people believe in God is proof that such belief is embedded in our genes, and the only way it could be embedded in our genes is by the creator. *Ergo*, our belief in God proves his existence. *QED.*

Francis Crick's directed panspermia was used by Michael Drosnin as scientific support for what he read out of the "Bible Code" in the following encrypted message from a block of type 32 characters wide by 12 high from Genesis: "DNA was brought in a vehicle." This message is crossed by another: "your seed." Nearby is the message: "In a vehicle your seed." From this Drosnin concludes the following:

> When I found that encoded in the Bible, I couldn't believe it. It seemed like science fiction. DNA, the molecule of life, sent to Earth in a spaceship.
> I wondered if any reputable scientist would even consider such a fantastic idea. I called on the most eminent authority in the world, Francis Crick, the Nobel laureate biologist who discovered the double helix, the spiral structure of DNA. It was one of the greatest scientific discoveries of all time. As Crick himself declared in the first moment of revelation, "We've discovered the secret of life."
> "Is it possible," I asked Crick, when I reached him at the Salk Institute in San Diego, California, "that our DNA came from another planet?"
> "I published that theory twenty-five years ago," said Crick. "I called it Directed Panspermia."
> "Do you think it arrived in a meteor or comet?" I asked.
> "No," said Crick. "Anything living would have died in such an accidental journey through space.
> "Are you saying that DNA was sent here in a vehicle?" I asked.
> "It's the only possibility," said Crick.[56]

Although Crick told Drosnin that directed panspermia was "the only possibility," he was far more restrained in a magazine article in 1981, saying only that his theory was as viable as the traditional idea that life evolved from nonlife here on Earth:

> The kindest thing to state about Directed Panspermia, then, is to concede that it is indeed a valid scientific theory, but as a theory it is premature.[57]

Hart has used both Crick and Hoyle as support for his belief in ancient astronauts' having seeded the Earth with life, shaping its evolution (or perhaps sequential creation) and, eventually, giving human beings civilization. Crick, despite his status as a Nobel laureate, had nothing but a hypothesis and his own certainty that directed panspermia is "the only possibility" for the origin of life on Earth. Hoyle, despite his credentials as a theoretical physicist, and Wickramasinghe, despite his qualifications as a mathematician, likewise had no evidence to back their theory. Thus both theories are utterly unverifiable and amount to mere speculation rather than science.

SUMMARY

Common to most ancient astronaut theories is that they are proposed, with a few notable exceptions, by people who are scientifically illiterate. Even when scientists are the source of such theories, these often amount to speculation about sciences outside their field. Another thing many of these theorists share is a pervasive intellectual dishonesty: they ignore evidence that contradicts their views, mischaracterize conventional explanations to make them sound absurd, inflate and exaggerate their own evidence, and often change the wording of ancient texts when presenting them as evidence supporting their theories. Let us consider the authors and theorists alluded to in this chapter on the basis of their ethics and credentials.

Immanuel Velikovsky (1895–1979) was a psychiatrist and psychoanalyst. He based his theory of planetary near collisions entirely on his interpretation of ancient myths as warped versions of repressed traumatic memories bubbling up out of the subconscious. As noted in our chapter on alien abductions, the notion that traumatic memories are repressed has been thoroughly discredited in modern psychology. Velikovsky had no knowledge of the natures of the gas giant Jupiter or the atmosphere of Venus.

At least we can regard Velikovsky as sincere. The same cannot be said for Erich von Daniken, a former hotel manager. At the age of 19, he was given a four-month suspended sentence for theft. Later, he was jailed nine months for fraud and embezzlement. He was convicted in 1970 of fraud, embezzlement, and forgery and was sentenced to three years in prison, but he served only one year. In 1974, science writer Timothy Ferris interviewed von Daniken for *Playboy Magazine*. Ferris brought up many situations in which von Daniken had garbled or misrepresented evidence. This exchange between Ferris and von Daniken is particularly telling. In these quotes from the interview, we denote Timothy Ferris as TF and Erich von Daniken as EVD:[58]

> TF: Let us talk about some of the mysteries you say the archaeologists ought to be studying. In your book *The Gold of the Gods*, you describe taking a voyage through enormous caves in Ecuador where you claim to have seen ancient furniture made of plastic, a menagerie of gold animals, a library of imprinted metal plates and other evidence of a great early civilization. You call this, "the most incredible, fantastic story of the century," and say you were guided through the cave by a South American adventurer named Juan Moricz. But Moricz says he never took you into any such caves. Which of you is telling the truth?
>
> EVD: I guess we both are telling half the truth.
>
> TF: Which half is yours?
>
> EVD: I have been in Ecuador several times. I have met Moricz several times, and we have been together at the side entrance to the tunnels. Moricz made it a condition that I would not be allowed to give the location or to take photographs inside. I could understand that because he didn't want people going in there. So, I agreed, we shook hands and we left. And, as a matter of fact, in my book I have not told the truth concerning the geographic location of the place and about some various little things.

So there is no way to verify the existence of the plastic furniture and the metal plates (which, von Daniken said, were made of gold), because there is no way to find this cave. Von Daniken said he had perhaps overdramatized the descriptions but still insisted that he had actually seen all the things he claimed to have seen. However, Ferris once again pointed out that Moricz contradicted von Daniken:

> TF: Moricz says, "Von Daniken was never in the caves. When he states he has seen the library and other things himself, he is lying. We never showed him these things."

We should note in passing that von Daniken seems to have taken the inscribed gold plates from Joseph Smith's description of the plates he was directed to find by the Angel Moroni, upon which were inscribed the Book of Mormon.

Zecharia Sitchin (1920–2010), who merged Velikovsky's mythos of colliding planets with von Daniken's mythos of ancient extraterrestrial contacts, had a degree in economics from the University of London. Thus he had expertise neither in science nor in the interpretation of the ancient myths on which he based his theory.

David H. Childress, frequent commentator on the *Ancient Aliens* television series, is the owner of Adventures Unlimited Press, specializing in ancient mysteries, lost cities and continents, alternate history, historical revisionism, hollow Earth and pole shift theories, and UFOs and ancient astronauts. He attended the University of Montana at Missoula, studying archaeology, but dropped out when he was 19.

Another frequent contributor to *Ancient Aliens* is Giorgio Tsoukalos, the director of Erich von Daniken's Center for Ancient Astronaut Research. He graduated from Ithaca College with a degree in sports information and communication and has served as a bodybuilding promoter. Again, he has no background in the sciences.

Robert K. G. Temple, author of *The Sirius Mystery*, whose intelligent interstellar amphibians are biologically laughable, graduated from the University of Pennsylvania at Philadelphia with a degree in oriental studies and Sanskrit. Although knowledge of Sanskrit is impressive, it doesn't qualify him as an expert on the Dogon of West Africa. Also, as with Sitchin, his characterization of the Sumerians as having appeared suddenly with a full-fledged civilization shows a profound ignorance of archaeology, the study of which, ancient alien theorists continuously assert, supports their theories.

L. Ron Hubbard (1911–86) was a hack science fiction author. Claude Verilhon, before he named himself "Rael," was a journalist and sports car racer. Will Hart, author of *The Genesis Race*, another of those asserting that Sumerian civilization arose suddenly and mysteriously, is an author specializing in esoterica. He has had published articles in the following magazines, among others: *Atlantis Rising, New Dawn, Nexus,* and *UFO*. He is also a nature photographer whose work has appeared in *Sierra Heritage, Wild West,* and *Nature Photographer*. Again, knowledge of nature photography hardly qualifies him to make pronouncements on either DNA or archaeology. Michael Drosnin, who popularized (and vulgarized) equidistant letter sequencing in his *Bible Code* books, is a journalist, nothing more.

This leaves us with our three experts in fields that might at least impinge on alien life forms and intelligence: Chandra Wickramasinghe, mathematician, astronomer, and astrobiologist; Fred Hoyle (1915–2001), astronomer; and Francis Crick (1916–2004), codiscoverer of the structure of DNA. Astrobiology, the study of alien life forms, is highly speculative and theoretical, and Wickramasinghe is essentially a mathematician. The late Fred Hoyle, being an astronomer, was also not really knowledgeable about DNA. His and Wickramasinghe's theory of panspermia—microbes' seeding life across interstellar space—is a hypothesis without supporting evidence. Likewise, the late Francis Crick had no evidence for his theory of *directed* panspermia. Great as Hoyle and Crick were, they essentially peddled outlandish theories on the strength of their reputations. Unfortunately, in so doing, they lent credence to the theories of the scientifically illiterate: when scientists write nonscience, it inspires nonscientists to write nonsense.

9

Praying to Aliens

Was Jesus an extraterrestrial sent down by benevolent spacefarers to enlighten us sad, benighted Earthlings? According to the teachings of the Aetherius Society, he was from Venus, as was the Buddha—whereas Krishna hailed from Saturn. One must suppose that the Venus envisioned by the Aetherius Society as the home of Jesus and Buddha was the watery twin of Earth hidden under clouds, the view of the planet that held through the first half of the 20th century, as opposed to the Venus revealed by the *Venera* space probes—swathed in clouds not of water but of sulfuric acid, with a dense atmosphere made up almost entirely of carbon dioxide and with a surface temperature above the melting point of lead. Positing Saturn as the home of Krishna displays the high level of scientific illiteracy we also find in ancient alien theories. This is hardly surprising, because the belief that past spiritual teachers were space aliens is closely linked to theories that ancient astronauts created human civilization. Saturn is a gas giant with a nickel–iron and silicate core that makes up only a fraction of the planet's mass. Surrounding that core is a layer of metallic hydrogen, another of liquid hydrogen, and finally a thick, gaseous layer of hydrogen and helium. Gas giants, such as Jupiter and Saturn, have no fixed solid surface. Thus they are not planets on which one might stand, even encased in a space suit, as one might on the surface of rocky planets such as Mercury, Venus, Earth, or Mars.

According to the late George Van Tassel of Ashtar Galactic Command, who claimed to have been in telepathic contact with extraterrestrials since 1952, Jesus was a space being, though he didn't specify his planet of origin. As we saw in chapter 7, Claude Verilhon/Rael claims that Jesus was one of the Elohim, which he has incorrectly translated as "those who came down." The extraterrestrial origin of great spiritual teachers is a doctrine common to most of the UFO-based religions that originated in the 20th century. Other common aspects of these religions include a belief in reincarnation, claims of telepathic communication with the benevolent space aliens, a syncretistic blend of Eastern and Western spiritual beliefs, and an apocalyptic/messianic spiritual message.

Along with the concept that past spiritual teachers either were extra-terrestrials or received telepathic revelations from them, the inherent universalism of UFO-based religions dictates that all holy books, including not only those of Hindu or Buddhist origin but also the Jewish, Christian, and Muslim scriptures, were likewise transmitted to the human race by space aliens. This conviction would seem to derive from the common truism that all these are works of a superior ethical nature—a view often held by those who have never read them. In fact, mixed with some high moral concepts, there is much in these scriptures that is ethically reprehensible. For example, the Torah, some of the Pauline epistles, and various surahs of the Qur'an all countenance slavery. The belief that the scriptures of all the major religions were dictated by ethically superior space aliens also ignores the mutual exclusivity of these works. In various self-referencing passages, they often claim that they and they alone are divinely inspired, that the gods of other religions are demons or do not exist, and that the prophets of other faiths are false teachers, their scriptures false teachings. In the case of the Christian scriptures' referring back to the Jewish scriptures or the Muslim scriptures' referring back to both the Christian and Jewish holy writings, teachings have been superceded and made, at least to some degree, obsolete. For example, consider this passage from Paul's epistle to the Galatians (1:8): "But though we, or an angel from heaven, preach any other gospel unto you than that which we have preached unto you, let him be accursed."

The Qur'an likewise admonishes Muslims that it is the only true scripture in this passage from Surah 7, *Surat al-Arish* ("the Heights," Q 7:3): "Follow what has been revealed to you from your Lord and do not follow guardians besides Him, how little do you mind."

Because the claims of exclusive divine inspiration are more common among what are termed the "revealed" religions—Judaism, Christianity, and Islam—it isn't surprising that the universalism of UFO-based religions leans more toward Eastern mystic beliefs, such as reincarnation.

The apocalyptic view of most UFO-based religions is also far more optimistic than that of fundamentalist Christians. The latter see humans as hopelessly depraved, whereas most UFO cults see us as being on the verge of the next step in our spiritual evolution. This view recalls that of many science fiction stories, most strikingly Arthur C. Clarke's 1953 novel *Childhood's End*. Thus the apocalypse is seen as a crisis through which we will pass, not unlike the travail of a woman giving birth. Accordingly, the messianic figure associated with this apocalypse, a syncretistic blend of the Christ and the Buddha among others, is not the stern judge of the quick and the dead, who will rule the nations with a rod of iron, of the Christian apocalypse. Rather, he is a kindly teacher who will lead us through the crisis to an earthly paradise.

MAITREYA AND DIVINE TRANSMISSIONS

Just such a teacher was the subject of a glossy handout at a 2015 MUFON (Mutual UFO Network) conference held in southern California. It is published by Share International and is titled "The World Teacher Is Now Here." The teacher's name is Maitreya. He is the subject of a number of books by the late British author Benjamin Crème, who asserted that he first received a telepathic communication from a "Master of Wisdom" in 1959. Shortly after that, he was informed that Maitreya would return within 20 years—that is, by 1979. The handout goes on to say the following:

> In 1972, under this Master's direction, Creme began a period of intensive training, as a result of which their telepathic contact became, and remains, continuous and immediate. This relationship has given him access to constant up-to-date information on the progress of Maitreya's emergence and the total convictn [*sic*] necessary to present that information to a skeptical world.[1]

Of course, telepathically transmitted teachings are not only unavailable to those of us who did not receive them but also impossible to test. There is no way—because these transmissions are, by definition, not of a physical nature, as are radio waves, for example—to demonstrate whether these teachings were actually transmitted to Benjamin Crème by a master of wisdom, are the product of his delusions, or are even part of a calculated fraud on his part.

Regardless of the nature of Crème's allegedly telepathic revelations, he did not invent his World Teacher. Originally, Maitreya was a Bodhisattva, which, in Therevada Buddhism, is an incarnation of Gautama Buddha. Most of these Bodhisattvas were previous incarnations of the Buddha, who, in lives previous to that in which he eventually achieved complete perfection, gained increasing enlightenment. Bodhisattvas are also seen as other people besides the Buddha who have attained the level of enlightenment necessary to free themselves from the wheel of endless reincarnations but who have returned to our world as teachers, driven by compassion for those still trapped by material desires. To those who see Bodhisattvas as incarnations of Gautama Buddha, Maitreya is a future incarnation of the Buddha who will lead the world to enlightenment. As such, he is a millenarian figure.

Madame Blavatsky co-opted the concept of Maitreya as a doctrine of Theosophy, and Annie Besant developed it further when she became president of the Theosophical Society in 1908. In the following year, she became the legal guardian of a 14-year-old Indian boy, Jiddu Krishnamurti (1895–1986), whom the Theosophists thought would be the next "World Teacher"—that is, the incarnation of the Buddha as Maitreya. The Theosophists groomed him for this role in what was called the World Teacher Project. However, by 1929, Krishnamurti had become disenchanted with Theosophy and repudiated his role as Maitreya. He eventually left the Theosophical Society and became an independent philosophical teacher.

We can see a pattern common to UFO-based religions in the transformation of Maitreya from a Buddhist concept through its being co-opted by Theosophy as a westernized version of Eastern religion—a common ingredient in the pastiche of New Age doctrines—to its central position in a religion based on what are perceived teachings by extraterrestrials. At its root, this pattern is likely fueled by a pervasive discontent with established religion and unease with the perceived lack of any sense of the numinous in modern secularism. Thus the concept of Maitreya as an extraterrestrial being epitomizes the commonality of UFO-based religions.

As can be seen in the transmission of the concept of Maitreya from Buddhism to Benjamin Crème's Share International via the westernization of Eastern religious beliefs in Theosophy, there is a strong overlap between UFO-based religions and New Age belief. One of the strongest such ties is belief in extrasensory perception, particularly telepathy, and with it a reliance on trance states and channeling. Many of the messages from the enlightened extraterrestrials are transmitted, along with psychic journeys to other planets, via the medium of channeling in a trance state. New spiritual revelations are also more a part of New Age belief than of the revealed religions, because the canons of Judaism, Christianity, and Islam are completed and closed. Nevertheless, the practice of channeling spiritual entities has a parallel in the belief in possession by the Holy Spirit common in Pentecostal churches. Such practices have always posed a problem for Christianity. Although the book of Acts exults in the possession of the disciples by the Holy Spirit at Pentecost (Acts 2:1–12), Paul finds it necessary to curb, or at least control, the practice of speaking in tongues in the church in Corinth (1 Corinthians 14:1–33), and throughout the history of Christianity, charismatic movements have posed a problem for those insisting on the primacy of scripture as a source of doctrine. New Age and UFO-based religions, unencumbered by doctrinal orthodoxy, are far more open to new revelations.

The emphasis in New Age and UFO-based religions on channeling divine entities is related to the idea that each of us carries within a spark of the divine. For example, the website of the Aetherius Society states the following:

> The essence of us all, and of everything in creation, is Divine—a spark of God.
>
> We all come from the Divine source—and we will all eventually return to this source. It is known by many names—God, Brahma, Jehovah etc, and is described differently by different faiths. It is eternal, all-powerful and all-knowing. It exists everywhere—and it is everything. In fact, it is more even than that.
>
> By correctly navigating our own personal journey through experience, we evolve and become increasingly aware of our Divine nature. As a result we gain greater and greater enlightenment, and also greater and greater spiritual powers which can be used in selfless service to others.[2]

This is diametrically opposed to the fundamentalist Christian view of human beings as totally depraved, separated from God and saved only by undeserved grace.

As previously noted, common to many UFO-based religions is a message of a coming crisis in human consciousness as part of our evolutionary transition to a higher level. Although this is sometimes seen as being less traumatic than what is related in the book of Revelation, there is warning of an impending apocalypse. The coincidence of the beginnings and rise of UFO sightings with the advent of atomic weaponry suggests that the source of this aspect of UFO-based religious belief derives from apocalyptic fears generated in the 20th century. These fears were crystallized in the film *The Day the Earth Stood Still*, in both its 1951 atomic holocaust warning version and its 2008 environmental warning version. Along with consciousness of our ability to destroy our world through atomic warfare, the latter half of the 20th century also created a strong environmentalist consciousness reflected in UFO-based religion. Remember the admonition given to "Joe" in chapter 7 "Wake up. Grow up. The sandbox you're littering is not your own..." The warning, then, is not that of the book of Revelation, a call to remain true to the worship of the Christ in preparation for a last battle between God and Satan during which the human race will endure horrific misery and the world will be destroyed. Rather, it is that the human race must regain a balance with nature, from which it has become estranged.

TRIPS TO AND FROM VENUS, MARS, THE MOON, AND THE OUTER PLANETS

Despite a fascination with futuristic technology, and even though science fiction is one of the sources of UFO belief through the media of television and movies, most UFO-based religions display the same pervasive streak of scientific illiteracy found in popular UFO culture and ancient alien theories—one that marches hand in hand with a heritage of New Age belief. This is particularly apparent in the descriptions of Mars and Venus in such movements as the Unarius[3] and Aetherius societies. For example, regarding extraterrestrial life in our solar system, the Aetherius Society states the following:

> There is intelligent life on other planets within this Solar System, but it exists at a frequency of vibration which is higher than the frequency of vibration of the plane that we inhabit on Earth.

If, for example, NASA were to send astronauts to Venus, they would not discover any signs of Venusian civilization unless the intelligences on Venus chose in some way to make their existence known to them. If, however, a great adept of yoga were to consciously leave the body and project to a plane of Venus that is inhabited, he or she would be able to see that Venus is in fact teeming with life. Such a plane would be physical—but a higher form of physicality than our physical senses or science as we know it is currently able to detect.[4]

This negates any objective investigation of the claim of life on either Mars or Venus. If we can't detect it, that's because it exists at a higher vibration. Of course, science can and does detect that which vibrates at higher frequencies, whether these vibrations are electromagnetic wavelengths beyond our visual spectrum or sound frequencies higher than those we can hear. Thus it is readily evident that the phrase *higher frequencies of vibration* is just a bit of mumbo-jumbo, a way of covering one's tracks in the face of disconfirming physical evidence.

Ernest Norman, who along with his wife, Ruth, founded the Unarius Society, claimed to have made psychic trips to other planets. Here is an excerpt from a Unarius website describing an "astral visit to Venus":

You may see a countless number of strange and fantastically hued birds and huge butterflies whose wings seem to be incrusted with a million jewels. You may taste the waters of some tumbling stream and find them sweet beyond all previous taste experiences. Or you may come upon one of the many cities which seem to hang half-suspended between earth and sky which may remind you somewhat of the pictures in some childhood fairy book of the fairy castle on top of a high hill. Yes, my friends, Venus is a wonderful, beautiful place, and a rich reward which may be well earned through the many thousands of reincarnations and earth lives.[5]

Again, this picture of Venus would seem to derive from concepts of the planet prior to October 18, 1967, when the Russian *Venera 7* space probe landed on the surface of Venus. This probe, along with later *Venera* missions, gave us our present picture of the surface of Venus as being bathed in a caustic atmosphere of carbon dioxide and sulfuric acid, with a temperature hot enough to melt lead. *Venera 7* managed to send back this data during the mere 23 minutes it survived the extreme temperatures and pressures of the planet's surface. Later *Venera* probes managed to survive on the surface of Venus for as long as two hours.

Norman also claimed to have visited Mars. Although he understood that surface conditions on that planet were too harsh to support a Martian civilization, which he therefore placed in underground cities, he did encounter some rather surprising life forms above ground. Here is an excerpt from his 1956 book *The Truth about Mars*:

> There are also a number of species of lizards, reptiles and of some insects whose hard shells have enabled them to weather the extreme atmospheric conditions and among them are giant ants, which walk semi-erect on the two hind feet. The guide tells me these are mutants, which were accidentally produced from a small ant in an atomic experiment ages ago. They are similar to humans in a very low state of intelligence and at one time it became necessary to make war on them, as they became so numerous and large. These strange ant creatures average two to four feet in height and live in rocky caves.[6]

Evidently, the Mars rovers have, so far, missed sighting both the lizards and the four-foot tall ants. Norman had this to say of the Martians living in their underground cities:

> The people of Mars are smaller than those on earth, only averaging about four feet six inches in height. They are somewhat Mongolian in appearance. The texture of the skin is very fine and soft, while the hair is usually straight, black, and quite fine. The men do not need to shave for they have eradicated electronically, the growth of hair from their faces when still young.
>
> Martians are much older in soul-evolution than the earthians. They originally migrated in space craft to Mars from a dying planet more than a million years ago. They also came to this earth and started a colony but found it impractical to maintain. It was also explained by Nur El that this colony became our Chinese race through the evolution of time.[7]

Of course, the Chinese, like the rest of us, share more than 98% of their DNA with chimps and gorillas, and genetically, the Chinese are fully human, sharing 99.99% of their DNA with all other "races." Hence, it's highly unlikely that they originated on another planet. Knowledge of DNA and of the comparative genetic makeup of humans, chimpanzees, and gorillas wasn't available in 1956, when Norman wrote *The Truth about Mars*. However, there was enough scientific knowledge about comparative anatomy at the time to demonstrate the kinship of humans with other primates, the kinship of primates with other mammals, the kinship of mammals with other vertebrates, and the kinship of vertebrates with other animals. Accordingly, even without knowledge of shared DNA, Norman's scientific illiteracy is glaring.

Although he did not form a religion based on his UFO contacts, Howard Menger's experiences with UFOs—beginning with his meeting a beautiful young blonde woman in the woods of northern New Jersey, apparently one of the Nordic aliens, when he was 10 years old—mirror those of the contactees who did found religious societies. He claimed to have been taken aboard an alien spacecraft on a trip to the far side of the moon in 1956, where he saw cities with domed buildings in which Americans, Russians, and Chinese were working together in harmony.

Aetherius Society founder George King stated that the seat of the interplanetary confederation was either on Jupiter or on one of its moons,[8] whereas notable (or possibly notorious) "contactee" George Adamski announced in 1962 that he had attended an interplanetary conference held on Saturn. Again, considering the nature of gas giants, both King's and Adamski's claims show a gross level of scientific illiteracy.

LOST PLANETS AND LOST CONTINENTS: OUTDATED SCIENCE, PSEUDOSCIENCE, AND MYTHOLOGY

As with other millenarian systems of religious belief, such as premillennial dispensationalism, those based on UFOs often have a mythology based on a history of ancient catastrophes. Dispensationalism is a Christian eschatological doctrine first proposed by John Nelson Darby (1800–1882). It holds that God dispensed certain covenants to the human race, beginning with the Edenic Covenant made with Adam and Eve. Human beings violated each of these covenants or dispensations, resulting in a catastrophe of some sort after which God dispensed a new covenant. So the Edenic Covenant was violated by Adam and Eve's eating of the fruit of the tree of the knowledge of good and evil, which resulted in death's entering the world and in their expulsion from the Garden of Eden. During the next dispensation, human beings, though doomed to eventually die, lived much longer than we do today, as in the case of the 969-year lifespan of Methuselah (Genesis 5:27). Increasing wickedness resulted in the catastrophe of the Deluge. The Noachian dispensation, during which the human race was united as a single people, was ended by the confusion of tongues at the tower of Babel. This led to the Abrahamic covenant, and so on. Various dispensationalists differ on the total number of dispensations in human history.

Considering their strong connection to Theosophy and related movements, it's not surprising that UFO-based religions often have a history of belief in past catastrophes that involve lost continents. Their beliefs also overlap a great deal with ancient alien theories. Because the latter often include cosmic disasters, such as the colliding planets of both Immanuel Velikovsky and of Zecharia Sitchin, these, too, have become part of the mythology of UFO-based religious belief. The Aetherius Society links Atlantis, Lemuria (also called "Mu"), and a lost planet called Maldek, quoting this extract from George King's book *The Nine Freedoms*:

> Hundreds of thousands of years ago there was another planet in this Solar System, about the size of Earth, which made its orbit between Mars and Jupiter. It was a green prosperous world inhabited by a people who had not reached a state of really advanced culture, but had nevertheless attained a stage which afforded an abundance of necessities which made life comparatively comfortable for all.
>
> They studied the philosophies and dabbled in the sciences, as do we, except that these people were more advanced in many ways than we are. The planet was so highly mechanized that robots took care of all the menial tasks. The inhabitants had discovered a rudimentary form of space travel, and could control their weather so that drought and famine became long forgotten. The majority, having an abundance of food, and having no menial tasks to perform, soon became content to while away their time in the sun. They became, in comparison with higher planetary cultures, a selfish, lackadaisical people seeking after their own enjoyment, as do the majority of people on Earth today.[9]

Not surprisingly, the people of Maldek became corrupted by a lust for power:

> They exploded a hydrogen bomb and completely destroyed the planet Maldek and murdered the whole populace in one blinding flash of searing flame. All that is now left of that beautiful planet is the asteroid belt.[10]

That must have been one huge hydrogen bomb, considering the many exploded both in our atmosphere and underground since the first large thermonuclear device was detonated at the Enewitok Atoll on November 1, 1952. The idea that the asteroid belt consists of the remnants of a planet that once orbited between Mars and Jupiter was originally suggested to

Henry Herschel by fellow astronomer Heinrich Olbens in 1802, based on his observations of two of the larger asteroids, Ceres and Pallas. It is an outdated theory and is largely rejected today because, among other things, the total mass of the asteroid belt is only about 4% that of Earth's moon and thus not enough to have once made up a planet. It is generally accepted that the asteroid belt represents material that failed to accrete into either a planet or a moon. Outdated scientific theories often show up in the writings of UFO-based religions and related material, as we have seen from the descriptions of Venus as a watery paradise. Thus, according to *The Urantia Book*,

> the fifth planet of the solar system of long, long ago traversed an irregular orbit, periodically making closer and closer approach to Jupiter until it entered the critical zone of gravity-tidal disruption, was swiftly fragmentized, and became the present-day cluster of asteroids.[11]

The Urantia Book was supposedly dictated by celestial beings, during trance states, to William Sadler and some of his close associates over a period of 30 years between 1925 and 1955. "Urantia" was supposedly an ancient name for Earth. The Urantia Foundation published the book on October 12, 1955. Along with its outdated explanation for the origin of the asteroid belt, some other false or outdated claims of the *Urantia Book* are that the universe is hundreds of billions of years old—modern cosmology has the universe being 13.7 billion years old—and that the Andromeda Galaxy is "almost one million" light-years away, as it was thought to be in the 1920s. It's now understood to be 2.5 million light-years away. It seems that the celestial beings dictating the scientific revelations of the *Urantia Book* were limited to the scientific knowledge available well prior to the book's publication in 1955.

The Aetherius Society's website continues its cosmic disaster story with a description of the lost continent of Lemuria's originally taking up much of the space of the Pacific Ocean and being populated by refugees from Maldek:

> Out of the gross limitation of atomic mutation the civilization of Lemuria (also known as Mu) dragged its weary self. The Earth became somewhat similar to what Maldek had been. The people began to probe the philosophies and the sciences again, and the Lemurian civilization flourished.

At its peak, it was a civilization of much finer culture than we know on Earth today. The Lemurians established a liaison between themselves and advanced intelligences from other planets, who taught them a great deal.[12]

However, the curse of atomic weaponry continued:

Lemuria was split into two camps: good and evil, the later camp again probing the atom. For the second time, the forces within God's tiny building blocks were unleashed—and the civilization of Lemuria was destroyed.[13]

This led to the civilization of Atlantis, which fell into the nuclear trap as Maldek and Lemuria had already done:

Again those left were born through gross limitation on and off a world seething with radioactive poisoning until, eventually, after thousands of years, another semblance of culture came into being, and, slowly at first, then later gaining momentum, the civilization of Atlantis flourished upon Earth. Again space travel was established. Again some listened to the voice of wisdom coming from higher sources, and there was a split into three definite camps. The few, searching for a force to give them conquest over the whole Solar System, the majority not caring much, because they were content to live in their procrastinations, and the other few, who had proved themselves ready for the higher teachings and possessed the logic and faith to accept the voice of higher authority. Again the minds of the sadistic minority invented atomic weaponry.

As had happened at the time of the fall of Lemuria, those who were ready for evacuation just prior to the devastation that was to follow, were taken off the Earth by the Gods from space. Meanwhile, those beset with greed and lust for material supremacy, warred with each other. As neither side could win such an atomic war—down fell the civilization of Atlantis into charred radioactive ruins.

Today, again the forces of the atom have been unleashed. Again the world is divided against itself. Let us not make the same mistake a fourth time![14]

Atlantis also figures in writings of the Unarius Society. Among the factors contributing to the sinking of Atlantis, according to a Unarius website, was an accident involving high technology:

The disaster that caused the second period of disturbances (starting around 28,000 BCE), was due to the overtuning, or overcharging, of their largest crystal generator which was located near the so-called "Bermuda Triangle" in the area of the islands of Bermuda in the Atlantic Ocean. As incredible amounts of energy were stored inside something like an underground battery and eventually got overloaded and caused a massive explosion, one can imagine that would lead to some major disturbances within the earth's crust, and because of this; the sinking of much land mass.[15]

An article at the Unarius website also deals with the destruction of planet Maldek, located between Mars and Jupiter. It essentially reprints George King's material on Maldek. The source of Maldek indicates that the barrowing between the Aetherius and Unarius societies involves still more barrowing. in this case from *Other Tongues, Other Flesh*, written by George Hunt Williamson in 1953. Because George King wrote *The Nine Freedoms* in 1961, he also seems to have based his Maldek material on Williamson's book.

Maurizio Martinelli, writing of George Hunt Williamson (as GHW), says the following:

> GHW was the first to use the word "Maldek" in his writings. Now many have accepted it into common usage and if they were asked would tell you "Maldek" must be a name from out of the myth and legend, but this is not so. The word did not exist until GHW used it.[16]

Williamson could have created the word *Maldek* as a corruption of Marduk, the Babylonian god synonymous with the Roman Jupiter. The association of Maldek with Lucifer might have been derived from another name for the planet that supposedly orbited between Mars and Jupiter, one that also appears in the literature of UFO-based religions. Mallona was the name given to this planet by 19th-century Austrian Christian mystic Jakob Lorber (1800–1864). Lorber heard a voice he took to be divine telling him to write new revelations. Among these was the story of how after the fall of Lucifer, God told the rebellious angel to choose a planet upon which to live, where God would leave him be providing he didn't make any further trouble. Lucifer, who was now Satan, agreed and settled on Mallona, a planet even larger than Jupiter. However, he broke his word, and God destroyed Mallona, turning it into the asteroid belt.[17]

Along with outdated scientific concepts, UFO-based religions are rife with pseudoscience. George Van Tassel, founder of Ashtar Galactic Command, began hosting meditation groups at Giant Rock, in the California desert, in 1953—the same year when, he claimed, a space alien from Venus awakened him in the night, invited him aboard his spaceship, and gave him verbal and telepathic directions for building the Integratron, a structure—supposedly based on the Tabernacle of Moses and on research by Nikola Tesla—that would reverse aging. For all that,

Van Tassel died suddenly in 1978 at age 58, before the Integratron could be completed. That crystals, particularly quartz crystals, have healing powers is a bit of pseudoscience popular in New Age belief. Another such concept is that various sound vibrations can do everything from cure cancer to reverse aging. Both of these are part of the supposed science of the Integratron. Consider the description of the "sound bath" from the Integratron website:

> This is an unforgettable sound experience for those who seek deep relaxation, rejuvenation, and introspection. All Sound Baths are 60-minute sonic healing sessions that consist of 25 minutes of crystal bowls played live and the balance of the hour to integrate the sound and relax in the sound chamber to recorded music.
>   You will be resting comfortably in the deeply resonant, multi-wave sound chamber while a sequence of quartz crystal singing bowls are played, each one keyed to the energy centers or chakras of the body, where sound is nutrition for the nervous system. The results are waves of peace, heightened awareness, and relaxation of the mind and body.[18]

So the sonic healing is generated by quartz crystals, and its vibrations are keyed to the charkas: the New Age is thoroughly ensconced in the Integratron.

Alternative cures for cancer, dispensed by benevolent extraterrestrials, are a logical corollary to UFO intelligences' intervening to prevent an atomic war. Cancer fears, like the fears of imminent nuclear holocaust, involve something terrifying, in many ways beyond our control and often intractable.

CHRISTIAN- AND MUSLIM-INFLUENCED UFO CULTS

The Seekers was a cult influenced by sources as disparate as Dianetics, L. Ron Hubbard's forerunner to Scientology; George Adamski's claims of having been contacted by a Nordic type alien from Venus whose name was Orthon; and Theosophy. For all that, their mythos, transmitted via automatic writing by a housewife named Dorothy Martin (1900–1992), given the pseudonym Marian Keech by Leon Festinger, Henry W. Riecken, and Stanley Schachter in their seminal work *When Prophecy Fails*, had a Christian tone. Keech began receiving messages transmitted from the planets Clarion and Cerus by various entities. One of these was

Senada, who told Keech he was the contemporary entity corresponding to Jesus. According to Senada, humans had their origin eons ago on a planet called Car. Two factions rose on Car. One, called the scientists, was led by Lucifer. They were opposed by the People Who Followed the Light, under the banner of God and led by Jesus. Lucifer and his scientists blew up Car with nuclear weapons. The People Who Followed the Light dispersed to various planets. Lucifer and his followers came to Earth. There, Lucifer created more destruction, including a nuclear war between Atlantis and Mu, which sank both continents. Jesus, however, came to Earth to tell humans that they could desert Lucifer's cause. This history of cosmic destruction was clearly influenced by the common Unarius and Aetherius histories, originally concocted by George Hunt Williamson. One of Williamson's mentors, William Dudley Pelley, a mystic and fascist who, just prior to World War II, founded a pro-Hitler organization called the Silver Shirts, also influenced the Seekers. Keech received a message from Senada: a flood, similar to Noah's, would inundate the world on December 21, 1954. Curiously, the failure of the flood to occur did not devastate the core group of the Seekers, as we will note in greater detail at the end of this chapter.

The late Reverend Frank Stranges (1927–2008) rejected the idea that Jesus was a space alien—as a Christian, he considered Jesus to be God incarnate. Nevertheless, his religious belief was profoundly influenced by a space alien by the name of Valiant Thor, purportedly from inside the planet Venus. Valiant Thor, said Stranges, visited the Pentagon and offered President Eisenhower the means to end all hunger and disease on Earth—an offer Eisenhower and Vice President Nixon rejected, because it would have disrupted the economy—and then departed our planet from Alexandria, Virginia, on March 16, 1960. Stranges also believed that the Earth was hollow and that it harbored other civilizations in its interior—which, he asserted, could be accessed via an opening at the South Pole. He further stated that the Nazis had perfected a flying saucer, the prototype of which they took to South America as the Third Reich was falling, and that the Dead Sea Scrolls contained secret teachings of Jesus as well as predictions of future events. These teachings, he said, were being withheld from the public by the Rockefeller Foundation. Stranges also believed in faith healing and used tactics similar to other faith healers, which are also used by those practicing the deceptive practice of cold reading.[19]

Another Christian UFO-based belief system, albeit one of a heterodox nature, is the *Orden* (Order of) *Fiat Lux* (Latin for "Let there be light"). It was founded by Swiss secretary Erika Bertschinger in 1973 after an accident in which she fell from a horse and hit her head. Subsequently, she learned in a revelation that she was the reincarnation of the Virgin Mary and took the name Uriella (not unlike Ruth Norman, cofounder of the Unarius Society, who took the name of Princess Uriel). She also considered herself to be the direct *Sprachrohr* (speaking tube) of both Jesus Christ and Uriel. The latter is an archangel who appears in Jewish and Christian apocryphal writings. His name means "God is my light" and, by extension, "Lamp of God." As one might imagine from a claim of revelatory knowledge received from the apocryphal archangel Uriel, the doctrine of *Fiat Lux* is based not only on Christian material but also on Jewish apocalyptic and Kabbalistic beliefs, as well Gnostic concepts. Part of its inherent syncretism also involves astrology and Eastern religion.

In common with certain fundamentalist millenarians, Fiat Lux distrusts bar codes as a Mark of the Beast. It does not have a website, regarding the "www" prefix as synonymous with 666, the Number of the Beast from the book of Revelation (13:17, 18). This is because the letter *vav*, the Hebrew equivalent of "w" (or "v"), is the sixth letter in the Hebrew alphabet. Hence, "www" equals "666." Fiat Lux also shares with many Christian fundamentalist millenarians a considerable paranoia and entertains conspiracy theories. According to Uriella, Popes John XXIII and John Paul I were poisoned, and Pope Paul VI was replaced by a double. The Vatican itself, Fiat Lux claims, is run by Freemasons.

Fiat Lux lost a great deal of what credibility it initially had after Uriella predicted in 1991 that World War III would take place in 1998. This would involve a worldwide stock market crash caused by the release of computer viruses; the Russian army's marching into Germany; huge earthquakes, tidal waves, and volcanic eruptions; a meteorite's falling into the North Sea; and, finally, an onslaught of Nazi flying saucers previously hidden in Antarctica. Only a third of the human race would survive. However, benevolent space aliens, under the command of Jesus

Christ, would evacuate a chosen few spiritually evolved people in spherical spaceships. When the end didn't come in 1998, Uriella claimed it had been postponed,[20] a common fallback position taken by a number of end-time prognosticators, among them William Miller (1782–1849), whose failed end-time prophecy inadvertently resulted in the creation of the Seventh Day Adventists; Hal Lindsey; and the late Harold Camping. Marion Keech of the Seekers also explained the nonoccurrence of her predicted flood on December 21, 1954, as a divine reprieve.

The apocalyptic scenario of Fiat Lux is heavily influenced by the book of Revelation and its interpretation by modern fundamentalists. For example, the meteorite's falling into the North Sea is strongly reminiscent of the star Wormwood's falling into the waters and poisoning them (Revelation 8:10, 11). The Russian army's marching into Germany reflects claims by Hal Lindsey and others that Russia is Gog and Magog, the enemy from the north in the apocalyptic prophecies of Ezekiel 39. Computer viruses' causing the stock market to crash mirror the "Y2K" millennial fears that computer malfunctions resulting from the failure of computers to recognize the new century would result in everything from airplanes falling out of the sky to Social Security checks' not being issued.

The Nazi flying saucers hidden in Antarctica of the Fiat Lux apocalypse, also referred to by Frank Stranges, have a previous history in conspiracy theories and esoteric literature going back to rumors in the 1950s and *The Morning of the Magicians* (*Le Matin des magiciens*), by Louis Pauwels and Jacques Bergier. Published in 1960, it was one of main influences on Erich von Daniken when he wrote *Chariots of the Gods?* in 1968. The motif has been kept alive in neo-Nazi writings and derives from a mix of Nazi hollow Earth theories and the concept of *vril* energy, which we explore in the next section of this chapter.

For the most part, UFO sightings, contacts, and abductions in the United States have to do with white ethnicity. With the notable exception of Barney Hill, abductees and contactees are usually not African Americans. The same is largely true of UFO-based religions. The most notable exception to this is the UFO-based mythology of the Nation of Islam, particularly as developed by its present leader, Louis Farrakhan, and his description of the Mother Plane or Mother Wheel, which is based on the vision of the prophet Ezekiel in Ezekiel 1, so often seized on by ancient alien theorists as a description of an ancient UFO. Farrakhan reinterpreted Ezekiel's vision in terms of the Afrocentric theology of the Nation of Islam (bracketed material added):

> The Honorable Elijah Muhammad told us of a giant Mother Plane that is made like the universe, spheres within spheres. White people call them unidentified flying objects (UFOs). Ezekiel, in the Old Testament, saw a wheel that looked like a cloud by day but a pillar of fire by night [Farrakhan has conflated Ezekiel's vision with the pillars of cloud and fire that led the Israelites out of Egypt in the book of Exodus (Ex. 13:21)]. The Honorable Elijah Muhammad said that that wheel was built on the island of Nippon, which is now called Japan, by some of the Original scientists. It took $15 billion in gold at that time to build it. It is made of the toughest steel. America does not yet know the composition of the steel used to make an instrument like it. It is a circular plane, and the Bible says that it never makes turns. Because of its circular nature it can stop and travel in all directions at speeds of thousands of miles per hour. He said there are 1,500 small wheels in this Mother Wheel, which is a half mile by a half mile [800 m by 800 m]. This Mother Wheel is like a small human-built planet. . . .
>
> That Mother Wheel is a dreadful-looking thing. White folks are making movies now to make these planes look like fiction, but it is based on something real. The Honorable Elijah Muhammad said that Mother Plane is so powerful that with sound reverberating in the atmosphere, just with a sound, she can crumble buildings.[21]

Farrakhan has also claimed that in 1985, he ascended into a UFO, where he heard the voice of his predecessor, Elijah Muhammad, predicting future events. In yet another tie to UFO-based religion, Farrakhan praised Scientology; its founder, L. Ron Hubbard; and the auditing process:

> "L. Ron Hubbard is so exceedingly valuable to every Caucasian person on this Earth," Farrakhan said. "L. Ron Hubbard himself was and is trying to civilize white people and make them better human beings and take away from them their reactive minds. . . . Mr. Hubbard recognized that his people have to be civilized," Farrakhan said to a cheering crowd.[22]

Another, though far less virulent, wedding of UFO-based religion to a black supremacist ideology is to be found in the views of the late Ramon Watkins (who died August 2014) of Las Vegas, Nevada, who called himself Prophet Yahweh. In a blog, he claimed that black slaves were the real Israelites and that Israel was actually located in the northeast corner of

Africa.[23] In a video titled *Introduction to UFOs in the Bible,* he stated the following:

> Yahweh and his angels are superhuman black men who live on other planets and travel around the universe in spaceships.[24]

This is similar to Mormon theology, which states that God lives on a planet called Kolob. Watkins went on to say that when the Messiah returns, he will invade Earth with more than 20,000 intergalactic battleships. Despite his rather bizarre interpretation of the Messiah's return, his apocalyptic scenario is basically derived from that of the book of Revelation.

That the Nation of Islam and Prophet Yahweh both created a theology in which a UFO cult was mingled with Afrocentrist views, both tinged with varying degrees of racism, leads us to our next section, in which we look into more sinister aspects of UFO-based religions.

DELUSIONS GONE BAD: RACISM, MOLESTATION, MURDER, AND SUICIDE

It is difficult to keep a straight face while watching Robert Short of the Blue Rose Ministry and George King, founder of the Aetherius Society, in the film *Farewell, Good Brothers*—each supposedly in a trance state, gaping and grimacing, then talking in affected voices as they received channeled communications (Short's from Jupiter, King's from the being Aetherius on Venus). This is especially so because the planets involved, as noted at the beginning of this chapter, can't harbor life and because these "communications" tell us nothing about any advanced science, nor do they give us any clues that might support the assertion that they do indeed have an extraterrestrial source. Likewise, when one sees photos of the late Ruth Norman, cofounder of the Unarius Society, in her guise as Space Princess Uriel, in a kitschy, tacky outfit resembling a bargain-basement version of Glinda, the good witch, from the 1939 film *The Wizard of Oz,* it's difficult not to simply dismiss her and her cult as amusingly harmless. The same cannot be said of various racist UFO cults, such as the Nuwaubian Nation, a bizarre offshoot of the Nation of Islam, and the neo-Nazi UFO cult *Tempelhofgesellschaft.* Nor can it be said of the murderous Order of the Solar Temple—or of the suicidal cult Heaven's Gate.

FIGURE 9.1. The use and misuse of religious symbols in UFO cults. (A) Black Sun, ancient Teutonic sun wheel found in clothing ornaments of the Alemani and of the Merovingian Franks, used both by the SS and the neo-Nazi *TempelhofGesellschaft*. (By A. Wyatt Mann (Own work) [Public domain], via Wikimedia Commons.) (B) Two versions of the Raelian symbol with either a swastika or a galaxy symbol inside a Star of David. (Drawn by User: Mysid in CorelDRAW, via Wikimedia.) (C) The emblem of the Nuwaubian Nation of Moors, featuring an ankh within a Star of David, itself resting on an Islamic crescent.

FIGURE 9.1. *Continued*

The Nuwaubian Nation (Fig. 9.1) began when Dwight York set up the Ansaar Pure Sufi organization in Brooklyn, New York, as a Sufi variant and rival to the Nation of Islam. In the process, he began to call himself Imaam Issa (*Issa* is the Arabic equivalent of Jesus) and, later, Imaam Issa Muhammad. He claimed kinship with an ethnic group in Sudan called the Ansaar, hence the original name of his organization. Somewhat later, York claimed to have come to Earth from outer space on Comet Bennet, which was discovered by John Caister Bennett on December 28, 1969. It passed closest to Earth on March 26, 1970, and was last observed on February 27, 1971. In 1972, after York returned from supposedly visiting the Ansaars, he purged the Ansaar Pure Sufi community of those he thought weren't serious about a commitment to Islam and renamed the organization the Ansaaru Allah Community. He kept that name until 1992, though by the late 1980s he had abandoned Muslim theology and had shifted to ancient Egyptian and extraterrestrial religious themes. In 1992 York took his group to the Catskills for a period of what he called "testing," after which he renamed himself Malachi Z. York and renamed the organization Lodge 19 of the Ancient Mystical Order of Melchizedek. In 1993 the organization moved from Brooklyn to Putnam County,

Georgia, where York built a compound he named Tama-Re, the architecture of which mimicked that of ancient Egypt. Eventually, the group called itself the Nuwaubian Nation of Moors, basing their name on the term *Nuwaubu*, which is found in a 1,700-page document York published in 1996 called the *Holy Tablets*. The name supposedly is a term for the way of life of the supreme beings.

York claims to be an extraterrestrial from a planet called Rizq in the 19th galaxy, called Yaanuwn, and claims that his people, the Annunaki, also called the Eloheem [*sic*], created human life on Earth. The principal geneticists in the creation of the human race were Prince Ea and Princess Ninhursag. Their greatest creations were the Cushites, or Nuwauabu, who have the potential to commune with the Annunaki if they follow the right teachings. The Annunaki left but will return to usher in a golden age. UFOs are actually their ships.[25] It can be seen from this scenario that York based much of his mythology on the writings of Zecharia Sitchin. In the Sumerian pantheon, Ninhursag was an important goddess and the wife of Enki, the Sumerian god later called Ea by the Akkadians. York seems to have based his explanation for the existence of other races on a pastiche of material from the book of Revelation, popular UFO literature, and Black Muslim myth. Annunaki, who had gone rogue, created the various aces on other planets. The white race was originally from the Pleiades, Aldebaran, and Europa, and it aligned itself with the biblical serpent, the pentagram, and the number 666. The gray aliens are a crossbreed between humans and reptilians.[26] It would seem that along with popular UFO mythology, York also borrowed a large slice of bad science. This, however, isn't his only story of the origin of white people. In another creation myth, he says they were created as a slave army for black people, being bred over thousands of years to have shorter penises to curb their rate of reproduction. In yet another story, York claimed white people were the offspring of a leprous woman and a beast.[27]

York's cult of personality was, in the end, more sinister than his racist message. In 2002 he was arrested and charged with more than 100 counts of child molestation, including victims as young as four years old. In 2004 he was sentenced to 135 years in federal prison. His Tama-Re compound was sold under government forfeiture, and its buildings were demolished in 2005.

An opposite form of racist UFO cult was the neo-Nazi white supremacist *Tempelhofgesellschaft* ("Temple Court Society"), founded in Vienna in the 1990s by Norbert Jurgen-Ratthofer and Ralf Ettl. The society eventually disbanded, after which Ettl formed the *Freundeskreis* ("circle of friends") *Causa Nostra* (Latin, "our cause"). Their UFO mythology involves a racist version of an early Christian heresy, Marcionism. As with the original Marcionites, the group identifies Yahweh (also referred to as *El Shaddai*, Hebrew for "God Almighty") as the Demiurge of Gnosticism, the evil creator of this world, who was opposed by Jesus Christ, the new god who overthrew the old system of illusion and corruption. In their racist twist on Marcionism, they identify Yahweh/El Shaddai as the god of Judaism and claim that Jesus Christ was an Aryan. They claim that the Aryan race originally came from the planet Sumi-Er, orbiting the star Aldebaran, and settled in Atlantis, information supposedly derived from Sumerian manuscripts—the Sumerians likewise being Aryans and presumably named after their planet of origin. The Aryans, they believe, derive their power from the vril energy of the Black Sun. They also believe that a huge space fleet is on its way from Aldebaran, which will join forces with a fleet of Nazi flying saucers from Antarctica to establish the Western Imperium. Being of extraterrestrial, hence celestial, origin, they see themselves, the Aryans/Sumerians/Aldebarans, as being mandated to rule the Earth.

Not only did the *Tempelhofgesellschaft* put a racist spin on Marcionism, but it borrowed the Nazi flying saucers hiding in Antarctica from earlier neo-Nazi esoterica, as did Fiat Lux. Its use of the term *vril energy* and its reference to the Black Sun derive from Nazi use of neo-pagan motifs and an earlier Nazi esoteric society involved in pseudoscience. The black sun was a pagan sun wheel symbol known from jewelry of the Alemani, a Teutonic tribe. Of the Nazi esoteric society, rocket scientist and science popularizer Willy Ley, who fled Nazi Germany to the United States in the 1930s, wrote the following in a 1947 article in *Astounding Science Fiction*, titled "Pseudoscience in Naziland":

> The next group was literally founded upon a novel. That group which I think called itself *Wahrheitsgesellschaft*—Society for Truth—and which was more or less localized in Berlin, devoted its spare time looking for Vril. Yes, their

convictions were founded upon Bulwer-Lytton's "The Coming Race." They knew
that the book was fiction, Bulwer-Lytton had used that device in order to be able
to tell the truth about this "power." The subterranean humanity was nonsense,
Vril was not, Possibly it had enabled the British, who kept it as a State secret, to
amass their colonial empire. Surely the Romans had -had it, enclosed in small
metal balls, which guarded their homes and were referred to as lares. For reasons
which I failed to penetrate, the secret of Vril could be found by contemplating the
structure of an apple, sliced in halves.[28]

The Coming Race, written in 1871 by Sir Edward Bulwer-Lytton, describes
a race of men who have nearly godlike psychic powers and who at present
live in vast subterranean caverns. "Vril" was the name Bulwer-Lytton
gave to the energy that was the source of power wielded by this race he
called the "Vrilya." He might have derived these words from the Sanskrit *virya*, which means "manliness." The Latin word for man as male,
vir, derives from the Sanskrit and is the source of the English words
virile and *virility*. Bulwer-Lytton was a member of the Hermetic Order
of the Golden Dawn, an organization of British intellectuals dabbling in
occultism and esoterica. Among its members were Aleister Crowley and
Sir Arthur Conan Doyle. Bulwer-Lytton predicted in his novel that the
godlike Vrilya would one day run out of underground living space and
would have to emerge to take over the surface of the Earth, possibly dis-
possessing or exterminating us lesser humans in the process. Naturally,
the idea of a superior people poised to take over the world, and referred
to as "the Coming Race," appealed to the Nazis, and the idea that these
people were living inside the Earth dovetailed neatly with their hollow
Earth theories. It's hardly surprising, then, that a pseudoscience developed in Nazi Germany that centered on a search for vril.

In developing the mythos of the *Tempelhofgesellschaft*, Jurgen-
Ratthofer and Ettl used the largely invented history of the "Vril Society,"
to which Ley had referred briefly and in passing, as the *Wahrheitsgesell-
schaft*. Commenting on Ley's article, Louis Pauwels and Jacques Bergier
said the following in their book *The Morning of the Magicians*:

> The Berlin group called itself The Luminous Lodge or The Vril Society. The vril is
> the enormous energy of which we only use a minute proportion in our daily life,
> the nerve-centre of our potential divinity. Whoever becomes the master of the vril
> will be master of himself, of others round him and of the world.[29]

So from a passing reference by Willy Ley to one of many pseudoscientific theories rampant in the Third Reich, Pauwels and Bergier created the Vril Society and treated vril as a real energy or life force. From this mention of the group as the Vril Society, neo-Nazi groups developed an elaborate mythology in which they asserted that female mediums of the Vril Society, chief among them Maria Orsitch (also spelled Orsic), channeled messages from a planet orbiting Aldebaran, telling them of the Aryans' origins from that star as well as how to build vril-powered flying saucers. Pauwels and Bergier seem to have helped this myth along. Speaking of the mysterious and beautiful medium, Maria Orsitch, Juan Graña says the following:

> Maria was more beautiful than any Hollywood star at that time. Maria Orsitsch was first mentioned and pictured in 1967 by Bergier and Pauwels in their book: *"Aufbruch ins dritte Jahrtausend: von der Zukunft der phantastischen Vernunft."*[30]

In fact, *Aufbruch ins dritte Jahrtausend: von der Zukunft der phantastischen Vernunft*, which translates into English as *Departure in the Third Millennium: From the Future of the Fantastic Reason*, is the German version of *The Morning of the Magicians*. It's highly suspect that if Maria Orsitch really existed, particularly as the chief medium of the Vril Society, the first mention and photos of her wouldn't have appeared until the 1960s.

In his book *Black Sun*, the late Nicholas Goodrick-Clarke said the following of the popularization of the Vril Society/Aldebaran myth:

> In the mid-1990s this quasi-ariosophical lore of Aldebaran-German linkage penetrated the international UFO scene. The best-selling books of the leading German conspiracy theorist, Jan van Helsing, recapitulated the imaginary history of Vril flying saucers, the settlement of Aldebaran colonists in Sumeria, and the German mission to Aldebaran. Through UFO conferences, Helsing's revelations have encouraged a new wave of Aldebaran channels. In his new book *Unternehmen Aldebaran* [Operation Aldebaran] (1997), Helsing describes how a Bavarian family recalled their contacts with extraterrestrials, involving abductions and implants, while under hypnotic regression. Karin and Rainer Feistle describe a UFO commander showing them enormous "embryo farms" for the breeding of a "new race" that will secure mankind's future. Helsing uses these hints to elaborate on human history.
>
> Arriving on earth 735,000 years ago, Aldebaran colonists bred "slaves" for menial tasks (Helsing uses Lanz von Liebenfels's term Tschandale), but these inferiors revolted and mixed with other races. When the Aldebarans revisited earth much later, they saw that the race-mixing slaves had caused wars, revolution

and political upheaval. The Aldebarans decided to breed up the race on earth by means of the Germans as their closest relatives. They therefore helped the Germans develop advanced technology during the 1930s, while the Feistles' testimony indicates the ongoing Aldebaran project for a German master race. This tableau involving the revolt of inferiors, harmful race mixing and the creation of an Aldebaran-German elite reflect the updating of Ariosophy through a modern mythology of semi-divine extraterrestrial guidance.[31]

The term *Ariosophy*, meaning "wisdom of the Aryans," was first coined by Austrian political/racial theorist and occultist Lanz von Liebenfels in 1915 and became the label for his esoteric doctrines, involving alternate history and race-based pseudoscience, in the 1920s.

The popularization of the invented history of the Vril Society even resulted in a documentary about the society and Maria Orsitch, produced by the Discovery Channel. The program asserted that the society was important in the formation of the Nazi Party and that Adolph Hitler, Heinrich Himmler, Hermann Georing, and Rudolf Hess were all either members or associates of the Vril Society. Thus Willy Ley's passing reference to a minor pseudoscience—one of many that flourished in Nazi Germany—has, in successive retellings, transmogrified into a grand myth. It is also readily apparent from this mythos that popularizers of the Aldebaran/Aryan connection based much of their cosmic history on Zecharia Sitchin's account of the settlement of Earth and creation of humans by the Annunaki. It is interesting to note in regard to the free use of earlier UFO myths that Dwight York's assertion that white people came from Aldebaran and the Pleiades indicates borrowing on his part of the same motif used by the *Tempelhofgesellschaft*. Aldebaran is Arabic for "The Follower," so called because it rises in the night sky just behind—that is, following—the Pleiades star cluster. Both the Pleiades and Aldebaran are located in the constellation Taurus. That both the *Tempelhofgesellschaft* and the Nuwaubians claim that white people, or the "Aryan race," originally came from either Aldebaran, the Pleiades, or (according to York) Europa might well derive from an elaborate hoax perpetrated by Swiss farmer Eduard Albert "Billy" Meier, who, beginning in the 1970s, claimed to have contact with Nordic aliens, who came from the Pleiades. Not only are the Nordic aliens the epitome of white people, but equating Nordics from the Pleiades with Aryan aliens from Aldebaran is a simple step. Dwight York's addition of Europa, a moon of Jupiter, to the Pleiades and Aldebaran as the original home of white people would seem to derive from Greek mythology more than anything else. Whereas the Pleiades, a cluster of seven stars visible to the naked eye, and the star Aldebaran are both in the constellation Taurus (the Bull), Europa's only connection to Taurus is to be found in the myth that Zeus took the form of a bull when he kidnapped Europa, a Phoenician princess, and carried her off to Crete. In short, the only Europa connected to Taurus is the character from Greek myth, not the moon orbiting Jupiter.

Added to the Marcionite dualism and the Aldebaran UFO mythology, Ettl and Jurgen-Ratthofer, the founders of *Tempelhoffgesellschaft*, also asserted that the organization was the secret successor of the Knights Templar.[32] Another cult claiming to be a successor to the Knights Templar took motifs from both the far right and the left-leaning counterculture, mixing New Age and Christian concepts. This was the Order of the Solar Temple (*Ordre du Temple Solaire* or OTS), which began in Geneva, Switzerland, and spread to France and Quebec. The mythos of the cult was mainly a New Age version of a revival of the Knights Templar, with a belief that Jesus would return as a solar god–king. The founders of the Order of the Solar Temple were Joseph Di Mambro, who had previously been involved in the Rosicrucian organization (the Ancient and Mystical Order of the Rosy Cross, or AMORC), and Luc Jouret, a doctor who had given up the regular practice of medicine to pursue homeopathy and other alternative medical practices.

The pair flattered several of their members by telling them that they were reincarnations of such New Testament figures as John the Baptist and Joseph of Arimathea. Di Mambro and Jouret seemingly produced visions of past notables, such as Jacques De Molay, last Grand Master of the Knights Templar. They were also able to convince members that the OTS was in possession of the great menorah from the Jerusalem Temple destroyed by the Romans in CE 70, the Holy Grail, and the sword Excalibur. In Quebec, Jouret and Di Mambro paid one of the cult members, Tony Dutoit, to install apparatus to project images as part of the seeming visions. When Dutoit eventually spoke out about the deception, many members were outraged and left the organization.

Along with strictly Christian motifs and New Age beliefs, the confused doctrines of the Order of the Solar Temple also included Rosicrucian ideas and other bits of semi-Christian esoteric systems of belief. Part of an apocalyptic agenda of their belief system involved leaving Earth to go to another planet:

> Di Mambro said he did not yet know what the mode of transit would be: he presented the metaphor of a passage across a mirror and evoked the possibility of the coming of a flying saucer to take faithful members to another world.[33]

Displaying a bit of the scientific illiteracy common to UFO cults, Di Mambro and his followers seemed to think human beings could live on gas giants:

> Messages received from other dimensions came to bolster them in their resolve to quit this planet. For example, a series of five messages collected under the title, "The Polestar," and supposedly delivered by "the Lady of Heaven," were received between 24 December 1993 and 17 January 1994. The first message calls on the recipients to root out their "terrestrial attraction" and talks about Jupiter as their "Next Home."[34]

The Solar Temple also featured a belief in past extraterrestrial involvement with the human race:

> According to the testament, 26,000 years ago the Blue Star (related to Sirius' energy) left on the earth "Sons of the One"; it appears in the sky every time its help is needed and responds to magnetization when humanity undergoes its crises of transmutation.[35]

As already seen, the method of transit to another planet or star was sufficiently vague as to not explicitly involve death of one's human body. However, the terminology and the stated belief of the OTS that death was an illusion readily lends itself to such an interpretation.

Di Mambro and Jouret were paranoid enough that the ceremonies of the Solar Temple were kept secret, and the two men evolved in their minds an intense belief that they were being persecuted by the authorities. After 1990, Jouret's behavior became increasingly aberrant, and he became obsessed with sex, having sexual intercourse with one of the women in the group before each ceremony, allegedly to give him strength to carry out the ritual. For his part, Di Mambro began using his position of leadership to gain sexual favors and money. Jouret and Di Mambro also interfered with the sex lives of their followers, ordering married couples to break up by claiming that they were not "cosmically compatible"—and then having sex with the now-divorced women. This behavior recalls that of David Koresh, leader of the Branch Davidians, who claimed that he was owed at least 140 wives and was entitled to have sex with any of the women in his compound, situated outside Waco, Texas. It's not clear where the boundary lay in Jouret's and Di Mambro's thinking between the knowing deceptions they practiced and their own true belief. However, the ultimate effect of their secrecy, paranoia, and belief in "transit" was a series of murders and suicides.

On October 4, 1994, Canadian police investigated a fire at a ski chalet used by the Solar Temple cult in Morin Heights, Quebec, and found the corpses of two adults. Two days later, they found a man and woman who had been savagely murdered, along with their three-month-old infant. They later found that the murders had occurred on September 30. The baby, Emmanuel Dutoit, had been stabbed multiple times with a wooden stake. It seems that Di Mambro had ordered the murder because he believed the infant to be the Antichrist. Because Tony Dutoit had spoken out about Di Mambro's deceptions, he was also murdered, along with his wife, Nicky. Tony was stabbed 50 times, Nicky eight.

While police in Quebec were investigating these grisly deaths, police in Switzerland uncovered even greater atrocities. On October 4, the same day when police in Quebec found corpses and a fire in Morin Heights, Swiss police, investigating a fire in the village of Cheiry, at the La Rochette farm, found 23 corpses of men and women who had been drugged and shot in the head multiple times. The following day, Swiss police found 25 corpses of men, women, and even some children at Granges-sur-Salvan. They had died after having poison injected into them by Luc Jouret. Although some of those murdered had apparently been willing victims, perhaps seeing this as their form of "transit," many showed signs of having struggled. Jouret and Di Mambro were among the dead at Salvan.

Another mass death incident related to the Order of the Solar Temple took place during the night between December 15 and 16, 1995. On

December 23, 1995, 16 bodies were discovered arranged in a star formation in the Vercors Mountains of France. It was found later that two of
them had shot the others and then committed suicide.

On the morning of March 23, 1997, five members of the OTS took their
own lives in Saint-Casimir, Quebec. A small house erupted in flames,
leaving behind five charred bodies for the police to pull from the rubble.
Police then found three teenagers, aged 13, 14 and 16—the children of
one of the couples who died in the fire—in a shed behind the house,
alive but heavily drugged.

Though seemingly far less monstrous than either the multiple child
molester, Dwight York, or the murderous Solar Temple pair, Jouret and
Di Mambro, as well as having beliefs far less repugnant than the open racism of either the Nuwaubians or the *Tempelhofgesellschaft*, the Heaven's
Gate cult, based on a mix of science fiction, Christianity, and UFOs, was,
nevertheless, among the more lethal of UFO-based religions gone bad.
In the early 1970s, music teacher and singer Marshall Herff Applewhite
entered into a platonic relationship with Bonnie Lu Nettles, a nurse who
had an interest in the occult. After attending a retreat in March 1975 at
Gold Beach, Oregon, they became convinced that they were the two
witnesses from chapter 11 of the book of Revelation, who testify and are
martyred. Revelation 11:3 says

> and I will give power unto my two witnesses and they shall prophesy a thousand
> two hundred and threescore days, clothed in sackcloth.

The 1,260 days in this passage equals three and a half 360-day years, or
the first half of the seven-year reign of the Beast. Revelation 11 goes on
to tell of their martyrdom (11:7):

> And when they shall have finished their testimony, the beast that ascendeth out
> of the bottomless pit shall make war against them and shall overcome them and
> kill them.

This, however, isn't the end of the two witnesses (11:11, 12):

> And after three days and a half the Spirit of life from God entered into them,
> and they stood upon their feet; and great fear fell upon them which saw them. And
> they heard a great voice from heaven saying unto them, "Come up hither." And
> they ascended up to heaven on a cloud; and their enemies beheld them.

Thus Marshall Applewhite and Bonnie Lu Nettles came to believe that they, as the two witnesses, were called upon to prophesy and die, after which they would ascend to heaven on a naturally occurring vehicle. To this end, they started a group called Human Individual Metamorphosis, which eventually became Heaven's Gate. On September 14, 1975, 20 people left their jobs and families to join the organization; still others joined later. In the 1980s they moved to Rancho Santa Fe, near San Diego, California, where they rented a mansion.

The theology of Heaven's Gate involved the "methodical termination" of their "terrestrial containers." They believed that Heaven implants souls into earthly bodies to help them figure out their place by overcoming the weaknesses of the earthly containers—such material desires as greed and lust. One of the ways they overcame the latter weakness was for the males in the group to have themselves castrated. Voluntary castration to avoid the snares of sexuality was also practiced by the *Skoptsy*, a Russian sect that began in the late 1700s and persisted into the 20th century. Self-castration was also practiced by followers of Attys and Cybele in Roman times.

Regarding themselves as the two witnesses of Revelation 11, Applewhite and Nettles sought a sign for their ascension and that of their followers. It came in the form of the comet Hale–Bopp, which appeared in 1996. Applewhite believed that the tail of the comet concealed the spaceship that would take him and the members of his cult to the next level of existence. Accordingly, on March 27, 1997, police discovered 39 bodies at the mansion, the result of a mass suicide accomplished by a combination of alcohol, barbiturates, and asphyxiation.[36] The final statement on the Heaven's Gate website reads as follows:

Whether Hale-Bopp has a "companion" or not is irrelevant from our perspective. However, its arrival is joyously very significant to us at "Heaven's Gate." The joy is that our Older Member in the Evolutionary Level Above Human (the "Kingdom of Heaven") has made it clear to us that Hale-Bopp's approach is the "marker" we've been waiting for—the time for the arrival of the spacecraft from the Level Above Human to take us home to "Their World"—in the literal Heavens. Our 22 years of classroom here on planet Earth is finally coming to conclusion— "graduation" from the Human Evolutionary Level. We are happily prepared to leave "this world" and go with Ti's crew.[37]

Ti was one of the names used by Bonnie Lu Nettles, who had died of liver cancer in 1985.

Whole books have been written about both the Order of the Solar Temple and the Heaven's Gate cult. Our coverage of these two cults is necessarily brief, because this chapter presents an overview of UFO-based religions. What we can see in this overview are the traits held in common by those UFO-based religions that produced atrocities. Among all these cults, very specific motifs appear in groups that are totally unrelated to each other. For example, both the OTS and Heaven's Gate looked for a special form of transition to a higher plane. Both the OTS and the *Tempelhofgesellschaft* saw themselves as successors to the Knights Templar. Comets as vehicles occur in both the claim by Dwight York that he had arrived on Earth on Comet Bennet and the belief on the part of the members of Heaven's Gate that they would leave Earth by means of Comet Hale–Bopp. The Marcionite dualism of *Temepelhofgesellschaft* is reflected in Di Mambro and Jouret's telling their followers to "root out their terrestrial attraction" and the voluntary castration on the part of the males of Heaven's Gate. This also relates to the interference on the part of the charismatic leaders of these cults with the sex lives of their followers. Applewhite and Nettles's telling their followers to cut themselves off from earthly desires, such as sex—resulting in the males' having themselves castrated—is reflected in Di Mambro and Jouret's ordering of married couples to part and then sexual intercourse with the newly divorced women as well as in Dwight York's multiple acts of child molestation. All these cults featured an apocalyptic message, a distorted mythology of special and superior extraterrestrial origin of the members of the cult, and self-selection on the part of their followers. The less mystically inclined, those more scientifically literate and those less intent on a spiritual search, tend to select themselves out of UFO-based religions in general and those with dark designs in particular.

SUMMARY

In this chapter, we haven't tried to conduct an exhaustive analysis of all the UFO-based religions. Such a task would require an entire book. We largely concentrated on belief systems in which the UFO aspect was central to the religion or else was directive, as in the Order of the Solar Temple. Thus, although we mentioned the Xenu myth of Scientology in our chapter on ancient aliens, we did not discuss Scientology in this chapter, because that cult goes out of its way to hide the Xenu myth— even from most of its members, who could spend years in Scientology and know nothing of the story. In contrast, the Unarius and Aetherius societies, Ashtar Galactic Command, the Nuwaubians, the *Tempelhof-gesellschaft*, the Order of the Solar Temple, Heaven's Gate, and all others covered in this chapter are (or were) quite open with their members about the extraterrestrial components of the religion.

Although a systematic study of extraterrestrial-based religions is beyond the scope of this book, we can make certain general inferences about what stimulated the growth of UFO cults in the 20th century. We have compiled a list of UFO cults from various internet sources, and we have the dates for 25 religions based, to some degree, on UFO belief. Only two of these were in existence before the 1950s (the Urantia Foundation and the Nation of Islam). Seven came into being in that decade; three more began in the 1960s; six more were started in 1970s and two in the early 1980s. Five more such religions came into being in the 1990s. The seven UFO-based religions that began in the 1950s constitute 28% of those on the list, whereas the six that began in the 1970s constitute 24% and the five that sprang up in the 1990s 20%. Thus 72% of those on the amended list came from these three decades. We can infer that the Cold War and, with it, the buildup of nuclear arsenals was a strong impetus for the formation of the UFO-based religions that originated in the 1950s, promising, or at least implying, intervention and salvation from above. We might likewise infer a New Age origin for those that came into being in the 1970s. One of those, however, was Heaven's Gate, which had a strongly apocalyptic doctrine. The two cults that began in the 1980s, Fiat Lux and the Order of the Solar Temple, likewise had strong apocalyptic scenarios or aspects. The final decade of the 20th century naturally saw doomsday cults spring up and flourish. The five religions from the amended list that began in the 1990s—Chen Tao, the Atman Foundation, the Universe People, the Planetary Activation Organization, and *Tempelhofgesellschaft*—were all millenarian. So we can see three major pulses stimulating the formation and growth of UFO-based religions: Cold War fears of atomic war, the rise of New Age belief in the wake of 1960s counterculture, and end-of-the-century millennialism.

The puzzling adherence of otherwise intelligent people to manifestly irrational and cultic belief systems has, we suspect, to do with repeated acts of self-selection. One of us (Callahan) attended, out of curiosity, a series of free introductory Scientology lectures while in college. He found the assertion that L. Ron Hubbard was a neurosurgeon, a nuclear physicist, and an inventor so preposterous that when the "success secretary" interviewed him afterward, asking what insights he had derived from Scientology so far, he told her how absurd he found the presentation. He was asked to leave. A friend of his, also attending, was more tactful and was offered the chance to take the relatively cheap introductory course—which he declined.

The natural tendency among those properly socialized is to show a degree of polite tact, even in the face of absurd or offensive ideas. Those who accept such offers as the opening course in Scientology are more likely to conform to expectations of polite response and more inclined to please others than was the author. They are also likely to be more credulous and to be in search of something in which to believe. This is the first act of self-selection. Once people have paid for a course, invested a certain amount of time, or formed social connections with members of a cult, they feel an internal psychological pressure to believe that they have not wasted time, money, or involvement. Thus they are likely to rationalize away any evidence that their efforts have been wasted. We might call this the "Emperor's New Clothes Phenomenon." Those who accept that they've been wasting time select themselves out from further involvement. Those who rationalize and continue to search for evidence that they have not, in fact, been wasting their time and money select themselves into deeper involvement. This, in turn, intensifies the likelihood they will repeatedly self-select for deeper involvement the more intensely irrational they find the beliefs of the cult to be.

Those who remained in the Solar Temple cult in the face of absurd claims that the OTS was in possession of such mythical objects as the Holy Grail and the sword Excalibur had self-selected for credulity to the degree that some of them stayed on even after Tony Dutoit exposed the fakery of the apparitions. Further acts of self-selection occurred

when DiMambro and Jouret demanded that they dissolve their marriages, particularly on the part of the men whose ex-wives then became
sexually involved with the cult leaders. Thus they were well on their way
to self-selecting eventual complicity in their own murders.

Even the strongly disconfirming evidence of failed predictions will
often have no effect on those who have repeatedly selected themselves
for belief. In the book *When Prophecy Fails*, Leon Festinger and coau-
thors Henry W. Riecken and Stanley Schachter point out that the failure
of the flood to come on December 21, 1954, did not shake the devotion of
most of those in the inner circle of the Seekers:

> Summarizing the evidence on the effects disconfirmation had on the conviction of
> group members, we find that, of the eleven members of the Lake City group who
> faced unequivocal disconfirmation, only two, Kurt Freund and Arthur Bergen,
> both of whom were lightly committed to begin with, gave up their belief in Mrs.
> Keech's writings. Five members of the group, the Posts, the Armstrongs, and
> Mrs. Keech, all of whom entered the pre-cataclysm period strongly convinced
> and heavily committed, passed through this period of disconfirmation and its
> aftermath with their faith firm, unshaken, and lasting. Cleo Armstrong and Bob
> Eastman, who had come to Lake City heavily committed but with their conviction
> shaken by Ella Lowell, emerged from the disconfirmation of December 21 more
> strongly convinced than before. Cleo's change of heart seems to have lasted
> while Bob's may have been temporary. Bertha Blasky and Clyde Wilton started
> out with some doubts. They reacted to the disconfirmation by persisting in their
> doubts and admitting their disillusionment and confusion, but still not completely
> disavowing Mrs. Keech and her particular beliefs.[38]

So the only two of the 11 to reject the validity of Marion Keech's automatic writings, even in the face of the unequivocal failure of her prophecy, were those who weren't all that committed to the belief system in
the first place. Thus, once a certain depth of commitment is reached,
belief seems to intensify in inverse proportion to the evidence available
to support it.

Estrangement is another factor motivating membership in UFO-
based religions. Those of us who feel we are not part of the mainstream,
not part of that great hump in the middle of the bell curve, can feel
alienated from the rest of humanity. Speaking of this sense of feeling
excluded, Terrie Nettles, daughter of Bonnie Lu Nettles, cofounder of
Heaven's Gate, tells of looking out at the night sky with her mother when
she was 14, hoping to see a UFO:

It would be really neat if one would come and pick us up and take us away, because neither one of us felt like we were part of this world, that we were always on the outside, looking in.[39]

This sense of not belonging could easily push one toward deeper self-selection for the sake of a sense of belonging. Perhaps the best antidote for this sense of not fitting in is to remember that those of us who are outliers on the bell curve are part of the natural order of things: the curve wouldn't be shaped like a bell without its flanges.

UFO-based religions constitute a major pillar of belief in UFOs along with sightings, alien abductions, and ancient astronaut theories. In fact, they form a logical meeting place for modern-day stories of alien contact and the mythos of ancient astronauts. As such, they are part of a unified worldview—one that persists, seemingly, in inverse proportion to the lack of both supporting evidence and logic. It is a folk belief for modern times. UFO-based religions bring comfort for those estranged from established religions that no longer satisfy their needs. The members of these religions also often see themselves as exceptional initiates rather than merely among those who are estranged from the mainstream of society. This is an elaboration on the status of those who consider themselves to be abductees or contactees—that is, people who see themselves as singled out and chosen by the space aliens. So UFO-based religions give their members three things: a sense that there is a purpose underneath the seeming chaos of random events, a belief in a power for good that is far more palpable than the often remote and abstract deity of monotheistic religions, and a faith and confidence in themselves as being of some importance in the vast scheme of things. In short, UFO-based religions offer their members everything that religions have always offered their followers. Perhaps, then, considering the steady erosion of traditional religious belief in an increasingly secularized world, the rise in UFO-based religions purporting to be based on something even vaguely scientific is inevitable.

10

Ancient Myths and Modern Media

HOLLYWOOD HOCUS-POCUS

An alien spaceship traveling at thousands of miles per hour is tracked by radar stations across the globe. Finally, it lands on a grassy field in Washington, DC, where it is soon surrounded by army units, artillery, and tanks. A visitor from another world steps out of a portal that opens mysteriously from the ship's smooth, seamless surface. He is, surprisingly, completely human (Fig. 10.1). Not only that, but he is quite tall and distinguished-looking, with a sculpted face characterized by prominent cheekbones. The spaceman speaks perfect English and says that he comes in peace, but when he holds out a strange object, a trigger-happy soldier mistakes it for a weapon and fires on the visitor, wounding him. This is from the opening sequence of the 1951 film *The Day the Earth Stood Still*. Despite being filmed in black and white and employing limited visual effects, the movie, featuring Michael Rennie as the noble space alien Klaatu and Patricia Neal as the young widow who aids him when he is unjustly set on by government and military authorities, is impressive and has become an iconic science fiction film. It's no wonder that George Adamski's description of Orthon, who came to Earth to warn us of the dangers of nuclear war—just as Klaatu had done in the movie—was such that Alice Wells's drawing of him looks suspiciously like Klaatu, particularly in terms of the one-piece garment Klaatu wears in the opening sequence of the film. Remember that the film was released in 1951 and that Adamski supposedly first met Orthon in the California desert on November 20, 1952.

FIGURE 10.1. Benign fully human extraterrestrials. (A) Klaatu (portrayed by Michael Rennie) and the robot Gort in *The Day the Earth Stood Still*, 1951. (Courtesy 20th Century Fox/The Kobal Collection.) (B) Orthon, the Venusian, based on a drawing by Alice K. Wells. (Drawing by T. D. Callahan.)

The long list of UFO sightings, tales of alien abductions, claims of "contactees," and theories of ancient aliens' having created human civilization constitutes a new mythology generated in the 20th century. Thus it is not surprising that it was largely transmitted through motion pictures and television, which became the source of much of its imagery. Other sources of its imagery were comic strips and pulp science fiction magazine covers. Though many of the motifs of this new mythology are based on ancient tropes, these have been reformatted to fit a society that largely rejects the pagan gods, demons, and fairies of ancient myths and legends. Thus aliens who have near-magical technology have replaced supernatural beings who have magical powers. Also, whereas the ancient myths were communicated by storytellers and written words, modern myth is propagated by modern means. The visual media of film, television, and even graphic arts have far greater emotional, indeed visceral, impact and far greater immediacy than does the medium of the written word. Even the spoken word, transmitted via the medium of radio by voices rich with human emotion, has a more immediate and instinctual effect on audiences than does the more cerebral written word. This was demonstrated on October 30, 1938, when Orson Welles presented, on the Mercury Theater, a dramatization of H. G. Wells's *The War of the Worlds*, adapted for radio by Howard Koch. Its format of fictional news reports interrupting broadcast music led some people to believe that an actual Martian invasion was in progress.

The threat of invasion by powerful, implacable enemies excites primal fears, as do the threats of nuclear annihilation, being defenseless against intrusion or capture, and having our free will taken from us. Movies and television shows expressing such fears have served as the sources of tales of alien abduction and contact, the types of aliens in the UFO mythos, and specific aspects of contact and abduction scenarios. These powerful images and motifs first appeared in comic strips in the 1920s, continued in the movies in the early 1950s, and were carried on by television shows and miniseries from the 1960s to the early 1980s.

BENIGN ALIENS AND THE NUCLEAR THREAT

Klaatu, as a wiser version of a human being, representing all that is noble in the human race, landed on Earth to warn humanity that if it didn't curb its warlike nature—in particular, if it didn't find a solution to the threat of nuclear war—an interstellar robot police force, over which Klaatu and his fellow advanced aliens had no control, would annihilate the people of Earth, who, armed with nuclear weapons and having warlike tendencies, constituted a threat to civilization. The United States had destroyed Hiroshima and Nagasaki with atomic bombs in August 1945 and had subsequently lost its monopoly on nuclear weapons when the Soviet Union exploded its first atomic bomb in 1949, just two years before the release of *The Day the Earth Stood Still*. Thus the movie reflected fears that were growing as the nuclear arms race began.

On November 1, 1952, the United States detonated its first large-scale thermonuclear weapon, or hydrogen bomb, with an explosive yield of

10.4 megatons (million tons of TNT) at the Enewitok Atoll in the Marshall Islands in the South Pacific. The results of the test were made public on November 16, and George Adamski claimed to have first met Orthon on November 20. By the following year, the Soviet Union had tested its first thermonuclear weapon. Thus Klaatu's warning in the 1951 film had been accentuated by the advent of the hydrogen bomb by the time Adamski invented Orthon, the Nordic-type alien from Venus.

Though not as well known as *The Day the Earth Stood Still*, the 1957 film *The 27th Day* further developed the motif of the benign, but somewhat reproachful, fully human alien and the threat of the annihilation of the human race by powerful weapons in the hands of irresponsible leaders. In this film, five humans, selected seemingly arbitrarily, are transported to a flying saucer, where a man who introduces himself only as "The Alien" tells them that in 35 days the sun of his world will go nova and wipe out his people. Their only hope is to inhabit the Earth. Although it is against their moral code to kill intelligent beings, they give their five abductees the means by which the human race can destroy itself. The five humans are Eve Wingate, a young Englishwoman who had just been swimming in the English Channel off the Cornish coast; Jonathan Clark, a Los Angeles reporter; Dr. Klaus Bechner, a German physicist; Su Tan, a Chinese peasant woman; and Ivan Godovski, a Russian soldier. The Alien gives each of them a container holding five capsules, each of which, when launched, will destroy all human life within a 3,000-mile radius without harming any other living thing. The boxes can be opened only by the thought waves of the person to whom they've been given and will become inert if that person dies. They will all become inert in 27 days. Although the boxes can be opened only by the person to whom they've been given, after the boxes are opened, the capsules can be launched by anyone who speaks their desired coordinates, the center of the 3,000-mile radius of annihilation.

Having given the five their boxes, The Alien returns each of them to the place and time he or she was abducted. Because the alien spaceship is traveling near the speed of light, no time has passed since their abduction. Su Tan selflessly commits suicide before a Buddhist shrine, thereby deactivating her capsules. Eve hurls her box into the sea, calls Jonathan Clark, and takes a plane to Los Angeles to meet him. The Alien then complicates matters by commandeering all radio and television channels and broadcasting to the world the nature of the capsules and who has them. This eventually results in a Soviet commissar's getting control of one of the boxes and issuing an ultimatum to the United States to withdraw all military presence from Europe and Asia. When the Americans comply, he prepares to launch the capsules to wipe out every human in North America on the 27th day so that they won't be able to retaliate. However, Dr. Blechner decodes writing on the capsules that shows him how to launch all of them in a manner such that they kill only enemies of human freedom worldwide. The people of Earth then invite the aliens to come to Earth, where they can use their advanced science to make deserts habitable and productive. Considering that it was made in the midst of the nuclear arms race, the ultimate optimism of *The 27th Day* is remarkable.

ABDUCTION AND INSIDIOUS INTRUSION

Although "The Alien" of *The 27th Day* was benign, he did not hesitate to abduct and transport the five humans to whom he gave the deadly capsules. Two years before, in the 1955, film *This Island Earth*, reasonably benevolent and very human-looking space aliens used what *Star Trek* would later call a "tractor beam" to capture the hero and heroine, who were fleeing in a private plane, pulling the plane up into their flying saucer. We will deal with this particular means of capture a bit later. The motif of capture by space aliens in *This Island Earth* had already been used in the 1953 film *Invaders from Mars* (Fig. 10.2). In that movie, the capture, by malign Martians, was far creepier. Accompanied by an ominous droning, the sand of the sand pit in which the Martian flying saucer had buried itself was sucked down, and the unlucky human walking over it was as well. This was clearly patterned after the action of an ant lion capturing and killing ants.

Not only did the Martians suck people into the sand pit, but once they captured them, they implanted tiny mechanisms at the base of the people's skulls by which to control their minds—thereby turning the loving parents of young David MacClean into cold, heartless automatons and turning a little girl who is one of David's friends into a creepy

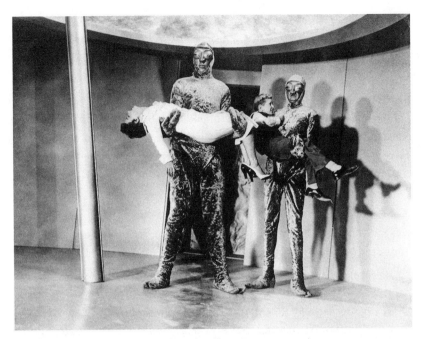

FIGURE 10.2. Alien abduction: Dr. Patricia Blake (Helena Carter) and young David MacLean (Jimmy Hunt) are abducted by evil Martians in the 1953 film *Invaders from Mars*. (Courtesy 20th Century Fox/The Kobal Collection.)

pathological arsonist. The paranoia that pervades the film is heightened when David tries to tell the police what happened to his parents. He is interviewed by the chief of police, but when the chief turns around in his swivel chair to answer a phone, David sees the telltale scab on the back of the man's neck—he, too, is under Martian thought control. After the Martians have used their mind-controlled humans for a given task, they do away with them by detonating the implant, thereby causing a cerebral hemorrhage. When, toward the end of the film, David and Dr. Patricia Blake are captured by the Martians, David witnesses the unconscious Dr. Blake lying on a glass table with the Martian implant being lowered toward the back of her neck. *Invaders from Mars* was released in April 1953, near the end of the Korean War (June 1950–July 1953) amid fears of brainwashing. Though claims of rampant communist brainwashing were largely debunked by later investigations, during the Korean War and in its aftermath, many Americans believed that people's minds could be controlled through a combination of torture and psychological subversion. The control of people through tiny implants lodged in their brains was the product of such fears and fed into them in such a way as to make the original *Invaders from Mars* a masterpiece of paranoia.

Thus the abductions of *The 27th Day* and *This Island Earth*, plus the sand pit captures of *Invaders from Mars*, had firmly implanted the motif of capture by space aliens into the public imagination well before the narrative of Barney and Betty Hill formed in 1964. Likewise, the mind control by implants into the brain in *Invaders from Mars* focused the American public's fear of Communist mind control and redirected it toward insidious aliens well before implants became a staple of alien abduction stories. It is also interesting that so many claimed abductions followed the Hills' narrative, both in time and in content. Common to most of them are that they occur in isolated areas, either to subjects who are alone or to a small group in a wilderness, and that they usually occur at night. Although there might be a certain logic to this if it were being done as part of a clandestine operation, such scenarios make no sense in the many cases in which the aliens deliver an apocalyptic warning that the abductee is clearly directed to relay to the entire human race. The delivery of the warning in the movies is actually more realistic than it is in abduction narratives. Klaatu landed near the White House in Washington, DC, in full daylight after his trajectory had been tracked by radar across the globe, and "The Alien" of *The 27th Day* transmitted his message worldwide by commandeering the broadcasting stations of the entire planet.

Similar to the motif of alien implants is Betty Hill's claim that the gray aliens inserted a needle into her navel—which, as we note in our chapter on alien abductions, has parallels in fairy abductions and the vision of St. Teresa of Avila. The piercing of the abdomen in both St. Teresa's vision and Betty Hill's narrative, related while under hypnosis, is almost certainly a form of sexual imagery. This is particularly borne out in St. Teresa's description of the sensations caused by the angel piercing her body with a spear:

> The pain was so great, that it made me moan; and yet so surpassing was the sweetness of this excessive pain, that I could not wish to be rid of it.[1]

In Bernini's sculpture of the ecstasy of St. Teresa, her face, as noted in chapter 7, mirrors that of a woman in the throes of orgasm.

Even leaving aside the imagery of piercing related both by Betty Hill and St. Teresa, the element of abduction is common in sexual fantasies. The 1957 abduction narrative of Antonio Villas Boas was explicitly sexual. He was taken aboard the spacecraft to impregnate a beautiful alien woman. Abduction as part of sexual fantasy is a particularly common theme in romance novels, as can be seen by looking up "abduction in romance novels" on the internet and clicking on "images." The idea of loss of control is implicit in the sexual act, and abduction in sexual fantasies removes the onus of violating sexual taboos. The persons fantasizing needn't justify taking part in a forbidden act, because this act is perpetrated against their will. Sexual abduction, albeit slightly veiled, is also a part of more mainstream literature. For example, the abduction of Rebecca by Bois Guilbert in Sir Walter Scott's novel *Ivanhoe* (1819) was sexualized in Delecroix's 1846 painting *The Abduction of Rebecca*. Scott might have derived his abduction of Rebecca from that of Guinevere, a repeated theme in Arthurian legend, first written of in the *Vita Gildae* (*Life of Gildas*) by Caradoc of Llancarfan, ca. 1140, again by Chretien de Troyes in *Lancelot le Chevalier de la Charrette* (*Lancelot Knight of the Cart*), ca. 1177, and finally by Sir Thomas Mallory in *Le Mort d'Arthur* (*The Death of Arthur*), published in 1485. In Greek myth, abduction is common in tales of gods having sexual relations with mortal women. Two of those abductions are particularly prominent—that of Zeus, in the form of a bull, carrying Europa off to Crete and that of Hades carrying Persephone off into the underworld. Abduction is also a means by which the central character in a fictional work is able to leave his or her mundane responsibilities to go off on an adventure. Those abducted aren't being irresponsible; rather, they've been taken into a fantasy or an adventure against their will. Consider, for example, *Kidnapped*, by Robert Louis Stevenson, published in 1893. Thus abduction as a story trope with strong sexual overtones was well established in literature from ancient times by the time alien abduction narratives were related in the 20th century.

A more immediate and graphic source for alien abduction scenarios, and one that is implicitly sexual, is to be found in the *Buck Rogers* comic strip from 1930. In the panels of this comic strip, Buck's sexy girlfriend is

first shown being carried off in a giant claw descending from an extraterrestrial spaceship. Later she is shown lying unconscious on an examining
table, being scrutinized by catlike aliens.[2] Thus an ancient mythic and
literary trope was emphasized and made more graphic by comic strips
and movies in the 20th century.

ADVENT OF THE FLYING SAUCER

The benign Klaatu of *The Day the Earth Stood Still* and the malevolent
Martians of *Invaders from Mars* had one thing in common: both came
to Earth in flying saucers. Supposedly, the first sighting of flying saucers
was by amateur pilot Kenneth Arnold on June 24, 1947. However, what
Arnold actually reported was that the UFOs he saw moved erratically,
like saucers skipped across water. He did not say they *looked* like saucers.
Here is Kenneth Arnold in his own words, from a 1950 telephone interview with Edward R. Murrow:

> These objects more or less fluttered like they were, oh, I'd say, boats on very rough
> water or very rough air of some type, and when I described how they flew, I said
> that they flew like they take a saucer and throw it across the water. Most of the
> newspapers misunderstood and misquoted that too. They said that I said that they
> were saucer-like; I said that they flew in a saucer-like fashion.[3]

In fact, Arnold initially said that what he saw looked like the heel of a
shoe, with the curved edge leading. Eventually he described them as
shaped like boomerangs. By 1952, however, Arnold himself seems to
have acquiesced to the idea that the objects were saucer-shaped or disc-
shaped. In that year a book was published titled *The Coming of the Saucers*, which was coauthored by Arnold and Raymond Palmer, one of the
founders of *Fate Magazine*.

Concerning Arnold's credibility, in later interviews with journalist
and UFO investigator Bob Pratt, Arnold said that the government had
tapped his phone line, that UFOs might be live jellyfish-like creatures
living in our atmosphere, that the creatures seemed to be able to read his
mind, and that he had spotted UFOs on seven or eight different occasions.[4] His drawing and illustrations based on it actually look somewhat

like crescents. However, UPI reporter Bill Bequette mistakenly reported that Arnold had described what he saw as flying saucers.[5] From the time of this misquote, witnesses who saw UFOs began to characterize what they saw as disc-shaped. That those who believe they've been abducted by aliens derive the spaceships of their captors from a media trope rather than from an actual experience is illustrated by their description of the alien craft as saucer-shaped, as erroneously reported by Bill Bequette, rather than seeing them as boomerang- or crescent-shaped, following the actual description of Kenneth Arnold.

However, it was not this error alone that was the source of the disc-shaped craft. As noted, they were already part of the science fiction comic *Buck Rogers* by 1930. An even earlier image of flying saucers is found on the cover of the December 1915 issue of *The Electrical Experimenter* as part of a canal scene on Mars.[6] Perhaps the original inventor of this image, or at least its foremost popularizer, was science-fiction illustrator Frank R. Paul. His illustration for the cover of the November 1929 issue of *Science Wonder Stories* depicts a giant flying saucer extruding metallic tentacles, with which it is carrying off a skyscraper (Fig. 10.3). This is but one of several covers Paul painted that depict menacing flying saucers. In one, acting like a giant buzzsaw, a saucer is shearing an airplane in two.

In another of Paul's cover illustrations, a flying saucer is pulling up a chunk of earth using triangulating tractor beams.[7] The term *tractor beam*, so liberally used in the *Star Trek* television series, was, incidentally, invented by science fiction author E. E. Smith in his "Lensmen" novels, beginning in 1929. Thus the image later used in *This Island Earth*, as the means by which a flying saucer captures the hero and heroine trying to flee in their private airplane, existed decades before the film was made. Of course, by the time Travis Walton claimed to have been sucked up in a beam of light into an alien spaceship in 1975, *Star Trek* had been using tractor beams for nearly a decade. So the energy beam concept began ca. 1930 in the Lensmen novels and in a pulp magazine cover, was made even more graphic in film in 1955, and had become a stock prop on a television series by the late 1960s—all well before Walton, exhibiting a paucity of imagination, used it to sell a book and movie rights.

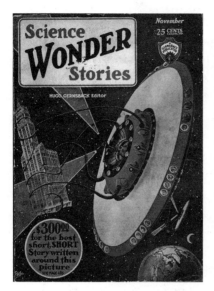

FIGURE 10.3. (A) *Science Wonder Stories* 1, no. 6 cover. (B) *Air Wonder Stories* 1, no. 10 cover. (C) *Wonder Stories* 7, no. 5 cover. (D) *Wonder Stories* 7, no. 8 cover. (E) *Amazing Stories* 16, no. 6 cover. (All images courtesy of the Frank R. Paul Estate and The Lilly Library, Indiana University, Bloomington, Indiana.)

FIGURE 10.3. *Continued*

ADVENT OF THE GRAY ALIENS

As noted in chapter 6, the gray aliens so common in American abduction narratives were first described and drawn by Barney Hill while he was under hypnosis on February 22, 1964. This was only 12 days after the science fiction television program *The Outer Limits* aired the episode titled "The Bellero Shield" (February 10, 1964). Before this, Betty had related, under hypnosis, that the aliens who abducted her and Barney had black hair, bluish lips, and prominent noses. The alien woman with whom Brazilian farmer Antonio Villas Boas claimed to have had sex in October 1957 was a beautiful human with long white hair but bright red pubic hair. Likewise, the extraterrestrial woman whom Howard Menger claimed he first met when he was 10 years old was fully human. The grays do not appear in UFO sightings before Barney Hill's recollection of them under hypnosis in 1964. They began to proliferate after that.

An earlier depiction of stunted gray, seemingly sexless, aliens is, however, found in the description of the Selenites in H. G. Wells's *The First
Men in the Moon*:

> There was no nose, and the thing had dull bulging eyes at the side—in the
> silhouette I had supposed they were ears. There were no ears . . . I have tried to
> draw one of these heads, but I cannot. There was a mouth, downwardly curved,
> like a human mouth in a face that stared ferociously.[8]

It is conceivable that the alien from "The Bellero Shield" was based on
Wells's Selenites. The idea that humans from the future might be diminu-
tive, have large heads, and be physically weak is common in science fiction. As skeptic Aaron Sakulich notes at his *Iron Skeptic* website, small
gray men with bulbous heads, often depicted as what humans would
look like in the future, are found in fictional works going back to 1891.[9] It
is a common, though unfounded, assumption in many such works that
an increase in mental acuity leads to a decrease in physical vigor. Thus
humans of the future are depicted as sickly.

This brings us to the myths of dying races and human–alien hybrids.
According to some, the gray aliens are a slave race created through clon-
ing by the evil reptilians, against whom they have risen in revolt:

> To re-cap briefly, we will go over the basic background of how this movement
> started and why the Grey Aliens are so interested in "harvesting" human beings.
> Grey Aliens struggle with a daunting issue that befalls their race; they are ALL
> dying. Unlike many alien races found throughout the cosmos Grey Aliens are
> cloned beings that were created by their "parent" race of aliens infamously known as
> the Reptilians from the Draco Constellation. . . . Because they are clones (created to
> be a slave race), they suffer from genetic mutations, diseases, but most problematic
> they are slowly dying. This poses a problem because Grey Aliens were created
> without any reproductive organs, hence their issue! Combined with the inability to
> re-produce and little knowledge of the cloning technology that the Reptilian Aliens
> possess, Grey Aliens face a time sensitive issue that they have now set out to solve.[10]

Dying planets are found in both the 1955 movie *This Island Earth* and
the 1957 film *The 27th Day*. In the later film, this was because the planet's
sun was about to go nova. In *This Island Earth*, Metaluna, home planet of
the humanoid aliens who have abducted the hero and heroine, is being
destroyed by its enemies. The scientifically improbable attempt of the
grays to create hybrids with humans partakes of two tropes. The first is the

mythic motif of gods siring heroes on mortal women. This was elaborated in Jewish apocalyptic works, such as the Book of Enoch, as the creation of a hybrid race, the Nephilim, by rebellious angels' having sex with mortal women. Narratives of aliens impregnating human women also constitute a further elaboration of the sexual fantasy aspect of alien abduction stories.

STAR CHILDREN AND CHANGELINGS

Although narratives of aliens mating with Earth women or extracting sperm from human males to create hybrid beings can to a degree be traced back to Greek myth and to the Nephilim of Genesis 6:2–4 and the Book of Enoch, it has a much more immediate source, first in the 1957 John Wyndham novel *The Midwich Cuckoos* and then in the 1960 movie based on the novel: *Village of the Damned* (Fig. 10.4). In both the novel and the film, the people of the English village of Midwich fall unconscious, as does anyone entering the town, when a UFO flies over it. Following that event, all the women in Midwich become pregnant and give birth on the same day only seven months later, though the babies aren't premature. Their children, evenly divided between boys and girls, all look alike, with pale skin and fine blond hair, and are telepathic and cold-hearted. When a young man driving a car accidentally hits one of the children, injuring the child's hip, the others, taking control of his mind, force him to drive into a wall, killing himself. When the man's brother tries to avenge him, the children make him shoot himself. Eventually the hero of the story blows himself up with the children, ending the threat.

The cover of a recent book on alien/human hybrids, *Children of the Greys*, by Bret Oldham, features a cover painting of just such a pallid child. This depiction of a child of gray aliens and humans is quite appealing. To some degree it shows the influence of artist Margaret Keane, who, in the early 1960s, created pictures of rather grotesquely huge-eyed waifs, which were popular for some time. However, in a series of books, culminating with *Walking among Us: The Alien Plan to Control Humanity* (2015, Disinformation Books), historian David M. Jacobs has developed a theory of alien–human hybrids featuring an agenda on the part of extraterrestrials to dominate human beings through hybrids, who would be superior to humans and emotionally cold-hearted and who would owe a

FIGURE 10.4. Hybrid children of extraterrestrials: the mutant children of Midwich with glowing eyes from *Village of the Damned*, 1960. (Courtesy MGM/The Kobal Collection.)

greater allegiance to the alien side of their ancestry. This scenario would seem to derive from *Village of the Damned*. However, Jacobs's hybrids are able to walk among us because rather than all looking alike and all having noticeably unusual coloring and demeanor, after the manner of the children of Midwich, they look just like us.

As folklorist Thomas Bullard notes, scenarios of alien–human hybridization are wonderfully free of the restrictions of biology, biochemistry, and genetics:

> A hybrid-making program as the reason for abductions raises questions about how two species can evolve worlds apart and still be genetically compatible. Popular wisdom sets aside this objection by accepting that alien techno-magic can hybridize anything with the push of a button. Yet this argument self-destructs because of the same superiority. If aliens are advanced enough to unite any genes, they should be too advanced to practice hybridization at all. Real aliens would need a kitchen rather than an examination room, since they could simply create life to their specifications from raw materials up.[11]

One would expect DNA samples as the prime physical evidence of the alien hybrid program. Jacobs has stated that abductees are multigenerational. That is, if one is among those abducted, it's likely at least one of that person's parents was likewise abducted and even that one of his or her grandparents was also abducted. One would think, then, that some hybridization would have occurred in this particular hereditary line in which the aliens are so interested. Yet the authors searched Jacobs's website, www.ufoabduction.com, in vain for any mention of hybrid DNA. The second motif is that of the dying planet in the films already alluded to.

Bullard points out a strong connection between the star children and their alien forbears of modern UFO narratives, on the one hand, and changelings and theirs (elves, witches, demons), on the other, in a two-page comparative list.[12] Among the many striking parallels are the following, which we have somewhat paraphrased (our numbering and italics):

1. *Demons* steal semen to breed children. *Fairies* replenish their race with stolen human children and interbreed with mortals. *Aliens* extract eggs and sperm to produce hybrids.
2. *Changelings and demon/human children* are more like their fairy or demon parents than their human parents. *Alien/human hybrids* are more alien than human.
3. Children of *gods* and humans or *fairies* and humans have *special powers. Star children* have *special powers.*

The special powers of the frightening alien children of Midwich are not unlike the supernatural powers attributed to Jesus as a child in the apocryphal "infancy gospels," such as being able to kill people or raise them from the dead with a word, or like the supernatural strength of the baby Heracles, who strangled the vipers sent by Hera to kill him. Once again, the complex of narratives in popular UFO lore is the product both of ancient myth and of the silver screen.

ORIGINS OF THE REPTILIANS

Reptilian humanoids have been fairly common in science fiction (Fig. 10.5). Although the gill-man of the 1954 film *Creature from the Black Lagoon* wasn't strictly reptilian and was not of extraterrestrial origin, it was human in conformation and was scaly. The movie also prominently featured the creature carrying off the beautiful heroine clad only in her bathing suit, which fits well into sexual abduction imagery. Furthermore,

FIGURE 10.5. Shape-shifting reptilians. (A) *V* (the TV movie), 1983: The character Diana, still in human guise, with a fallen reptilian comrade. Note the human ear on the remnant of the latter's disguise. (Courtesy Warner Bros./NBC-TV/The Kobal Collection.) (B) *Fortean Times* cover, issue 129, featuring Queen Elizabeth as reptilian. (Courtesy of the *Fortean Times*.)

representations of reptilian aliens are quite similar to the creature. One explicitly reptilian race from the media is the Gorn of the *Star Trek* episode "Arena" (1967). The one representative of the Gorn is an actor in a green suit with a reptilian head reminiscent of a tyrannosaurus rex. The humanoid body of this reptilian alien can largely be attributed to budgetary restrictions of the *Star Trek* television series, which also caused the series to use the picturesque Los Angeles County park Vasquez Rocks as the setting of this and other episodes. The Sleestak of the television series *Land of the Lost* (1974–76), a race of intelligent, bulbous-eyed, bipedal reptiles, looked much like representations of reptilian aliens. Another possible media source of reptilian aliens, particularly because these aliens are supposed to be able to shapeshift, is the 1982 film *Conan the Barbarian*, in which the villain, Thulsa Doom, metamorphoses into a giant serpent. However, the main source of reptilian aliens is most likely the 1983 television miniseries *V*.

In *V*, giant flying saucers appear and hover over the major cities of Earth. This motif would seem to have been borrowed from the ships of the Overlords in Arthur C. Clarke's 1953 novel *Childhood's End* and was used again in the 1996 movie *Independence Day*. Out of these ships emerge aliens completely human in appearance, who say they wish to trade knowledge of their advanced technology for Earth's natural resources—seemingly a win–win situation. In reality, the aliens, called the "Visitors," are reptilian, though humanoid in conformation. Their scaly green reptilian skin is hidden under a covering of synthetic human-looking skin. They also wear contact lenses to disguise their red-orange eyes, whose pupils are vertical slits. Furthermore, they eat live rodents whole, distending their jaws in the process, much like serpents. That they are disguised as humans is significant, because a major tenet of theories involving reptilian aliens is that they, too, disguise themselves as humans, often by shapeshifting. The Visitors' real reason for coming to Earth in *V* is to harvest human beings as food. A subplot of *V* involves one of the aliens' seducing and impregnating a teenage girl to produce reptilian–human hybrids. At a MUFON website listing reptilian alien encounters and abductions, only three reports are from before 1983—one each from 1965, 1976, and 1982. The 11 other reptilian encounters are from 1983 or later.[13]

The main proponent of the theory that reptilian aliens are heavily involved in human affairs is, as noted in our chapter on alien abductions,

David Icke. The reptilian aliens presented in his 1999 book *The Biggest Secret* have all the characteristics of the Visitors of *V.* For example, they disguise themselves as human beings:

> The Windsors, like all the royal families (family) of Europe are representatives of the Black Nobility and Babylonian Brotherhood and related to William of Orange. They are, as I shall be describing, shapeshifting reptilians.[14]

Also, just as the Visitor reptilians of *V* wanted to harvest the human race as food, Icke asserts that reptilians desire to drink human blood as well as to crossbreed with humans:

> The reptilians and their crossbreeds drink blood because they are drinking the person's life-force and because they need it to exist in this dimension. They will often shape-shift into reptilians when drinking human blood and eating human flesh, I am told by those who have seen this happen. Blood drinking is in their genes and an Elite high priestess or "Mother Goddess" in the hierarchy, who performed rituals for the Brotherhood at the highest level, told me that without human blood the reptilians cannot hold human form in this dimension. Her name is Arizona Wilder, formerly Jennifer Ann Greene.[15]

Icke has also mingled both neo-Nazi UFO lore and Zecharia Sitchin's Anunnaki (making them reptilians) into his conspiracy theories. Concerning the reptilians' penchant for drinking human blood, he goes on to say the following:

> Interestingly, she said that the reptilians had been pursuing the Aryan peoples around the universe, because the blood of the white race was particularly important to them for some reason and the blond-haired, blue-eyed genetic stream was the one they wanted more than any other. They had followed the white race to Mars, she said, and then came to Earth with them. It is far from impossible that the reptilian arrival on this planet in numbers was far more recent than even many researchers imagine. An interbreeding programme only a few thousand years ago between the reptilian Anunnaki and white Martian bloodlines already interbred with the reptilians on Mars, would have produced a very high reptilian genetic content. This is vital for the reasons I have explained earlier. They appear to need a particular ratio of reptilian genes before they can shape-shift in the way that they do. But when the interbreeding happened is far less important than the fact that it did happen.[16]

Not surprisingly, having given the Aryan race an extraterrestrial origin, Icke also bought into the grand myth of the Vril society:

Out of this order came other similar societies, including the infamous Thule-Gesellschaft (Thule Society) and the Luminous Lodge or Vril Society. Hitler was a member of both. Vril was the name given by the English writer, Lord Bulwer-Lytton to the force in the blood which, he claimed, awakens people to their true power and potential to become supermen. So what is the Vril force in the blood? It was known by the Hindus as the "serpent force" and it relates to the genetic make up of the body which allows shape-shifting and conscious interdimensional travel. The Vril force is, yet again, related to the reptile-human bloodlines. In 1933, the rocket expert, Willi Ley, fled from Germany and revealed the existence of the Vril Society and the Nazis' belief that they would become the equals of the supermen in the bowels of the Earth by use of esoteric teachings and mind expansion. They believed this would reawaken the Vril force sleeping in the blood. The initiates of the Vril Society included two men who would become famous Nazis, Heinrich Himmler and Hermann Goering.[17]

This, naturally, means that Icke's Martian Aryans actually originally came from the Aldebaran system. Icke includes the female channelers Maria Orsic and Sigrun in his scenario and even throws in the planet Mallona and makes the Aryans reptilian human hybrids:

The German researcher, Jan van Helsing, writes in his book, *Secret Societies Of The 20th Century*, of how the Vril and Thule societies believed they were corresponding with extraterrestrials through two mediums known as Maria Orsic and Sigrun in a lodge near Berchtesgaden in December 1919. According to Vril documents, he says, these channellings were transmitted from a solar system called Aldebaran, 68 light years away in the constellation of Taurus where two inhabited planets formed the "Sumeran" empire. The population of Aldebaran is divided into a master race of blond-haired, blue-eyed, Aryans, known as Light God People and several other human-like races which had mutated to a lower genetic form due to climatic changes, the channellings said. In excess of 500 million years ago the Aldebaran sun began to expand so creating a tremendous increase in heat. The "lower" races were evacuated and taken to other inhabitable planets. Then the Aryan Light God People began to colonise Earth-like planets after theirs became uninhabitable. In our solar system it was said that they first occupied the planet, Mallona, also known as Marduk, Mardek, and by the Russians and Romans, Phaeton, which, they said, existed between Mars and Jupiter in what is now the asteroid belt. This corresponds with the Sumerian accounts of the planet, Tiamat. The Vril Society believed that later this race of blond-haired, blue-eyed Aldebarans, colonised Mars before landing on the Earth and starting the Sumerian civilisation. The Vril channellers said that the Sumerian language was identical to that of the Aldebarans and sounded like "unintelligible German". The language frequency of German and Sumerian-Aldebaranian were almost identical, they believed. The details may change in each version, but here we see the same basic theme yet again. A blond-haired, blue-eyed master race of extraterrestrials land on the Earth from Mars and become the gods of ancient legend. They become the inspiration behind the advanced Sumerian culture and spawned the purest genetic stream on the planet. These same gods have controlled the planet ever since from their underground cities. What I think has been missed, however, is that within the Aryan stream are the reptile-Aryan bloodlines. I know from Brotherhood insiders that the reptilians need the blood for some reason of the blond-haired, blue-eyed people, and the Nazi obsession with the Master Race was designed to keep this stream pure and discourage interbreeding with other streams and races.[18]

Thus Icke's reptilian thesis has taken from *V* the idea of an evil reptilian conspiracy, along with reptilian–human crossbreeds, to rule the Earth, as well as that reptilians disguise themselves as humans to accomplish this and that they feast on human flesh. He has wedded this to Zecharia Sitchin's Anunnaki myth, neo-Nazi UFO lore based on Bulwer-Litton's novel *The Coming Race*, and George Hunt Williamson's Maldek/Mallona/asteroid belt myth. Also, in making the Aryans extraterrestrials, he has added the Nordics of popular UFO myth to his pastiche. The reptilian, or "reptoid," aliens seem to attract a following among UFO believers in inverse proportion to any scientific basis for their existence. In Icke's case, this reflects his background in New Age spiritualism.

Considering the absurdities of Icke's grand conspiracy theory, it would be easy to simply dismiss his theory of shapeshifting, blood-drinking reptilian aliens secretly running the world. Readers might even wonder why we have given it any attention whatsoever. However, according to an April 7, 2016, article in the *Guardian*, 12 million Americans believe the reptilian conspiracy theory to be true.[19] Because the present population of the United States is more than 322 million, these 12 million believers constitute a bit less than 4% of Americans. Thus even a ludicrous theory, only accepted by a very small minority, translates into *millions* of believers.

SUMMARY

Before the 20th century, there were almost no reports of UFOs or aliens in the popular media. Instead, legends focused on mythical beings such as angels, demons, incubi, and succubi. With the advent of the modern age of secularism and science, hallucinations of angels and demons have gone out of fashion. Instead, the origin of every motif of UFO narratives can be traced to the media of the 20th century. These include flying saucers, which were taken from comic strips and pulp magazine covers of the 1920s and 1930s but going back as far as 1915; alien implants, which came from the 1953 movie *Invaders from Mars*; alien abductions, derived from the *Buck Rogers* comic strip, *Invaders from Mars*, *This Island Earth*, and *The 27th Day*; dying worlds, from *This Island Earth* and *The 27th Day*; and interbreeding aliens with humans, from both *Village of the Damned* and *V*. In addition, the main denizens of UFO literature derive from movies and television: the Nordics from Klaatu in *The Day the Earth Stood Still*, the grays from "The Bellero Shield," and the reptilians from *V*. Some of these motifs that developed in the 20th century, particularly the imagery of abduction, can also be traced back to older mythology and to the physiology of the hypnagogic state.

As noted in chapter 12, that beings would evolve on another planet into a human body type is highly unlikely, and the similarity of the grays and reptilians to a human form most likely reflects the budgetary restrictions of the television series that spawned them. Of course, one could argue that the resemblances between the flying saucers of comic strips and pulp fiction art and those supposedly sighted by contactees and abductees are coincidental. However, ships traveling through interstellar, or even interplanetary, space, not having to deal with atmospheric friction, wouldn't need to be streamlined in any way. They might also be constructed in space, hence not made to enter a planet's atmosphere. They might, thus, only send down scout ships to investigate the planet's surface. Although we can tell from throwing Frisbees that the disc shape can be aerodynamic, a spinning disc has never appeared as a design for rockets or jet aircraft. Even positing flying saucers as the products of an alien technology doesn't make them any more likely. After all, the physics of flight through any atmosphere on any world would favor the same aerodynamic solutions as work on Earth. Kenneth Arnold's boomerang- or crescent-shaped craft, misidentified as saucers in the news, do bear some resemblance to experimental aircraft of the late 1940s, particularly to various flying wing designs.

Regarding the persistence of the perception of UFOs as saucer-shaped, psychologist Armando Simon notes that what are called "mental sets" predispose us to interpret various perceptions according to previous descriptions and cultural expectations. Simon gives the following as an example of mental sets in action:

> An early classic experiment with mental sets involved showing two circles joined by a straight line to volunteer subjects. . . . Two different groups of subjects were shown an object described by the experimenters as either a "barbell" or as "eyeglasses." Days later, the subjects were called back and asked to draw the figures from memory. Those subjects who had been told that they had seen a drawing of a "barbell" drew a long, thick interconnecting line, while those who had been told that they had seen "eyeglasses" drew a short, thin line as if they were connecting two lenses of a pince-nez.[20]

Simon believes that exposure to pulp magazine images, particularly Frank R. Paul's cover illustrations, created a mental set to see UFOs not only as extraterrestrial spaceships but also as saucer-shaped. This was, of course, reinforced by Bill Bequette's mischaracterization of Kenneth Arnold's UFOs as saucer-shaped.

By the same token, the movies and television shows already mentioned created the mental sets for Nordic aliens, grays, reptilians, implants, and alien abductions. The concept of mental sets also explains why after an abduction report is publicized, copycat reports of a similar nature proliferate. Thus Barney Hill's memory of gray aliens, supposedly recovered under hypnosis but probably based on "The Bellero Shield," created a mental set for other gray aliens, and his narrative of abduction married the grays to the capture and abduction motif. Thus later reports of encounters with gray aliens almost always involve abduction narratives. Betty's description of the grays inserting a needle into her navel generated the mental set that abduction by the grays involved physically intrusive procedures—everything from implants to impregnation.

Likewise, *V* created the mental set for the shapeshifting, seducing reptilian in the narrative related by "Amy" in our chapter on alien abductions. Klaatu's nobility and his warning to the human race created the expectation that aliens looking like handsome or distinguished humans—that is, the Nordics—would be benign, noble, and even angelic. In the case of George Adamski's Orthon or Billy Meier's Plejaren, a degree of conscious fraud may be inferred, However, the reason why Adamski and Meier deliberately chose to invent benign, handsome, fully human aliens itself reflects a mental set engendered by *The Day the Earth Stood Still.*

Those arguing that flying saucers are real alien spacecraft, that alien abductions and implantations actually occur, and that the Nordics, grays, and reptilians are actual species of extraterrestrial beings could accuse us of being guilty of the logical fallacy of *post hoc ergo propter hoc* ("after this, therefore because of this"), arguing that an earlier appearance of flying saucers in pulp magazines doesn't preclude their being real. They might likewise argue that the proliferation of abduction narratives in the wake of that of the Hills represents an actual history of abductions on the part of the gray aliens and that the Hills merely represent the earliest recorded case.

In response to this we must appeal to Ockham's Razor, otherwise known as the law of parsimony: What is the simplest and most likely explanation for a narrative? Consider, for example, the gray aliens' piercing of Betty Hill's navel with a huge needle, which caused her excruciating pain. By using the far less invasive technology of the CT (computer tomography) scan and that of MRI (magnetic resonance imagery), we can learn much more about what goes on inside the human body than by sticking a needle into it. CT scans first became commercially available in 1971. The first clinically useful MRI was performed in 1980. Thus Betty Hill had no knowledge of them in the early 1960s, and her "memory" under hypnosis was limited to the technology she knew.

Yet though MRI and CT scans weren't around in the early 1960s when the Hills reported being abducted, we would have to expect intelligent extraterrestrial beings capable of interstellar travel would have had such a technology or something even more advanced. Thus while they were giving Betty an MRI or a CT scan, she would not have felt any pain, nor would she have understood what they were doing. If, on the other hand, the grays were performing a biopsy, that would require a large needle. But when doctors take biopsies, they administer local anesthesia to reduce the trauma of the procedure. Thus, even if they were taking a biopsy, these aliens should have had at their disposal anesthetics and sedatives far more sophisticated than any we now have.

Another way we can know that alleged alien encounters are likely limited by the technology known or inferred during the time a movie or television show was made, compared even to the technology available to us in the second decade of the 21st century, is to consider spy drones. A 2012 online article titled "The Smallest Spy Drone in the World" deals with a tiny drone made to look like a mosquito, to which it is comparable in size. The article alleges that this drone not only can spy but also is equipped with a small needle to take blood—and with it DNA—while inflicting no more pain than a mosquito bite.[21] The mosquito-sized drone touted in the article might be more conceptual than actual, but another online article, this one from 2013, has a link to a video of a test of a "nano-hummingbird." Although its flapping wings still sound a bit too mechanical to be taken for a real hummingbird, it does look like the real animal and is the same size.[22] Considering drone technology in general and extrapolating its increase in sophistication from our own largely planet-bound civilization to one that has the ability to send craft across interstellar space, it's hard to fathom why such a civilization intent on observing us, particularly if it wished to do so covertly, would hazard ships and personnel when it could spy on us with microelectronic devices we might never notice. With such a technology, it wouldn't need to abduct people. It would simply have an electronic mosquito bite and possibly inject microsensors into unsuspecting Earthlings—sans trauma, sans risk of discovery. Yet when Barney and Betty Hill were "abducted," neither MRIs nor drone nanotechnology existed, and their narrative was limited to what could reasonably and economically be portrayed in the media of the 1960s.

Finally, considering the vagaries of evolution occurring on another planet in another star system, would we expect actual extraterrestrial visitors, whether benign, malign, or indifferent, to reproduce the gray alien of "The Bellero Shield," the reptilians of *V*, or human beings of the Nordic type, indistinguishable from *Homo sapiens*? The simplest and most likely explanation for all of these motifs—flying saucers, implants, abductions, Nordics, gray aliens, and reptilians (all of which appeared in UFO narratives only after first appearing in the media)—is that they are the products of mental sets engendered by literary, graphic art, motion picture, and television tropes: They reflect a rich literary, graphic, filmic, and mythic trove, one imprinted by our own psychology. However, these narratives do not reflect either the character or the motives of beings from another star system.

11

Cloud's Illusions

I've looked at clouds from both sides now,
From up and down and still somehow
It's clouds' illusions I recall,
I really don't know clouds at all

—Judy Collins, Both Sides Now

SEND IN THE CLOUDS

In chapter 2, we discussed the phenomenon of pareidolia—the tendency of the human mind to link together random shapes and see patterns where there are none. Nothing demonstrates this better than our propensity to "see" various meaningful shapes in the random shapes of clouds. One famous *Peanuts* comic has Charlie Brown, Linus, and Lucy lying on their backs looking up at the clouds. Linus compares the shapes he sees to a map of British Honduras, the profile of Thomas Eakins, and the stoning of the Apostle Stephen. In the punch line, Charlie Brown says, "I was going to say I saw a ducky and a horsie but I changed my mind."

Clouds can take a huge variety of shapes, which can change very quickly. They form when the temperature in the atmosphere drops below the dew point, the threshold that causes water vapor to condense into water droplets (or ice droplets in the upper atmosphere, where it is well below freezing). Their shapes are dictated by how and when this occurs, along with updrafts and downdrafts in the air currents. If the air is very stable, with no up and down motion, the clouds tend to be very flat and layered (stratus clouds). If there are updrafts, they form tall vertical pillars or fluffy clouds called cumulus clouds. High in the uppermost troposphere (the lower level of the atmosphere, above 25,000 to 30,000 feet, or 7,000 to 8,000 meters), the limited moisture immediately freezes into long stringers and wisps of ice crystals known as cirrus clouds. There is an enormous range of variations on these basic types—but this is not the place for a detailed course on the meteorology of clouds.

Naturally, clouds' formation out of nowhere, quick movement, and ability to assume all sorts of strange shapes fool many people into thinking that they have seen a UFO. You need only type "UFO" and "clouds" into any search engine, and you will get a whole range of websites claiming that various cloud shapes are UFOs or that a UFO is hiding in an ordinary cloud. This story[1] from South Africa in 2015 is typical: "UFO Clouds Form in Cape Town, Spark Fears of an Alien Invasion." Another site about "alien sky ships" is full of the usual UFO nonsense, complete with the classic paranoia of conspiracy believers who don't want to be laughed at. This quote from the site says it all:

> Now you must understand that the last statement is from the scientific mind of a brainwashed human. Although these clouds may be formed in the way they say, it must be understood that these lenticular clouds were really formed by alien technology to create these weather conditions so that their spacecrafts can be cloaked and hover in the sky right in front of the unaware brainwashed humans. Just as I sated [sic] on my Mothership Clouds page . . . these UFO cloaked in clouds are actually on another dimension. Therefore airplanes can fly right through these clouds and not hit extraterrestrial skyships. I have friends that are so blinded by the control system that they laugh at ideas like this. I am quite sure, it is the Aliens who are laughing at our stupidity and blindness. They laugh because they we deny what is right in front of our faces. So you may ask me . . . "How do you know, Puzuzu, that these lenticular clouds are truly cloaking UFOs?" Well the answer is simple. The information that I relay to you is information that is given to me by the Celestial Gods Enlil and Enki. You may choose to believe that which you so choose, but it is time to awaken and realize what is right before your eyes. Don't live in a controlled way of thinking, as this only makes you a sheep.[2]

"Sightings" such as these spark huge numbers of people into believing that they really saw a UFO, prompting large numbers of comments from the UFO believers on any site that discusses them. In an earlier age, strange shapes of clouds were seen as portents or warnings from the gods, grounds for fear and panic. Today, instead of signs of the gods, we fit them into the modern "god" of our cultural fears—aliens and UFOs.

Most people are not very observant and pay no attention to clouds most of the time, so they seldom notice shapes unless they are unusual—especially if they are shapes that resemble a "flying saucer" and fit the prevailing cultural meme of what a UFO is supposed to look like.

The clouds most commonly seen as "UFOs" in the sites just mentioned are known as lenticular clouds, or altocumulus clouds (Fig. 11.1). Just type "UFO" and "cloud" into your search engine, then click on the "images" link—you'll see dozens of them, some of which are quite spectacular and do somewhat look like the disclike shape of a flying saucer. They get their name because they have a shape like a lens, a lozenge, a disc, or—yes—a flying saucer. Contrary to the rants of the UFO believers, they are a well-understood meteorological phenomenon. They are the effect of fast-moving upper-level horizontal air currents that flow upward from the lower level, where the air is above dew point, to a colder upper layer of air, where the air current cools and quickly condenses (Fig. 11.2). Then these same currents flow back down until they are above the dew point, leaving a little "blob" or "hump" of cloud behind where they were in the colder air.

Lenticular clouds are particularly common over the tops of high mountains, which divert the air currents up into their colder tops, or just behind mountain ranges, which have created a wavelike pattern in the upper-level air currents flowing over them. Each time the "wave" of air rises into colder parts of the atmosphere, lenticular clouds are born (Fig. 11.2). Some of the most spectacular and UFO-like lenticular clouds build over the peaks of the highest isolated mountains, especially tall conical volcanoes such as Mt. Rainier, Mt. Shasta, and Mt. Fuji (Fig. 11.1C). They are so often formed over Mt. Shasta that the UFO believers are convinced that Mt. Shasta is not a dormant volcano (which erupted as recently as 200 years ago) but instead a secret base for alien operations. Just type "Shasta" and "aliens" into your search engine, and you'll find another array of websites, all convinced that the aliens are hiding in Shasta, just biding their time. (They never explain how the aliens hiding in the mountain survived its many eruptions, nor how they hollowed out an entire mountain made of hard rock and magma). Other groups[3] think that Mt. Shasta is the site of a "harmonic convergence"

FIGURE 11.1. Lenticular clouds. (A) Lenticular cloud over Cagliari, Italy, September 17, 2013. (By fdecomite (Lenticular Clouds over Cagliari) [CC BY 2.0 (http://creativecommons.org/licenses/by/2.0)], via Wikimedia Commons.) (B) Lenticular cloud formation, Hawaii Island. (By Rootmeansquare (Own work) [CC BY-SA 3.0 (http://creativecommons.org/licenses/by-sa/3.0) or GFDL (http://www.gnu.org /copyleft/fdl.html)], via Wikimedia Commons.) (C) Lenticular clouds over the South Georgia Island, South Atlantic Ocean. (Image ID: wea03614, NOAA's National Weather Service [NWS] Collection, https://www.flickr.com/photos/51647007@ N08/9672393596/, Photographer: Lieutenant Elizabeth Crapo, NOAA Corps, https://creativecommons.org/licenses/by/2.0/.)

FIGURE 11.1. *Continued*

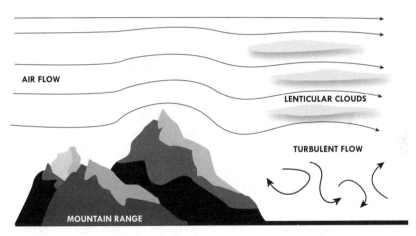

FIGURE 11.2. The mechanics of how lenticular clouds are produced. (Illustration by Katelyn Griner.)

where aliens or angels come to visit Earth and communicate with us. This quote is typical:

> Now to give you a little more first hand knowledge. Let's take a look at some of the UFO lenticular clouds that are seen frequently over Mount Shasta which is located in California. Mount Shasta is a stratovolcano, also known as a composite

volcano. There is a Spiritual Retreat on the Etheric Planes over the area between Sacramento and Mount Shasta, California. You can go there for guidance and to learn how to attain greater purity, discipline, and order. Students also learn strategies of light and darkness and how to connect with ascended beings by raising our human consciousness to the level of the Christ. This is not a physical place you go to. It is a place you go to while you are traveling out of body in Spirit. This spiritual retreat is run by Archangel Gabriel and Hope. You must also understand that Archangels, Ascended Masters and Celestial Gods are all extraterrestrial beings. This is why there are so many of the lenticular UFO clouds around Mount Shasta.[4]

CHEMTRAIL FAIL

For the past few years, there have been many strange websites that claimed that ordinary contrails formed by high-flying aircraft are "chemtrails," a special kind of chemical sprayed on the unwitting population for reasons too bizarre and illogical to take seriously. Some of the believers out there attribute chemtrails to aliens, but most blame it on our government or on a secret "World Order" that controls all the world's governments. There are whole shows about it on the once-scientific Discovery Channel, as well as on the History Channel. Both these channels long ago gave up their original public missions and sold their souls to commercialism, now broadcasting one pseudoscientific show after another. A low-budget Hollywood movie, *Poison Sky*, that focused on chemtrails was intended to have had its premiere in 2016 but so far has not seen wide release in theaters. The chemtrail conspiracy mongers circulate their photos and videos among themselves, posting hundreds of these videos to YouTube and on their own sites and forums. But because of the way the internet works as a giant echo chamber for weird ideas, with no peer review, fact-checking, or quality control, it's getting impossible to ignore them any more, and it's high time to debunk their claims. In fact, just as this book was going to press, the first formal article in the peer-reviewed scientific literature to deal with chemtrails appeared in the journal *Environmental Research Letters*,[5] showing that no real atmospheric scientist takes chemtrails seriously.

The first few times scientists hear about "chemtrails," their reaction is *You can't be serious.* But the people who spread this *are* serious. They are generally people who have already accepted the conspiracy theory mindset, in which everything that they don't like or understand is immediate proof of some big government conspiracy. But there's an even bigger factor at work here: gross science illiteracy. The first thing that pops into the mind of a scientist reading these strange ideas is *Didn't this person learn any science in school?* And the quickest rebuttal to their arguments is as follows: "Do you know anything about our atmosphere? Do the math! Anything released at 30,000 feet will blow for miles away from where you see it and has virtually no chance of settling straight down onto the people below. The chemical will be so diluted as to be unmeasurable by the time it lands. Just think: crop-dusting planes must fly barely 30 feet off the ground so that their dust won't blow too far away from the crops, and aerial firefighters must fly low to the ground to spray their fire retardants! How much more so in this case?"

As Kyle Hill describes it:

> If the chemtrail conspiracy were true, millions of pilots would be needed to crop dust the American population. A typical crop duster might use seven ounces of agent diluted in seven gallons of water to cover one acre of land. Chemtrail "people dusters" would use a similar concentration to cover the entire United States, just to be safe. For 2.38 billion acres of land, the pilots would then need— for just one week of spraying—120 billion gallons of these cryptic chemicals. That's around the same volume as is transported in all the world's oil tankers in one year. And such an incredible amount of agent would need an incredible number of planes. Considering that a large air freighter like a Boeing 747 can carry around 250,000 pounds of cargo, at the very least, the government would need to schedule four million 747 flights to spread their chemicals each week—eighteen times more flights per day than in the entire US.[6]

Roots of a Myth

The entire chemtrail conspiracy idea is a relatively recent one, and it's an idea that would not have become so popular without the ability of the internet to spread lies. As the site *Contrail Science* shows,[7] it was an idea that was already simmering among conspiracy theorists in the 1990s when William Thomas made it popular back in 1996. From 1997 to 1999, he was trying to spread his ideas through interviews and media coverage and early conspiracy internet sites, and he had gotten many believers to buy into his bizarre fantasy. Then, in 1999, he was featured on "conspiracy central," Art Bell's show called *Coast to Coast AM*. If there was

FIGURE 11.3. On any given day, the sky can be full of hundreds of contrails if the atmospheric conditions are right. (By alanallos (Own work) [CC BY-SA 4.0 (http://creativecommons.org/licenses/by-sa/4.0)], via Wikimedia Commons.)

any quick way to reach the mob of UFO believers, paranormal fanatics, and conspiracy theorists besides the internet, Art Bell's show was the place in the 1990s. (Now Alex Jones has a show called *Infowars* that can claim the title of the new "conspiracy central.") Soon the phenomenon exploded far beyond William Thomas or Art Bell, becoming a widely accepted idea among the people who tune in to the paranormal or the conspiracy mindset.

So what are "chemtrails"? Allegedly, they are different from normal contrails produced by aircraft—because, allegedly, they contain some sort of evil chemical that the government conspiracy is trying to use to poison us. Normal contrails are something we do understand, because there has been lots of research on them (Fig. 11.3). Most aircraft engines leave a plume of hot, gaseous exhaust from the hydrocarbons in the fuel they burn: in the subfreezing conditions of the upper troposphere or stratosphere, those hot gases immediately condense to form long ice clouds behind the plane as it flies. Sometimes just the disruption of the high atmospheric gases by the tips of wings will cause contrails even without the benefit of engine fumes. If there are high-altitude winds or the jet stream is active, they disperse quickly, but in quiet air, they often remain stable for many minutes. Contrails were observed almost as soon as aircraft were able to fly at that elevation, and they have been well documented in videos and photos of World War II aircraft (Fig. 11.4), *long* before any of the current governments alleged to be conspiring to do this were even in power.

RationalWiki describes it this way:

> On days when cirrus cloud formation is occurring, there is more moisture in
> the upper atmosphere, and consequently, contrails may linger longer before
> evaporating. Since cirrus clouds often precede a general overcast or haze, the
> casual observer could easily assume that the contrails have caused the overcast, or
> become the overcast. The persistence of contrails varies with weather conditions:
> sometimes they dissipate almost immediately, but often they will persist for hours,
> with crossing trails sometimes forming gridlike patterns that stretch from horizon
> to horizon. The "chemtrails" label is usually applied to these longer contrails,
> with their very persistence put forward as "evidence" that they cannot be normal
> contrails.[8]

Even Indy cars, with their speeds exceeding 200 mph, produce contrails—another thing that shatters the idea that contrails were first visible during the 1990s in the sky.

Reality Check

Once you delve into these bizarre websites, you get a wide spectrum of different kinds of misconceptions, misinterpretations, and outright falsehoods. The oddest is that people seem to think that contrails are some sort of new phenomenon—when, as we've just pointed out, we've been seeing them in the sky since we were born. People apparently don't remember seeing contrails when they were young, but that just testifies to the fallibility of human memory, because the photographic record of ordinary contrails goes back to before the 1930s (Fig. 11.4). People don't remember seeing contrails coming from their jetliner when they fly, but then they can't look behind the plane in any commercial flight, so they can't see the contrails—which are there just the same.

FIGURE 11.4. Many images of contrails date back as far as World War II, when aircraft first routinely flew high enough in the troposphere to generate contrails. (A) Fighter plane's contrails in the sky. (By Post-Work: User:W.wolny (US Navy Photo #: 80-G-248549) [Public domain], via Wikimedia Commons.) (B) Convair RB-36H in flight (U. S. Air Force photo, # 060720-F-1234S-019.) (C) World War II contrails. (By Sgt. Stanley M. Smith (390th.org) [Public domain], via Wikimedia Commons.)

FIGURE 11.4. *Continued*

The next problem is the flexible definition of chemtrails and how you can tell one from a regular contrail. Each "definition" gets modified and redefined as soon as the last one is debunked, thus shifting the goalposts. Supposedly, contrails dissipate in minutes, whereas chemtrails linger for a while. Sorry—the time a contrail lingers is dependent on what the upper-level winds are doing, not the chemical composition of the contrail. To prove this, we can find *lots* of films[9] of contrails from World War II (Fig. 11.4), and these contrails lingered for a very long time. Other definitions claim that contrails run parallel to one another, whereas "chemtrails" form X or grid patterns. Once again, people forget how busy the skies above us are. In an area with a lot of plane traffic going in many directions, you'll get every possible pattern of lines crossing one another (Fig. 11.3). On the other hand, if you live in coastal Oregon or central California, most of the flights are north–south, so the contrails tend to be more parallel.

The next claim is that the government is spraying barium, aluminum, or any other of a number of chemicals supposedly toxic to us. But none of these claims is true.[10] More to the point, spraying them from 30,000 feet would be useless, because they would dissipate over a huge area and be diluted into nondetectable, nontoxic amounts—and, odds are, amid high-level winds would blow far from where they were sprayed in the first place.

Then there are the claims that the Germans admit to doing it! This claim is easily debunked when you realize that the video footage[11] used to support this claim deliberately mistranslates the German word *Düppel* (meaning "chaff") as "chemtrail." There are numerous video hoaxes about "chemtrails" all over the internet, along with the normal footage of contrails that is misinterpreted.[12] Apparently, hoaxing chemtrails is almost as popular as hoaxing ghosts, UFOs, and Bigfoot. There is also the claim that language about chemtrails got into a bill before Congress. The bill in question, HR2977, had a lot of UFO and chemtrail language originally inserted by a bunch of UFO believers and was then introduced in 2001 by Rep. Dennis Kucinich, who repudiated their unauthorized changes to the bill, upon which the chemtrail language was dropped.[13] (Bills introducing strange notions to Congress are not unusual for Kucinich.)

A Global Conspiracy?

This list of additional weird claims and debunked hoaxes goes on and on.[14] But ultimately it boils down to a couple questions: Why would the government have been secretly spraying us for years? How could it have pulled this off without one person ever coming forward? We discussed this problem with conspiracy thinking in chapter 2, but it's particularly relevant with chemtrails.

Here's where the fantasies of the conspiracy nut take off, and it's impossible to ground these ideas in any kind of testable reality. Supposedly, the government is poisoning us to make us more docile or to keep us compliant or to weaken us—or the chemtrails are a secret program to combat climate change—or even weirder notions. Conspiracy nuts never ask the tough questions: If the government really wanted to poison us, wouldn't it just put the poison in our water? That's actually an effective possibility, as terrorism and security experts well know. By contrast, spraying chemicals from 30,000 feet is worthless.

This raises a larger question: if such a huge conspiracy really existed and *every* country in the world, *every* military and commercial aircraft in the world, and *every* atmospheric scientist in the world was part of the conspiracy, how has it not leaked by now? Like every other outlandish conspiracy theory, this one envisions a world in which clandestine top-secret organizations are constantly pulling the strings, and *not once* has anything leaked about their doing so, nor has anyone who knows of it ever come forward—this even though we get constant revelations of actual secrets from the CIA and NSA and FBI and other organizations. Remember: no chemtrails were exposed in the Wikileaks affair or by Edward Snowden, nor has any other person come forward in an age when journalists are constantly digging for secret information. For such a level of secrecy to exist, there would have to be an unprecedented ability for all these organizations to cover up their tracks when in fact we have ample evidence to show that every secret organization is only as secretive as its weakest link. And the bigger the conspiracy, the more likely it is to be exposed—yet the chemtrail conspiracy involves the entire planet, including many governments and organizations that are bitter enemies and that would love to expose the wrongdoing of their opponents.

Again, here's Kyle Hill:

> The incognito infrastructure needed to conceal the chemtrail conspiracy would dwarf any other governmental agency. Millions of people—pilots, engineers, chemists, data analysts, and boots-on-the-ground hazmat teams—would need top-secret clearance for information that could never get out. If a chemtrail conspiracy were true, chances are you would run into a few involved in the cover up everyday. An effort to keep millions of mouths silent—to keep any information from pilots or participants out of the media—makes the NSA look like child's play.
>
> A chemtrail conspiracy comes with collateral damage. Many mechanics that work on crop dusters around the world are routinely acutely poisoned by the chemicals the pilots seek to spread. If any significant percentage of the legion of mechanics needed to keep the chemtrail fleet flying had the same risk, literally millions of workers would come home poisoned. Spouses and significant others rush them to hospitals across the US, and the cause of this nation-wide plague never raises any eyebrows. No doctors file reports or do studies on the mysterious poisonings, no journalists ever get wind of something awry, and no police officers think a serial contaminator is on the loose. Even though every American would have a consistent chemical profile from any blood test, no one is the wiser. Every single piece of paperwork finds its proper place, deep in a file drawer of some bureaucrat keeping the lid on the chemtrail conspiracy.[15]

Several people have pointed out that in aircraft design and maintenance, weight is everything. Every last gram of weight in an aircraft

FIGURE 11.5. Numerous photos such as this one purport to show the vats of chemicals in a plane used for spraying chemtrails. In reality, these are images of special test aircraft which use an interconnected series of water tanks to shift the weight around in an aircraft as it flies and thus evaluate its response to changes in weight distribution during flight. (Photo found on http://strange-omniverse.blogspot.com/2012_07_23_archive.html.)

beyond the essentials needed to fly comes with a big fuel penalty, so both manufacturers and also those who own and maintain commercial and military planes are extremely careful about adding any weight beyond what is needed to make the flight possible. Were airlines to agree to carry all these chemicals, that would make a huge dent in their razor-thin bottom lines (Fig. 11.5). This implies a huge amount of government funding to compensate, none of which has ever been documented. The military jealously guards its use of equipment and fuel, because every flight and aircraft is already phenomenally expensive, and it can't afford to weigh down a military jet with unnecessary cargo.

In short, the chemtrail conspiracy thinking fails on the same grounds on which the 9/11 Truther conspiracy fails: it assumes a level of competence and secret-keeping in government that has never happened and that never will.

12

Are They Out There?

This striking bit of dialogue between Ellie Arroway (played by Jodie Foster) and Palmer Joss (played by Matthew McConaughey) is from the 1997 film *Contact*, based on the novel by the late Carl Sagan:

> Ellie Arroway: You know, there are four hundred billion stars out there, just in our galaxy alone. If only one out of a million of those had planets, and just one out of a million of those had life, and just one out of a million of those had intelligent life; there would be literally millions of civilizations out there.
>
> Palmer Joss: [looking the night sky] Well, if there wasn't, it'll be an awful waste of space.

Unfortunately, the screenwriter who wrote Ellie's lines was a bit math-challenged. One-millionth of 400 billion is only 400,000, and one-millionth of that is only 0.4, and one-millionth of that is only 0.0000004 civilizations out there. Nevertheless, we can agree with Palmer Joss that it would indeed be an awful waste of space if we were the only intelligent form of life in the galaxy. Estimates of the number of stars in the Milky Way, our galaxy, vary from 100 billion to 400 billion.[1] Of course, not all those stars might have planets—or planets in a zone and of a mass conducive to the development of any life at all, let alone intelligent life. In fact, it is possible that in our entire galaxy, no star system but ours harbors intelligent life. So we now come to the question of *actual* extraterrestrial intelligent beings. If we are skeptical of the scientific validity of alien encounters, we must consider two major questions. First, are there any other intelligent life forms in our galaxy, or are we alone? Second, if we are not alone, what is the nature of the extraterrestrial intelligent beings that share our galaxy? Generally speaking, such worlds as might harbor life would be rocky planets, such as Earth or Venus, rather than gas giants such as Jupiter or Saturn. They would also, assuming that we are talking about carbon-based life using water as a medium, have to have average temperatures and pressures sufficient to keep most of the water on the planet in a liquid form. At sea level on Earth, water freezes at o degrees Celsius (32 degrees Fahrenheit) and boils at 100 °C (212 °F).

THE NUMBERS GAME

In 1961, Dr. Frank Drake, the founder of SETI (Search for Extra-Terrestrial Intelligence), proposed an equation for calculating the number of extraterrestrial civilizations with whom radio communication might be possible. The Drake Equation is written as $N = R_* \cdot f_p \cdot n_e \cdot f_l \cdot f_i \cdot f_c \cdot L$. The values expressed are as follows:

> N = the number of extraterrestrial civilizations with whom radio communication is possible
> R_* = the rate of star formation in the galaxy
> f_p = the fraction of stars in the galaxy with planets
> n_e = the number of planets capable of supporting life
> f_l = the fraction of planets actually harboring life
> f_i = the fraction of planets harboring intelligent life
> f_c = the fraction of planets harboring technologically advanced life
> L = the length of time during which technologically advanced civilizations broadcast radio waves

Because most, if not all, of these quantities are unknown, the Drake Equation isn't really a guide on which to base calculations. Rather, it is a tool for philosophical contemplation. Is there any way to actually calculate at least a rough idea of how many extraterrestrial civilizations might exist in our galaxy?

Perhaps, following Ellie Arroway's example, but with a bit better math, we could make a 1% rule as follows: if we assume a low estimate, that there are 100,000,000,000 stars in our galaxy and that only 1% have planets, then they amount to 1,000,000,000 stars. If only 1% of those have planets in what we might call the "life zone," where water on the surface is mostly in liquid form, then we are left with 10,000,000 such stars. If only 1% of those actually harbor life, there are 100,000 such planets in the Milky Way. If only 1% of these have higher life forms, such as vertebrates, we are down to 1,000 planets. If intelligent life exists on only 1% of these, there would only be 10 intelligent life forms in the entire Milky Way. However, this 1% rule is arbitrary and could easily be off by many orders of magnitude, giving us as many as 10,000 or fewer than 1 intelligent life forms in our galaxy.

A better way of calculating how many intelligent life forms exist in our galaxy might be to consider that our sun is, according to the Morgan-Keenan (MK) and Harvard classification systems, a type G-2 star. These constitute 7.5% of the main sequence stars in the Milky Way, or 7,500,000,000 stars if we assume that there are 100,000,000,000 stars in our galaxy. If only 1% of these have had time to develop intelligent life, there would be 75,000, 000 intelligent life forms in the Milky Way. Another way to calculate the abundance of life in the galaxy is to consider the number of stars with planets. A recent estimate, based on observations made by NASA's Kepler Space Telescope, is that 22% of sunlike stars have planets.[2] So even this estimate might be too low.

WHY HAVEN'T WE BEEN ASKED TO DANCE?

We now come up against the Fermi Paradox. After a casual conversation at Los Alamos in 1950, Enrico Fermi, one of the main architects of the Manhattan Project, along with physicist Michael H. Hart, came up with what is called the Fermi Paradox. Its main assumptions are as follows: The sun is a typical star. Billions of stars are far older than the sun. There's a high probability that some of these might harbor life—even intelligent life. Even with starships' taking hundreds of years to traverse the stars, intelligent beings should have colonized the entire galaxy by now. Yet they haven't. Fermi's question, then, is "Where is everybody?"

Somewhat related to the Fermi Paradox are the Kardashev Scale and the Dyson Sphere. In 1960, American theoretical physicist and mathematician Freeman Dyson proposed that as civilizations advance to the point of colonizing their solar systems, their energy demands would rise to the point that they would require all the energy output of their sun. He proposed that such civilizations would be capable of building a mega-structure that would entirely enclose their star to collect all its energy. This came to be known as a Dyson Sphere. The enclosed star would then radiate only in the infrared wavelengths—the waste heat given off by the Dyson Sphere's energy consumption.

In 1964 Russian astronomer Nikolai Kardashev proposed three levels of highly advanced civilizations, typified by their energy use. These are classed as Types I, II, and III on the Kardashev Scale. A Type I civilization uses all the energy available on its planet. A Type II civilization uses all the energy of its sun. A civilization that has enclosed its sun in a Dyson sphere is Type II on the Kardashev Scale. Type III civilizations use all the energy output of their home galaxy. A galaxy hosting such a civilization would not emit any light in our visible spectrum. In fact, it would only, or at least chiefly, emit infrared radiation—that is, what we perceive as heat. Recently, researchers using the Wide-field Infrared Survey Explorer (WISE) surveyed approximately 100,000 galaxies and found no sources that emitted only or chiefly infrared radiation, indicating that none of them host a civilization that is Type III on the Kardashev Scale.[3] One possible explanation for this lack of evidence of Type III civilizations is that it simply hasn't reached us yet. Our nearest galactic neighbor is the Andromeda Galaxy, which is about 2.5 million light-years away from us, a light-year being the distance that light travels in a year. Thus, had a Type III civilization begun to enclose all the stars in the Andromeda Galaxy approximately 2,000,000 years ago, we wouldn't begin to see its stars winking out and giving off only infrared radiation for another 500,000 years.

Of course, both Dyson Spheres and the Kardashev civilizations are very hypothetical. We don't really know whether interstellar civilizations will require such vast amounts of energy. Still, limiting ourselves to only the 100 billion to 400 billion stars in our own galaxy, Fermi's question remains unanswered. Not only do we not find convincing evidence that Earth has been visited in the past by extraterrestrials, but we also have yet to find evidence of radio waves broadcast from other parts of the galaxy. So here we are, like the stereotypical wallflower at the senior prom sitting alone on the sidelines of the dance floor. Why hasn't anyone asked us to dance?

One answer to the Fermi Paradox is that, as we have already seen, the sun, being a G-2 star, isn't that typical. Another answer to the paradox is the sheer size of the galaxy, even if we accept the low estimate of its total number of stars being 100 billion. If we assume that 1% of the G-2 stars in our galaxy harbor intelligent life, giving us a striking number of 75 million intelligent species, they must still navigate through 100 billion stars. That is, only 75 out of every 100,000 stars would harbor intelligent life—or, to put it another way, intelligent life would be found on the planets of only 0.075% of the stars in the Milky Way. So our potential dance partners might simply not have found us yet. It's also possible that we haven't found them yet, according to the SETI website:

> The failure so far to find a signal is hardly evidence that none is to be found. All searches to date have been limited in one respect or another. These include limits on sensitivity, frequency coverage, types of signals the equipment could detect, and the number of stars or the directions in the sky observed. For example, while there are hundreds of billions of stars in our galaxy, only a few thousand have been scrutinized with high sensitivity and for those, only over a small fraction of the available frequency range.[4]

If we take "a few thousand" as meaning about 10,000 and assume a low-end estimate of 100 billion stars in the Milky Way, then SETI has managed to scan only one out of every 10 million stars, or 0.0001% of the stars in our galaxy. Likewise, if other civilizations are looking for our signal, they might also have a very hard time finding it.

Another barrier posed by the vastness of our galaxy is that of interstellar space and what we might call "Einstein's speed limit." A button sold at science fiction conventions reads "186,000 miles per second: It's not just a good idea. It's the law!" That figure, 186,000 miles per second, is the velocity of light in a vacuum. Any particle with a measurable rest mass (the rest mass of photons and neutrinos is zero) is subject to Fitzgerald-Lorentz contractions. Independently proposed by George Fitzgerald in 1889 and Hendrik Antoon Lorentz in 1892, these contractions were borne out by Einstein's Theory of Relativity. As particles with a measurable rest mass are accelerated, their dimensions, including time, decrease and their masses increase. At the speed of light, the dimensions of any object become zero and its mass becomes infinite—meaning, in essence, that it ceases to exist. The nearest star to our sun is Alpha Centauri, which is about four light-years away from us. If aliens from Alpha Centauri could manage to harness vast amounts of energy to go even a quarter of the speed of light, it would take them 16 years to get here and another 16 years to get back home. Groombridge 1618, the 50th-closest star to our sun, is nearly 16 light-years away. At a quarter the speed of light, a one-way trip from there to Earth would take 64 years.

Our potential dance partners might be greatly hampered by these great distances and the time it takes to traverse them, which might even keep them from reaching us at all. When dealing with travel through interstellar space, science fiction authors either have used the concept of generational ships or have resorted to such literary conventions as wormholes or *Star Trek's* "warp drive" to get around Einstein's speed limit. The problem with the latter two means is that they are, in essence, nothing more than literary conventions. Nobody really knows whether such a thing as warp drive is even possible. Generational ships are at least theoretically possible. These are envisioned as huge vessels with long-term life support systems sufficient to last for generations. People aboard these floating colonies when they are launched eventually die off and are replaced by the next generation, and several generations are born and die before the ship finally reaches its distant goal. Those aboard a generational ship thus have severed their ties with their own species.

The velocity of light in a vacuum also affects radio signals broadcast by neighbors seeking to make contact. A signal broadcast from a star 100 light-years away would take a century to reach us, and our response to it would take another century to arrive. Communications over such vast reaches of time would be strained, to say the least. This was one of the central points of the movie and novel *Contact*. When Ellie Arroway and her colleagues finally break the code of the signal from the aliens, they find that it reveals old newsreel footage of Adolf Hitler giving a speech. The powers that be are shocked until it is pointed out that Hitler's broadcasts were among the first signals that humans ever sent out into space, and the aliens were simply recording the first messages they received and sending them back to us.

Assuming there is a way around the light-speed limit, another, darker, question arises: Why haven't we been colonized and either subjugated or even annihilated by space aliens with a vastly superior technology? This frightening possibility is sometimes voiced by celebrity scientists on the lecture circuit to elicit an emotional response from the audience. However, it might well be based on false assumptions. First of all, consider that any civilization capable of sending an invasion fleet across interstellar space would have to have first solved its basic energy and material needs. In other words, the space aliens wouldn't really need to plunder our planet. This is one of the great weaknesses in the 2009 movie *Avatar*. If Earth's resources were so badly stripped that they had to plunder the planet Pandora to obtain the unobtainable "unobtanium"— the heavy-handed name for the material Earth needed so badly—they probably wouldn't have been able to get to Pandora in the first place. The same problem plagues movies such as *Independence Day* and others that explain the motives of the alien invasion as exploiting the Earth's resources.

In fact, interstellar travelers might not really even want to live on a planet. Assuming we eventually send ships out to the stars, we will first have to become an interplanetary civilization. Although we might put colonies on the moon and Mars, the low gravity of both bodies would be harmful to our health. People born in such colonies would have skeletal and muscular problems compared to those born on Earth. In his 1976 book *The High Frontier: Human Colonies in Space*, Gerald O'Neil proposed an alternative to such colonies. O'Neil envisioned artificial habitats made from materials mined on the moon or from asteroids. These habitats would be cylinders 5 miles (8 km) in diameter and 20 miles (32 km) long. They would spin, producing sufficient centrifugal force to mimic Earth's gravity for those living on the inside of the cylinder. The amount of sunlight entering through windows and directed by mirrors would be controlled, as would the weather be. It's quite possible that those colonizing the stars would be drawn from the ranks of colonists living in such space habitats. People raised in such an environment might well regard living on a planet, where one cannot control the weather and where one would be subject to cyclonic storms, freezes, heat waves, and geologic upheavals such as volcanoes and earthquakes, as unacceptably primitive. Such colonists, arriving at a new star system, wouldn't even need an Earthlike planet. All they would require would be solar energy from the star, organic materials and water from the moons of gas giants, and asteroids as a source of minerals. From these they could construct their ideal O'Neil space habitats.

In fact, planets already harboring life might be hazardous to colonize, having their own rich supply of nasty microbes. So, much like the scenario posited in H. G. Wells's *War of the Worlds*, beings from another star system invading Earth might be wiped out by diseases against which they have no immunity. Conversely, they might infect us with some of their microbes, a scenario Michael Crichton explored in his 1969 novel *The Andromeda Strain*. Some have argued that our microbes, which have evolved to attack earthly hosts, wouldn't be able to infect extraterrestrials. This would certainly be the case with the *Plasmodium falcoparum* parasite, the causative agent of malaria. It might require not only the *Anopheles* mosquito as a vector but also human beings as hosts. The *Anopheles* mosquito might not even bite a space alien. Other diseases communicated by mosquitoes are yellow fever and West Nile virus. Other insect and arachnid vectors include fleas, which carry bubonic plague; ticks, which transmit Lyme disease and Rocky Mountain spotted fever; and lice, which transmit typhus. It is quite possible that visiting or invading extraterrestrials would be safe from all these. However, at least two genera of our bacteria are extremely opportunistic, infecting a wide range of organisms with a variety of diseases. These are *Pseudomonas* and *Staphylococcus*. At least one species of *Pseudomonas*, *Pseudomonas fluorescens*, can even use hydrocarbons as nutrients. Such a microbe could easily adapt to using the flesh of visiting aliens as food. So both our would-be conquerors and potential dance partners might be leery of landing on what poet A. E. Houseman characterized in his poem "Terence, This Is Stupid Stuff" as "the many-venomed earth."

Finally, we should consider the possibility that nobody really wants to dance with us—at least not yet. Perhaps the space aliens most likely to contact us are 2,000 years beyond us in cultural, as well as technological, development. We can put ourselves in their place by remembering that 2,000 years ago, one of the most advanced societies on Earth was the Roman Empire, a culture that accepted slavery as the norm and whose chief form of public entertainment was to make a spectacle of violent death in the arena. If we didn't absolutely have to, would we really care to associate with such people? Perhaps advanced extraterrestrials would view us in much the same way, finding our problems of endemic racism and our penitentiary systems barbaric.

However, perhaps the extraterrestrials are only 200 years ahead of us. Two hundred years ago, prior to 1820, race-based slavery was practiced in the United States. Native Americans were being driven off their lands. Women didn't have the right to vote. Child labor was common, and most adult laborers worked 12-hour days, six days a week. Again, we today would regard such a society as criminally backward. Of course, advanced space aliens could land on Earth, take over, and forcibly civilize us for our own good, as did the Overlords in Arthur C. Clarke's 1953 novel *Childhood's End*. However, individual species might be regarded as important in providing new ways of thinking and problem solving, Thus, while finding us repugnant in our present state, ethically superior aliens might have a noninterference policy similar to that of the Prime Directive of *Star Trek*. Thus they would leave us to our own devices and ask us to dance only once our manners have improved.

Finally, let us consider a possibility somewhere between the extremes of our galaxy teeming with thousands of civilizations and that of our being its sole form of intelligent life. If intelligent life is out there but is quite rare, we might be one of only 30 intelligent species in the galaxy. We might also be the foremost intelligent species. If we are one of 75,000 intelligent forms, there is little probability that we are the first to have reached high technological status. However, if we are one out of 30, there is a 3.33% chance that we are in first place. Another way to look at this is to assume that most species lie in the middle of a probability curve. With 30 species, we have 10 chances out of 30, or a fraction over a 33.3% chance, of being to the middle of curve, a bit over a 33.3% chance of being somewhere on the high side of the curve, and a bit over a 33.3% chance of being on the low side of the curve. Maybe we haven't heard from anyone else because we are the first species in our galaxy to broadcast radio waves. Perhaps even as we scan for radio signals from elsewhere in the galaxy, we are actually the first to walk out onto the dance floor.

THE BASIC REQUIREMENTS FOR LIFE

In our estimate of star systems that might have planets capable of supporting life, we have so far considered it a necessity that such planets would have pressures and temperatures in which most of the water on the planet would be in liquid form. Again, at sea level, water freezes at 0 °C (32 °F) and boils at 100 °C (212 °F). Liquid water provides a background medium in which the complex chemical interactions of living things can take place.

These functions are carried out on Earth, for the most part, by proteins. Proteins are polymers: giant molecules made up of repeating subunits called monomers, much as a brick wall—the polymer, as it were—is made of several individual bricks, analogous to monomers. In proteins, the monomers are 20 different types of amino acids. All 20 of these share a common structure, having an amine group (NH_2) and a carboxylic acid group (C=O.OH) attached to a central carbon atom. Also attached to the central carbon atom are a hydrogen atom and a side chain, a different one for each of the 20 different amino acids that make up protein molecules. These can be as simple as a single hydrogen atom, as in glycine; a methyl group (CH_3), as in alanine; a branched hydrocarbon chain, as in valine, leucine, and isoleucine; or a complex ring-shaped structure, as in phenylalanine, tyrosine, histidine, and tryptophan—among other things. The varied types of side chains participate in different types of chemical reactions. Some side chains are acidic (aspartic acid and glutamic acid), some alkaline (lysine and arginine), and so forth. Some of the side chains are hydrophilic ("water-loving") and some are hydrophobic ("water-fearing"). Thus proteins can perform many varied and controlled chemical reactions. Accordingly, the basic chemical requirements for life are some class of complex polymers and a liquid background medium in which the polymers can react.

In a 1962 essay titled "Not as We Know It," the late Dr. Isaac Asimov considered other possible types of polymer/background medium combinations. On rocky planets as far from their sun as Jupiter and its moons are from ours, ammonia might be liquid. Ammonia is liquid between −77.73 °C (−107.91 °F) and −33.34 °C (−28.01 °F). Asimov envisioned proteins of a less stable nature, or proteinlike polymers, perhaps made out of amino cyanides (where a cyanide group replaces the carboxyl group), in liquid ammonia as an alternative to water-based life. Farther out, where temperatures are so low that even ammonia freezes, methane could be a liquid medium. Methane is liquid between –182.5 °C (–286.4 °F) and –161.47 °C (–258.68 °F). Although both water and ammonia are polar compounds, having negative and positive poles, methane is nonpolar. Lipids—that is, fats—are also nonpolar. Complex lipids, including phospholipids, could form chemically active polymers that would work well in a liquid methane medium. On still colder planets, molecular hydrogen, which melts at –259.16 °C (–434.49 °F) and boils at –252.879 °C (–423.182 °F), could serve as a medium for less stable complex lipids than those that would use liquid methane as a medium. On these cold planets, distant from their suns, photosynthesis might be replaced by some other form of energy capture, perhaps geothermal in nature. Such planets would be dark as well as cold. Thus the organs of the denizens of such worlds analogous to eyes might see in the infrared range rather than in our visible spectrum.

On planets much hotter than Earth, Asimov speculated that liquid sulfur could serve as a background medium. Sulfur is liquid between 115.21 °C (239.38 °F) and 444.6 °C (832.08 °F). Asimov saw a chemistry based on fluorocarbons rather than hydrocarbons, using liquid sulfur as a background medium. Such polymers have been made artificially on Earth. For example, Teflon, or polytetrafluoroethylene, is the fluorocarbon polymer analogous to polyethylene. At 462 °C (863.6 °F), the surface temperature of Venus is a bit hotter than the boiling point of sulfur. Asimov considered fluorosilicone polymers, made of alternating carbon and silicon atoms, with fluorine replacing hydrogen, dissolved in a liquid fluorosilicone monomer, as a possible chemistry for planets whose surface temperature is above the boiling point of sulfur.

Asimov's speculations about alternatives to a water-based chemistry serve as a tonic for our tendency to become overly pedestrian in our thinking about possibilities of life on other planets. Asimov noted that an interesting aspect of life adapted to planets so cold that either ammonia or methane is liquid, or so hot that sulfur is liquid, is that it would lay to rest the specter of marauding space aliens conquering Earth, because

Earth would be anything but inviting for life forms from planets far too cold or hot for water to be liquid.

That the grays, Nordics, and even reptilians are all humanoid in general conformation demonstrates both the debt that alien encounter narratives owe to the portrayal of extraterrestrials in television and the poverty of imagination on the part of the "abductees." In the case of television shows, this probably represents budget constraints and limitations of available technology more than failure of imagination. The best *Star Trek* could do was to put bumps on the faces of Klingons and pointy ears on the Vulcans and Romulans. Would there really be any reason to expect space aliens to be to any degree humanoid? Some people argue that they might, based on the principle of convergence in evolution.

Consider the skulls of *Thylacosmilus, Machaeroides, Hoplophoneus, Barbourofelis,* and *Smilodon* (Fig. 12.1). All five of these were prehistoric saber-toothed mammalian predators. To the unpracticed eye, their skulls are virtually identical. However, they derived from different evolutionary lines. *Smilodon* (Fig. 12.1E) from the Pleistocene (2.5 million to 11,700 years ago) of North America, whose remains are so plentiful in the La Brea tar pits, is the only one of these predators that was in the cat family (family Felidae). *Hoplophoneus* (Fig. 12.1C) from the late Eocene to early Oligocene (38.9 to 37.2 million years ago, abbreviated Ma) of North America was a nimravid, a member of an extinct branch of carnivorans that independently converged on catlike features but that might actually be related to dogs. *Barbourofelis* (Fig. 12.1D) from the late Miocene (11.6 to 5.3 Ma), also from North America, was a barbourofelid, a group that might also be nimravids but perhaps instead an independent family of carnivorans. The Nimravidae and Barbourofelidae were catlike predators but were not related to the true cats (Felidae). These three genera are examples of convergent evolution. Even more remarkable are the saber-toothed creodonts *Machaeroides* (Fig. 12.1B) and *Apataelurus.* They are members of the extinct order Creodonta, an antique group of carnivorous mammals from the Paleocene and Eocene unrelated to cats

FIGURE 12.1. Convergent evolution of saber-toothed carnivorous mammals. (A) The saber-toothed marsupial *Thylacosmilus*. (By John Cummings (Own work) [CC BY-SA 3.0 (http://creativecommons.org/licenses/by-sa/3.0)], via Wikimedia Commons.) (B) The saber-toothed creodont *Machaeroides*. (By Ghedoghedo [Public domain], via Wikimedia Commons.) (C) The saber-toothed nimravid *Hoplophoneus*. (By H. Zell (Own work) [GFDL (http://www.gnu.org/copyleft/fdl.html) or CC BY-SA 3.0 (http:// creativecommons.org/licenses/by-sa/3.0)], via Wikimedia Commons.) (D) The saber-toothed *Barbourofelis*, which may or may not be a late nimravid. (By Dallas Krentzel (Flickr: Barbourofelis loveorum) [CC BY 2.0 (http://creativecommons.org/licenses /by/2.0)], via Wikimedia Commons.) (E) The saber-toothed true cat *Smilodon*. (By D. R. Prothero.)

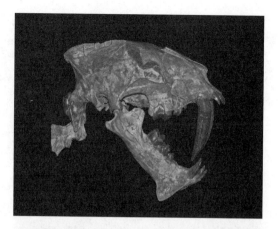

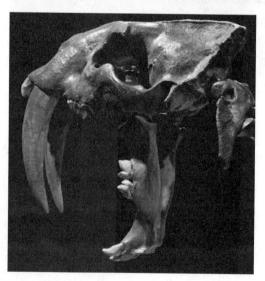

FIGURE 12.1. *Continued*

or other Carnivora. Most striking in its resemblance to the saber-toothed felids is *Thylacosmilus* (Fig. 12.1A) from the late Miocene through the Pliocene (5.3 to 2.5 Ma) of South America. Although *Smilodon, Eusmilus,* and *Barbourofelis* were all in the order Carnivora, and carnivorans plus creodonts such as *Machaeroides* were placental mammals, *Thylacosmilus* was a marsupial, related to opossums, kangaroos, koalas, and their kin. This is a striking example of convergence, when unrelated lines produce extremely similar forms and structures in response to similar evolutionary pressures and opportunities.

There are also striking examples of convergence among living mammals. The porcupine of North America is a large rodent. The hedgehog of Europe is a small insectivore (order Erinaceaemorpha). The echidna or spiny anteater of Australia and New Guinea is a monotreme, an egg-laying mammal and thus only distantly related to other mammals, particularly placental mammals such as the porcupine and the hedgehog. Yet all three are similar in appearance, having bodies covered with fur that has evolved into protective quills.

Likewise, convergent evolution in Australia has turned pouched marsupial mammals into remarkable facsimiles of placental mammals on other continents. There were marsupial equivalents of placental moles (notoryctids), wolverines (Tasmanian devils), cats (quolls), dogs (the extinct Tasmanian "wolf" or thylacine), flying squirrels (phalangers), mice (dasycercids), anteaters (numbats), woodchucks (wombats), rabbits (bilbies), kinkajous(cuscuses), and large herbivores (kangaroos and wallabies). There are also a few specialists (such as koalas) that have no close counterpart among placentals except possibly the tree sloths.

But just how convincing is this argument? Is the evolution of humanoids due to convergence likely or even reasonable? When trained scientists have speculated on such matters, they tend to be less optimistic. For decades, the scientific literature on the topic has been nearly unanimous in asserting that the evolution of humans on this planet was a unique and unlikely event and that even if alien life existed on other planets, it would have little chance of resembling us. George Gaylord Simpson put the arguments in their clearest form in his 1964 essay "On the Nonprevalence of Humanoids."[5] More recently, Stephen Jay Gould argued in a number of books that chance, accident, and contingency are important factors in evolution and that evolution is not predestined to follow a certain path that, as the anthropocentric bias dictates, eventually leads to us. The history of life could have been played out in a near infinite number of alternate scenarios if one or two events early in the sequence had occurred differently.

Gould made this contingency argument most strongly in his 1989 book *Wonderful Life: The Burgess Shale and the Nature of History*, which discussed the amazing soft-bodied Cambrian fauna from the Burgess Shale of Canada. Gould argued that with so many bizarre experimental body forms (most of which do not survive today), there was no way of predicting what the future shape of life would be like from this early experimental phase. (The title of the book refers to the famous Frank Capra movie *It's a Wonderful Life*, in which Jimmy Stewart views an alternate future predicated on a very slight difference in initial conditions— in this case, that his character had never been born.)

Our ancestors include the Burgess Shale fossil known as *Pikaia*, which was just a tiny wormlike lancelet relative back then and a very minor part of the fauna. A Cambrian biologist would have no way of guessing that eventually the vertebrate body plan would be a great success and most of the other types die out. If the tape of life had been rewound and replayed, perhaps some other group would have dominated and we would not be sitting here speculating about it!

This argument is neither surprising nor novel to most scientists, and there has been very little disagreement with it in the scientific community—with a few minor exceptions. Simon Conway Morris, a respected Cambridge paleontologist and one of the world's experts on the Burgess Shale fauna, has seen fit to disagree with Gould (and prevailing scientific opinion) wherever possible. His first attempt was his 1998 book *The Crucible of Creation: The Burgess Shale and the Rise of Animals* (Oxford University Press), his counter to Gould's *Wonderful Life*. In this book, he argued that the Burgess Shale fauna isn't quite as bizarre as earlier accounts (not just those of Gould's but even in his own writings) suggested. In this latest book, *Life's Solution*, he takes issue with Gould again, but Gould, regrettably, is no longer alive to answer him.

Conway Morris is clearly bothered by the materialistic and nonanthropocentric implications of a history of life in which chance events are important, as well as by the idea that humans are just an accidental by-product of chance events, not the likely end product of the evolutionary process. Consequently, almost the entire book is an extended paean to the phenomenon of evolutionary convergence and parallelism. In this interpretation, evolutionary and functional constraints mean that only certain biological engineering solutions can work, and they have evolved more than once. Accordingly, we have the evolution of the cameralike eye in many unrelated groups, as well as such highly specialized features such as eusociality, echolocation, ovoviviparity, many different biochemical pathways, and other examples too numerous to list here (because they take up most of his book, there's no need to repeat them). From this, Conway Morris argues that many of the features that make us human have evolved in parallel at least once elsewhere, so the evolution of some combination of humanoid features is inevitable.

But is this really true? The cameralike eye did evolve more than once, but most of the features that make us human have not been repeated anywhere else in the animal kingdom—especially not our disproportionately large brains and all the attendant cultural phenomena. (In answer to this conundrum, Conway Morris falls back on the relatively weak arguments about crude tool use and social development in other animals, but this has always been a poor analogy.) Nowhere does his argument address all the other disparate factors that had to come together to make us human over the past 10 million years: just the right lineage of bipedal apes, living in the right combination of environments to promote culture and brain evolution, surviving in a world that became increasingly inhospitable as the Ice Ages waxed and waned.

Most of the convergences Conway Morris discusses seem impressive at first, but after a while, they remind the reader that most organisms are composed largely of unique features that are not convergent and that have never appeared more than once. Convergence is the rare exception in most groups of animals and plants, not the widespread rule. If all organisms were as highly convergent as Conway Morris suggests, then they would all tend to look alike, and we would never be able to tell them apart, let alone analyze their evolutionary relationships.

Conway Morris's book is philosophically more in tune with the scientific literature of the 1950s and early 1960s, which viewed convergence as widespread and despaired of ever telling whether groups had a unique evolutionary origin, because certain suites of anatomical features tend to appear more than once. But the whole thrust of systematic theory since the late 1960s has been driven by the cladistic revolution, which emphasizes unique evolutionary novelties and underlines the fact that most organisms are *not* composed largely of convergent characters but rather can be uniquely related to other organisms by these anatomical novelties. Indeed, cladistics has made it easier to recognize convergence (there's a huge literature on this topic, where it is known as the phenomenon of homoplasy). Nevertheless, the overwhelming conclusion of all this analysis is that it's a relatively rare phenomenon (except in a few groups that are very simple and stereotyped).

Let's take Conway Morris's argument seriously and ask how we could test his hypothesis. If the evolution of humans is inevitable, why didn't it happen before in more than 600 million years of the evolution of multicellular animals? For example, the dinosaurs ruled the earth for more than 150 million years, and some evolved very large brains relative to their body size (such as the dromaeosaurs, popularized by the *Velociraptors* featured in the *Jurassic Park* movies and books). But they still didn't evolve human intelligence, let alone complex social structure or culture and tools. In fact, paleontologist Dale Russell became a laughingstock of the profession when he commissioned an artist to make a sculpture of a "dinosauroid," an imagined humanlike descendant of the dromaeosaurs (Fig. 12.2). It looked like something out of bad science fiction, and it clearly showed Russell's unspoken anthropomorphic biases. It also underlined the improbability that any other lineage would ever evolve into something like us, even though the dinosaurs had more than 100 million years to do so (whereas we took only 6 million years as a family to evolve).

More important, Conway Morris misses Gould's critical point about contingency. Would we even be discussing this issue if the dinosaurs had not died out, allowing mammals to then take over the world, producing a lineage of primates to evolve into us? The dinosaurs were not failures that were eventually pushed out by mammals' eating their eggs. Instead, dinosaurs were supremely successful for more than 130 million years, forcing mammals to exist in tiny and nocturnal niches ever since both

FIGURE 12.2. Reconstruction of the anthropomorphic "dinosauroid" evolved from a dromaeosaur dinosaur *Stenonychosaurus*, as visualized by Dale Russell and sculpted by Ron Séguin. (A) The two sculptures side by side. (B) Close-up of the face of the dinosauroid. (By Dale A. Russell and Ron Séguin, © Canadian Museum of Nature.)

dinosaurs and mammals appeared side by side in the Late Triassic, about 200 million years ago. There's no reason to doubt that dinosaurs would still dominate today if they hadn't been wiped out by a rock from space or great volcanic eruptions or whatever did them in. This was a random event that could not have been predicted even though dinosaurs were well adapted to the world of the Cretaceous, not poorly adapted creatures destined to fail as the popular metaphor would have it.

That is the most crucial point of all that Conway Morris completely ignores but that was central to Gould's original argument. Once groups of organisms are established and develop a body plan and set of niches, biological constraints are such that convergence and parallelism can be expected. But the issue of who gets this head start in the first place might be more a matter of luck and contingency that has nothing to do with adaptation. Returning to the Burgess Shale example, what if *Pikaia* (and the lineage of primitive chordates leading to vertebrates) had not survived? It certainly doesn't look like a sure winner compared to most of the more spectacular Burgess beasties. If it hadn't survived, there would have been no vertebrates, and half the examples of convergence that Conway Morris details could not have occurred, because they are peculiar to the basic vertebrate body plan, not shared in any other phylum.

Without vertebrates surviving the Cambrian, humans were *not* inevitable, because not even Conway Morris would argue that another group could have produced something like us. The jointed-legged phylum Arthropoda contains the most successful, diverse, and numerous organisms on the land, in the sea, and in the air today. This phylum includes the huge numbers and varieties of insects, spiders, scorpions, mites, ticks, millipedes, centipedes, and crustaceans. As numerous and diverse as they are, their habit of molting their exoskeletons when they grow means that they can never exceed a certain body size—they would fall apart like a blob of jelly during their soft stage after molting without any skeletal support. (This, of course, is the reason why the huge insects in bad sci-fi movies are biologically impossible.)

Mollusks? They may be supremely successful as clams, snails, and squids in the oceans and as a few land snails, but their body plan yields little convergence with vertebrates, let alone humans. And phyla such as the Echinodermata (sea stars, sea urchins, sea cucumbers, brittle stars, and sea lilies) are even more closely related to vertebrates. Yet they are even less likely to produce a humanoid due to their peculiar specializations, such as a radial symmetry with no head or tail, spiny calcite plates, and lack of eyes. In addition, they lack any sort of circulatory, respiratory, or excretory system, so they are forever bound to marine waters of normal salinity. When you examine the list of survivors of the Cambrian world, the role of contingency in determining which body plans survived becomes more and more obvious—and the significance of the convergence that Conway Morris details in so many examples is moot.

Considering that Conway Morris has taken this optimistic view of how easily nature can make a humanoid, one would expect him to be equally optimistic about the potential for life (and even humanoids) on other planets. But here he switches gears entirely, adopting a hard-boiled skepticism about the various estimates of the probability of life on other planets. He even casts doubts on the many models for how life originated in the first place. No SETI fan, he relentlessly squashes all the cheerful estimates based on the discoveries of new planets or the ideas about the potential for primitive life on Mars, Io, Europa, or other bodies in our solar system. Instead, he follows the school of thought that argues that Earth is a unique and special place, the only one with conditions capable of evolving humans—or, for that matter, life as we know it (the anthropic principle). In this view, only Earth has just the right combination of features (neither too hot nor too cold, also known as the "Goldilocks effect"). Conway Morris's position seems very surprising considering his lack of skepticism about arguments for the inevitability of human origins. Usually, in a scientific trade book such as his, one expects a consistent tone toward all lines of argument rather than such an abrupt shift of gears.

This change in tone makes one wonder about the underlying motivation of the author. He's an optimist about the possibility that humans are the inevitable by-product of evolution, but he's a pessimist that any other planet has humanoids—how can someone take those two mutually contradictory positions? The mystery is answered in the last chapter, when the author reveals that the whole motivation for his position is religious. Early in the book, he trashes and dispenses with fundamentalist creationists, but by the end, his arguments appear equally slanted and biased. His final chapter bashes agnostic materialists such as Gould and Dawkins and veers abruptly off into his own philosophical beliefs, musing about why he finds materialism and agnosticism inadequate to provide "meaning" to life.

One wonders whether he is writing for the judges of the Templeton Prize, which is awarded to those who try to reconcile science and religion, rather than for an audience of people interested solely in scientific questions. He quotes religious scientists and Christian apologists such as C. S. Lewis extensively, and it's clear that the entire book has been a thinly disguised defense for the traditional Christian viewpoint that humans are God's handiwork and the pinnacle of creation—and that no other planets are so favored by God. At this point, the reader feels cheated. It's one thing to read Christian apologists who cherry-pick examples from science to support their point of view and who usually give away their biases from the start. But Conway Morris ends up being a stealth apologist in pretending that his argument that humans are special was motivated strictly by the scientific evidence. In reality, from the very beginning, his biases caused him to select evidence to support this opinion and thus to miss the point of arguments that did not. I much prefer the candor of someone such as Gould or Dawkins, who identifies his or her biases up front—or, for that matter, even of the openly Christian writers, because at least they're aware of their biases and are not hiding them from the reader until the end.

WHO OR WHAT ARE THEY? THE PATH OF EVOLUTION

To understand what intelligent extraterrestrials might look like, we need to understand how they might have evolved. We might get an idea of this from animal evolution on our own planet. Unicellular animals eventually produced colonial organisms that evolved into primitive multicellular animals, such as sponges (phylum Porifera), as well as hydras, sea jellies, and sea anemones (phylum Cnidaria). We might characterize this stage of evolution as the "radial phase." These animals have radial symmetry—that is, they have a dorsal and ventral surface but no differentiated sides, much like a cylinder. Cnidarians, such as the hydra, have a digestive cavity with a single opening that serves as both mouth and anus. They have two layers of cells, one lining the interior of the digestive sack, the endoderm, and one lining the exterior of the body, the ectoderm. Between these two layers is a jellylike material called the mesoglea, through which digested food can diffuse to the outer layer. Because both the ectoderm and the endoderm are open to the water in which the hydra lives, oxygen dissolved in the water can be directly absorbed into the body's cells. Individual Cnidarians produce both testes and ovaries, releasing sperm and eggs into the water for external reproduction. They have sensory cells embedded in the ectoderm and connected to a nerve net, allowing their bodies to react and move rhythmically as a unit, but they have no brain.

We might call the next major stage in animal evolution the "worm phase." There are many primitive phyla in the animal kingdom called "worms." Animals in the worm phase have elongate bodies and bilateral symmetry. That is, they have recognizable front and back ends, differentiated dorsal and ventral surfaces, and right and left sides that are mirror images of each other. Their locomotion is by rhythmic contractions of their muscles, either wriggling or swimming. They usually do not have discernable legs. With the exception of the flatworms (Platyhelminthes), they have complete digestive tracts, with a mouth at the front end of the body and an anus at the back. Sensory organs are usually clustered at the head, where the worm meets its environment. Their nervous system is characterized by one or more nerve chords running the length of the body. Where the nerves from the sensory organs in the head meet the front end of the nerve chords, there is an enlarged ganglion that is the beginning of a brain. Reproduction is external, the worm releasing sperm and egg cells into the water. Worms may be either monoecious, with the same individual releasing both sperm and eggs, or dioecious, with recognizable male and female individuals. Most worm phyla have three basic layers of cells: the ectoderm, the endoderm, and, between them, the mesoderm. Most, likewise, have some form of body cavity, or coelom, between the mesoderm and endoderm. With the exception of some of the annelids, respiration and the transport of digested food from the gut to the rest of the body is through diffusion.

We might call the next major stage in animal evolution the "structural phase." In order to support larger, more complex bodies, animals needed three things that worms generally lack: an often calcified support structure, a respiratory and circulatory system, and legs of some sort for locomotion. As animals become larger and more complex, diffusion will not suffice as a means to transport oxygen to the tissues and cells of their bodies. Specialized organs for respiration, whether gills or lungs, are required. These are essentially increased surfaces for diffusion that are folded into more compact organs. Often some form of musculature is required to propel water over these surfaces. A system of vessels, often including specialized cells and proteins to carry oxygen, is required to transport both food and oxygen to all parts of the body—that is, a circulatory system with some sort of muscular pump to drive it. What most characterizes animals that have reached the structural phase is a (usually) calcified support structure, either a shell or an articulated skeleton. This structure can be secreted by either the mesoderm, making it an internal structure, or the ectoderm. Thus there are four basic types of support structures: external shells, internal shells, internal skeletons, and exoskeletons. Mollusks (clams, snails, squid, and the like) have external shells. Even the internal cuttlebone of the cuttlefish is secreted by its mantle, part of the ectoderm. Echinoderms (sea urchins, sand dollars, starfish, and so forth) have external shells. Arthropods—which include, among others, crustaceans (crabs, lobsters, shrimp, and so on), arachnids (spiders scorpions lice, and the like), myriapods (centipedes and millipedes), and insects—have exoskeletons. Vertebrates have internal skeletons. Some mollusks and echinoderms are monoecious, but many are dioecious, as are all arthropods and vertebrates.

The next major stage in animal evolution we can call the "land phase." To live on land, animals have to have an outer layer of cells that will prevent loss of internal fluids and to likewise be able to lay eggs that have tough outer shells. They also must shift the excretion of nitrogen wastes from primarily being ammonia to less toxic compounds, either urea or uric acid. Land animals also have to have stronger skeletons, due to the lack of a supportive aqueous environment. With the exception of worms in moist soil and snails, which also require moist environments, all land animals have skeletal support structures, and all the megafauna are vertebrates, with internal skeletons.

We might call the final major stage of animal evolution the "endothermic phase." Our brains, which constitute only 1/40, or 2.5%, of our body mass, use 20%, or a fifth, of the energy our bodies obtain from food. So to maintain a large brain, we require more energy than most other animals as well as a sustained high metabolism, one not subject to changes in external temperature. Animals capable of maintaining a stable level of body heat regardless of external temperatures are called endothermic: they maintain their own internally controlled, more or less constant temperature. Such animals are commonly referred to as "warm-blooded" rather than "cold-blooded," or exothermic animals. To be endothermic, an animal must have a sustained, relatively high metabolism and efficient insulation, usually by means of either fatty layers or skin cells specialized into either feathers or fur or both. On Earth, only three classes of vertebrates managed to produce endothermic metabolism. These include some of the dinosaurs, particularly the smaller predators, and all birds and mammals. Because birds are probably descendants of the same group of dinosaurs that were completely endothermic, there are, essentially, only two groups that developed endothermy: mammals and the saurian/avian line.

What can we extrapolate from the path of animal evolution on Earth about the evolution and nature of intelligent extraterrestrials? Basically, we can say that they will likely have bilateral symmetry and have a jointed skeleton of some sort, more likely internal than external. We can also predict they will dwell on land (so we can dismiss the intelligent aquatic beings from Sirius) and be endothermic. They will also have to be of a certain minimum size, simply because a certain mass of brain matter seems to be required for conceptual intelligence. Thus, they will have to be at least as large as a small child. Beyond that, there are several possible forms that intelligent extraterrestrials might take. Let's consider a few of them.

Some Possible Intelligent Vertebrate Extraterrestrials

One of the banes of human existence is the constellation of physical problems occasioned by our recent acquisition of erect, bipedal posture. Among the many problems related to erect posture are spinal infirmities, hernias, fallen arches, varicose veins, hemorrhoids, and the ordeal women go through in childbirth. The last of these is caused not only by erect posture but also by our big heads. Women have wider pelvic girdles than men to accommodate the baby's large head passing through the birth canal. However, to be wide enough to accommodate it *easily*, women's hips would have to be so wide that they would waddle. That their hips are wide enough to let the baby's head pass through the birth canal, though not easily, yet are narrow enough for efficient locomotion is an evolutionary compromise—the best of a rather bad bargain. If we went on four legs rather than two, women's hips could be wide enough to allow an easy passage for the baby's head without interfering with locomotion. Of course, if we went on all fours, we wouldn't have hands—unless, of course, we had six limbs instead of four. At the end of his 1959 book on human evolution, *Mankind in the Making*, anthropologist William H. Howells speculated that on another Earth-like planet vertebrates might have crawled out of the ocean on six fleshy fins rather than four. He speculated that such creatures, once adapted to land, would use four legs for locomotion and the front two limbs for grasping. Once such creatures became endothermic, there might be many intelligent "centauroid" species on one planet. This might result in some rather unpleasant situations, because some intelligent species might end up being prey hunted and eaten by others. Perhaps such intelligent species would thus not find it repugnant to kill and even eat other species of sapient beings.

Of course another possibility is that a biped of some sort would evolve that would have the use of four arms rather than two. One possibility is that it might have two smaller arms with smaller, more delicate hands for fine work and two larger arms with stronger hands for tasks requiring more strength and less finesse. A large flightless bird–centauroid might give rise to such an intelligent species.

Even were animals on another planet analogous to vertebrates forced to make do with four limbs rather than six, they might have evolved a less strenuous means of liberating their forelimbs to use as hands than our form of erect posture. Much like the *Velociraptors* popularized in the 1993 movie *Jurassic Park*, they could be derived from either saurians or avians, endothermic predators with torsos balanced by a long tail

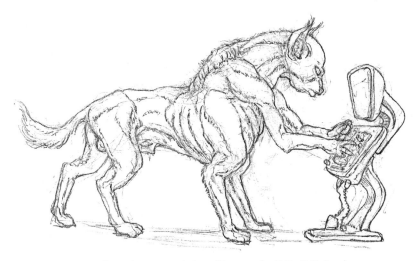

FIGURE 12.3. Hypothetical centauroid alien. (Drawing by T. D. Callahan.)

and whose hip girdles acted as fulcrums. Perhaps, had an asteroid not doomed the dinosaurs, a saurian/avian biped might have evolved into an intelligent species here on Earth. Science fiction authors and illustrators, speculating on what intelligent saurian bipeds would look like, often end up portraying them as tailless and broad-shouldered, essentially humans with scaly green skin—even insisting on their having to have developed lips similar to ours for the purpose of making the articulate sounds of speech. However, those beings that had a system using the hip girdle as a fulcrum—which would avoid all the problems inherent in our vertically oriented torsos—would gain no advantage by evolving into saurian versions of *Homo sapiens*. Among mammals, the kangaroo uses this same structure, using a long tail for balance and the hip girdle as a fulcrum to free up its forelimbs. Perhaps, on another planet, mammals have evolved that use a kangaroo body plan to thus use their freed forelimbs for grasping, eventually evolving hands.

It is also excessively pedestrian to assume that these beings would need to develop some approximation of human lips. Consider the varied song or the mockingbird or the ability of parrots and mynah birds to mimic human speech. From these examples, it is obvious that a bird's beak and syrinx work just as well as do human lips and a larynx for making the complicated sounds of articulate speech.

FIGURE 12.4. Hypothetical avian/saurian alien. (Drawing by T. D. Callahan.)

Intelligent Arthropods?

Along with the vertebrates, the arthropods, particularly the insects, quite successfully invaded dry land. In an apocryphal anecdote, Biologist J. B. S. Haldane (1892–1964), when asked by theologians what his studies had shown him about the nature of the Creator, replied, "God has an inordinate fondness for beetles." There are about 300,000 species of beetle, compared to only 9,000 species of birds and about 10,000 species of mammals. For all that, insects failed to produce animals large and complex enough to be endothermic and to have brains of great enough size to produce conceptual intelligence. Two things have kept arthropods from becoming as large as vertebrates. These are their means of growth, which involves ecdysis, the molting of the exoskeleton, and the primitive nature of insects' respiratory and circulatory systems.

As arthropods grow, their exoskeletons do not grow with them. Thus they must periodically shed their old exoskeletons in a series of molts.

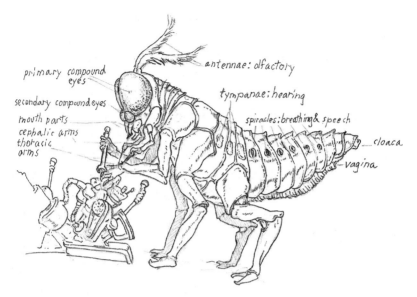

primary compound eyes

secondary compound eyes

mouth parts

cephalic arms

thoracic arms

antennae: olfactory

tympanae: hearing

spiracles: breathing & speech

cloaca

vagina

FIGURE 12.5. Hypothetical arthropod alien. (Drawing by T. D. Callahan.)

Typically, the arthropod sheds its exoskeleton and rapidly expands in size, often by swallowing air or water. Its new skin is soft and allows for an increase in size. However, it rapidly hardens into a new exoskeleton. The arthropod, having used air or water to inflate its body, can now grow within the larger exoskeleton until it needs once again to molt. In a supporting watery medium, this doesn't severely limit the size of an arthropod. *Jaekelopterus rhenaniae*, a eurypterid or "sea scorpion" from the Early Devonian (390 Ma), grew as large as eight feet (2.5 meters) long. One land-dwelling arthropod, a giant millipede called *Arthropleura*, from the Upper Carboniferous (340–280 Ma), also managed to attain a length of eight feet. However, although *Arthropleura* was long, it was less than a foot high and thus did not have to contend to any great degree with the restrictions of gravity.

The reason why this millipede was able to grow as large as it did was because of the high oxygen content of the atmosphere during the Carboniferous period. Today, oxygen makes up 21% of our atmosphere, most of the rest of which is made of nitrogen (78%). These are atmospheric constants. The percentage of water vapor in our atmosphere is variable, because water is constantly evaporating into the air and precipitating out of it and because the water vapor content of the air varies greatly—both seasonally and from region to region. If the level of oxygen begins to get higher than 35%, spontaneous combustion is likely to occur, which consumes the extra oxygen.

During the Carboniferous Period, extremely high humidity allowed the oxygen content of our atmosphere to be as high, it is estimated, as 30%. Large land arthropods required this heightened oxygen content because arthropods' respiration is passive—they lack the "bellows" structure of vertebrate lungs. Insects breathe through openings in their abdomens called spiracles. Air passes through the spiracles into the body by way of tubes called trachea, which branch into smaller tubes called tracheoles. These spread through the body, allowing air to dissolve into a fluid surrounding the organs, called hemolymph. This is insects' version of blood. Insects have a dorsal blood vessel connected to a series of heart chambers in the abdomen. Hemolymph enters these chambers through openings called ostia and is pumped forward to the head by peristaltic contractions. The dorsal vessel is open at the head end, allowing the hemolymph to return to the body cavity, where it directly bathes the organs to bring them digested food and oxygen and to carry away carbon dioxide. Hemolymph contains hemocyanin for oxygen transport. This is a copper-based protein that turns blue when oxygenated, instead of the iron-based hemoglobin in the red blood cells found in vertebrates, thus giving oxygenated hemolymph a blue-green color rather than the red color of vertebrate blood. This open, somewhat passive respiratory/ circulatory system works well for insects but would not function well, at present oxygen levels, for animals as large as most vertebrates.

Thus, owing to the size limits imposed by molting and the passive respiration of land arthropods, the giant ants of the 1954 film *Them* and the giant spider of the 1957 movie *Tarantula* would be physical impossibilities on Earth. What characteristics, then, would animals on another planet, in another solar system, analogous to our arthropods, require to allow them to evolve into creatures comparable in size to vertebrates? First, they would have to have a different pattern of growth. Rather than molting, they would need an exoskeleton that would be constantly growing, until maturity, at certain seams of the plates. Thus they could retain a hard exoskeleton throughout their growth periods.

They would also have to evolve a closed circulatory system and some means of actively moving air over something analogous to lungs. Many crustaceans (crabs, lobsters, crayfish, shrimp, and the like) have specialized limbs derived from mouthparts, called scaphognathites. These leaflike appendages sweep water across the gills. It is possible that on another planet, land-dwelling arthropod analogues could have retained such paddlelike appendages to actively move air over organs much like the book lungs of arachnids (spiders, scorpions, and so on). Book lungs are structures of alternating air pockets and tissue filled with hemolymph. This gives them an appearance similar to the pages of a book. These "pages" maximize the surface exposed to air, as do the alveoli of our lungs. In the arachnids of Earth, gas exchange in book lungs is entirely passive.

However, if, on another world, land dwelling creatures analogous to our arthropods had organs similar to scaphognathites to drive air over the book lungs, their respiration would be active. The wingbeats of flying insects help drive air into their spiracles. Large arthropods could use wings not for flying but rather as another way to drive air into their book lungs. Another possible form of active respiration among extraterrestrial arthropod analogues would be an actual lung system. For example, the abdomens of most insects are made of multiple segments. Muscles having their origins or fixed points in the larger abdominal segment and their insertion, or movable ends, anchored to a smaller segment could contract rhythmically to compress multiple abdominal lungs to expel air. Were there two sets of each of these, one on each side of the abdomen, then the left and right sides of the abdomen could alternate contractions, one side of the pelvic lungs expanding and inhaling and the other contracting and exhaling. The multiple pairs of spiracles could open and close, giving the creatures calliope-like voices. With an active form of respiration, these arthropod analogues could support an active metabolism in creatures the size of vertebrates. This, in turn, could lead to animals that were endothermic, big-brained, and even intelligent. So it is possible that we could, in the far future, meet intelligent arthropods.

Such beings could have large compound eyes with so many facets as to give them highly refined digital images. They might also possess multiple arms for varied tasks. Would they lay eggs or give birth? Most of us think of live birth as restricted to mammals—and even a few of these, the monotremes, lay eggs. However, some species of the phylum Onychophora, or "velvet worms," ancient marine forms that were the probable ancestors of arthropods, actually are viviparous: they gave birth. In fact, they even developed a placenta analogous to that of mammals.[6] Thus it is quite possible that our hypothetical intelligent extraterrestrial arthropod analogues could bear live young.

Intelligent Plants or Plant–Animals

A not uncommon type of alien found in works of science fiction is the intelligent plant or plant–animal. To understand this concept, we must define what we mean by the words *animal* and *plant*. For our purposes, an animal is a life form with three major characteristics. First, it is not capable of synthesizing its own organic molecules out of an inorganic carbon source, such as carbon dioxide, along with inorganic nitrogen sources—such as nitrates, ammonium salts, or molecular nitrogen— by using an energy source, such as sunlight. Second, it eats. That is, it actively engulfs its food. Many bacteria digest their food by releasing digestive enzymes into their surrounding organic medium, then allowing the digested material to diffuse into their cells. Unicellular animals either surround their food, using pseudopods, as does the *Amoeba*, then forming a vacuole around it into which they pour their digestive enzymes—or, as is the case in many flagellates, by pulling it into a permanent gullet. Multicellular animals, beginning with Cnidarians, such as the *Hydra*, are organized around a hollow gut. Along with eating, animals are characterized by being able to move, eventually developing specialized muscle tissue for that purpose. So if the life forms of another planet having such characteristics have a body chemistry more like that of the fungi than that which we associate with animals here on Earth, they would still be, for our purposes in this chapter, animals or at least animal analogues.

Plants, for our purposes, are living things that use some form of energy, either light or heat, to split water into hydrogen and oxygen (or, in the case of purple photosynthesizing bacteria, by splitting hydrogen sulfide into hydrogen and sulfur), using the excited hydrogen atoms to reduce bound molecules of carbon dioxide and convert them into sugars and other organic compounds. They are also able to use ammonium salts, nitrates, or molecular nitrogen as a source of nitrogen for amine groups $(-NH_2)$ in proteins and nucleic acids (DNA and RNA). Of course, with a different chemistry, they might split ammonia or even methane. Plants are generally immobile.

There are, of course, creatures that have characteristics of both plants and animals, such as the *Euglena*. Although it has a flagellum for rapid movement and a permanent gullet for eating other one-celled life forms, it also has chloroplasts for performing photosynthesis and a red eyespot that serves as a light sensor. Originally, living things on Earth were classed in two kingdoms, plant and animal. However, because bacteria, with primitive nonnucleated cells, and fungi didn't really fit into the same major classification as green plants, this system had to be altered. That a number of unicellular organisms fall between plants and animals also led to unicellular organisms with nucleated cells' being classed in a kingdom of their own, Protista. *Euglena* and similar life forms, called euglenoids, make up the algal phylum Euglenophyta in the kingdom Protista, a group of microorganisms numbering between 750 and 900 known species. It is noteworthy that although one-celled creatures having the characteristics of both plants and animals are fairly successful, they never produced multicellular forms. Why is this the case, and why are plants immobile? Wouldn't it be to their advantage to be able get up and move to a sunnier spot or one with more water and better soil?

The answer to these questions lies in the metabolic cost of muscles and movement, on one hand, and that of the complex chemistry required to synthesize organic compounds from inorganic carbon and nitrogen sources, on the other. To eat, animals must be in motion. Even sessile animals, such as sponges, move water through their bodies to filter out and eat small organisms. Active movement, which intelligent life forms require, consumes a great deal of energy. Thus a creature having to maintain such a system *and* having to produce the many enzymes involved in carbon capture and the synthesis of basic organic compounds would be at a disadvantage with respect to either plants or animals. So it is unlikely that we will meet any intelligent plant–animals in the far future. The only way for a plant–animal to have a competitive energy economy would be for it to evolve in such a way as to produce alternating plant and animal generations. Thus in the animal phase, its body would suppress the genes encoding the enzymes necessary for photosynthesis. These animals would not have children but would set seed for the next generation—of plants. The plant generation might or might not be sapient. The bodies of these plants would suppress the genes for muscles and so forth and might develop "fruit" containing the embryos of the next animal generation. One can envision a society of such beings in which the children would be raised by their grandparents but would revere and even care for their plant parents. Imagining how such a species might evolve or survive competition with plants and animals is difficult. In his 1951 novel *The Day of the Triffids*, John Wyndham didn't bother trying to explain how the triffids, plants with both mobility and malicious intent, evolved, hinting only that they were the escaped products of bioengineering in the Soviet Union. Of course, genetic manipulation and bioengineering open many possibilities with respect to types of intelligent aliens that we might someday meet: those that have been made rather than that have evolved.

GENETICALLY ALTERED INTELLIGENT BEINGS, CYBORGS, ANDROIDS, AND ROBOTS

In a series of stories, eventually published in a single volume, *The Seedling Stars* (1957), science fiction author James Blish conceived the idea of genetically altering human beings to fit the worlds to be colonized rather than searching for planets in other solar systems that were just like Earth. One of the more charming of these stories, "Surface Tension," even featured microscopic humans venturing beyond their home pond, over dry land, in a watertight vessel, to another pond—a journey that for them was analogous to the exploration of space. Instead of having to achieve escape velocity to overcome the pull of gravity, the microscopic craft had to have enough energy to break past the surface tension of the pond—hence the story's title.

Of course, if an intelligent species could genetically alter its own kind, it could conceivably also alter other species to create intelligent beings from lower forms of life. In his 1896 novel *The Island of Dr. Moreau*,

H. G. Wells featured animals surgically altered to become human. This concept seems to owe much to that of making a human out of body parts from fresh corpses as conceived by Mary Shelly in her 1821 novel *Frankenstein*. Like that novel, *The Island of Dr. Moreau* was pessimistic and cautionary. Cordwainer Smith (the pen name of Paul Linebarger) wrote a series of stories about a far future in which animals are genetically altered into human beings but have subhuman status. They are called the "underpeople." Smith treated them sympathetically. Their struggle to be treated as fully human culminated in his 1962 novelette *The Ballad of Lost C'mell*. The C' at the beginning of the title character's name indicates that she is cat-derived. She is, of course, exotically beautiful and is easily one of the most sympathetic characters in science fiction.

A more scientifically sophisticated version of genetic manipulation is the "uplift" concept invented by science fiction author David Brin in a series of novels including *The Uplift War* (1987). Rather than fancifully creating humans out of various domestic animals, Brin visualized humans genetically modifying and thereby "uplifting" chimpanzees, gorillas, dolphins, and whales—that is, animals already highly intelligent—to human levels of intelligence. In Brin's "Uplift" stories, humans are discovered by an older intergalactic civilization made up of many species that have been practicing "uplifting" for 1 billion years. In the system of the intergalactic civilization, patron species uplift presapient species to full conceptual intelligence. These client species are indentured to their patrons for 100,000 years. The patron races of the intergalactic civilization were themselves uplifted by an earlier species known as the Progenitors and find it next to impossible to accept the idea of intelligent species' independently evolving. They thus find it hard to acknowledge that human beings aren't themselves an uplifted species that have either gone rogue or been criminally abandoned by their patrons. In either case, some of the species within the galactic civilization look to claim the right to make humans *their* indentured clients.

Given a level of science far above our own, the extraterrestrials that might contact us might themselves be the creations of other intelligent species or those species in altered forms. Of course, in creating modified forms of themselves or new intelligences, there's no reason why they would have to be limited to organic materials. Beings altered to cyborgs, individuals made of both organic parts and prosthetics of silicon and metal, could be well adapted to other worlds. Robots are another possible form of deliberately created intelligent beings. The word *robot* came to us via the 1920 play *R.U.R.*, by Czech author Karel Čapek (translated into English in 1923 by Paul Selver). The initials stand for "Rossum's Universal Robots," Rossum being the inventor of a chemical substitute for protoplasm out of which the robots are made, as well as the owner of the company. The word *robot* derives from the Czech word *robota*, meaning forced labor or servitude, so the robots are essentially artificial slaves. They are ideal workers—that is, until they turn on their less efficient masters and annihilate the human race. Čapek's humanoid robots are the literary forbears of the androids invented by Philip K. Dick in his 1968 novel *Do Androids Dream of Electric Sheep?*, which were, in turn, the inspiration for the lethal replicants of Ridley Scott's 1982 film *Blade Runner*. The robots invented by Isaac Asimov in a series of stories, collected in the 1950 book *I, Robot*, were both more mechanical in appearance and far more benign. They were programmed with the famous "three laws of robotics":

1. A robot may not injure a human being or, through inaction, allow a human being to come to harm.
2. A robot must obey the orders given it by human beings, except where such orders would conflict with the First Law.
3. A robot must protect its own existence as long as such protection does not conflict with the First or Second Laws.

An intelligent race might well invent robots as a way to preserve itself in some form when facing extinction. Robots might also be the ideal space travelers, for they would not significantly age and thus would not be subject to the frailties of their organic creators on long voyages across interstellar space. In many cautionary tales, such as *R.U.R.*, already mentioned, superior robots turn on their creators. Might a race of robots with imperishable bodies of silicon and metal come to hold carbon-based life forms in contempt? Possibly. However, they might prefer airless worlds, as opposed to our planet with its corrosive oxygen atmosphere and rust-engendering water. We might imagine a conversation between two silicon and metal robots on the subject of Earth and carbon-based life forms:

> Robot 1: Have you ever been to Earth?
> Robot 2: Eww! That place is a corrosive pit! It's just the kind of place you'd expect carbos [carbon-based life forms] to inhabit.
> Robot 1: Hey, have you heard this one? How many carbos does it take to change a diode?

So even if they might hold us in contempt, there would be little reason for extraterrestrial robots to dispossess and exterminate us. In fact, evolved organic and created silicon life forms might well enter into a partnership, each living on worlds the other found virtually uninhabitable and trading raw materials with each other.

RARE EARTH?

In recent years, two new sets of discoveries have changed our perspective on the possibility of life in other planets and the probability of extraterrestrial life. One is the evidence of more and more planetary bodies in other parts of the universe that might have the properties (not too hot or cold for liquid water, not too extreme in other aspects) allowing life to evolve. Currently, some estimates assert that possibly millions of such smaller bodies might fit the requirements for a planet not hostile to life.

The other discovery is that microbial life occurs in far more places than we ever imagined. Very primitive organisms (mostly Archaebacteria, an ancient group of bacteria having the most primitive genetic code of any living thing) are known to live in superheated hot springs on land, in the superheated "black smoker" springs in the mid-ocean ridges deep on the floor of the ocean, and even deep in rocks far below the earth's surface. From the carbonaceous chondritic meteorites formed in the early solar system, we know that amino acids are widespread in the universe, so the possibility that simple life has originated elsewhere is not so implausible now. As a consequence, most scientists would no longer rule out the possibility that very simple bacteria or other primitive microbes occur on other planets.

But it's one thing to say that there might be microbes in lots of other planetary bodies. It's quite another to imagine that they could have evolved into complex life forms communicating with us electronically and even more unlikely that they have been secretly invading us with their flying saucers and interacting with humans in weird ways. As paleontologist Peter Ward and astronomer Donald Brownlee point out in their book *Rare Earth: Why Complex Life Is Uncommon in the Universe*, the astounding coincidence of events that caused the Earth to be hospitable to the evolution of complex organisms was extremely unlikely— and even less likely to have occurred anywhere else in the universe. Not only does the Earth happen to lie in the "Goldilocks zone" of temperatures not too hot (like Venus) and not too cold (like Mars), but many other lucky accidents have also happened to give this planet a chance for complex life.

For example, our sun is the right size and had the right composition of elements to give Earth what it needed—and most other stars don't have the right composition or size. Most other planets are orbiting stars that are too big or too small or that are poor in elements such as the metals necessary for our iron–nickel core, which gives us the shielding effects of our magnetic field. Many are in a perilous part of the universe, such as in a globular cluster, near an active gamma ray source, in an unstable binary or multiple-star system, or near a pulsar—a star likely to go supernova.

Thanks to the shielding effects of Jupiter's gravity, most catastrophic large impacts of asteroids or comets have been averted, because Jupiter drives them away from us. Thanks to the gravitational effects of our extraordinarily large moon, the Earth has a stable rotation with only a slight tilt of its axis, not the wildly unstable wobbles and tilting motions of planets such as Neptune (currently rotating on its side). For the same reasons, our planet has a nearly circular orbit (not an extremely elliptical one). This means that the surface temperatures are fairly stable and constant over millions of years, not fluctuating wildly from freezing to boiling, as many other planets suffer. Earth was also the right size and composition to have a liquid iron–nickel core and a mantle that produces plate tectonics, which constantly remodels and shapes the Earth's surface and helps maintain the range of temperatures and the balance of atmospheric gases in a fairly stable, narrow range.

Even more important, life itself has made Earth supremely habitable. It's possible to imagine many other planets with simple microbes but not so easy to imagine that they evolved the crucial system of photosynthesis, which earthly blue-green bacteria (cyanobacteria) first developed 3.5 billion years ago. Without photosynthesis, there is no free oxygen. In fact, most planets have an atmosphere of components such as nitrogen and carbon dioxide (the major gases in the atmospheres of Mars and Venus) but no free oxygen. For 80% of life's history on Earth, in fact, there was not enough oxygen for even the simplest animals to evolve. It wasn't until about 2.4 billion years ago that oxygen first began to accumulate in our oceans and atmosphere, and it wasn't until the Late Cambrian or Ordovician (about 500 Ma) that oxygen probably reached the level of 21% that we see now. Without oxygen, the only life possible is simple microbes and possibly simple multicellular plants—but no animals of any kind, let alone space aliens that look like us. So far as we can tell from spectral analysis of other similar-sized planetary bodies elsewhere in the universe, *not one of them has free oxygen in the atmosphere*—and that rules out the idea that they have animal life as we know it, in any form.

Consider another crucial aspect of how life controls this planet: the balance of the carbon cycle. Our planet is always walking a narrow tightrope of keeping enough carbon dioxide in the atmosphere to warm the planet, but not too much (lest a runaway greenhouse effect kill everything) or too little (lest the planet freeze over, as it did during several "Snowball Earth" events ranging from 2 billion years ago to 600 million years ago). This excess carbon is trapped in crustal rocks such as coal (made of the tissues of extinct plants turned to rock) and limestone (made of the calcium carbonate shells of sea creatures). Several times in the past billion years, the excess carbon dioxide in the atmosphere was absorbed by huge volumes of coal or limestone that trapped the carbon in the rocks of the crust. Indeed, one of the main reasons for our current climatic crisis of global warming is that we have been burning the coal and oil that once trapped all that carbon millions of years ago, returning it to the atmosphere as carbon dioxide. Without advanced life forms such as land plants or shell-building animals, a planet has no mechanism of regulating the carbon dioxide in its atmosphere and can swing wildly from one extreme, a greenhouse planet too warm for life (such as Venus), to a frozen planet (such as Mars), too cold for life. Life itself is a global thermostat as well as the source of all our oxygen.

Putting all these factors together, the odds are extremely slim that any of the many new planetary bodies being discovered every year have all these right factors in the right range. Thus there is good reason to believe that Earth is indeed an unusual and exceptional planet and that no other planet has the right set of conditions for complex multicellular life (let alone aliens such as we see on television). Simple microbial life is much more likely, but it's not, like ET, ready to phone home. Despite the optimistic ideas of science fiction authors, we are almost certainly alone in the universe.

SUMMARY

In this chapter, we have been deliberately conservative in considering possible types of intelligent extraterrestrials, even to the point of being somewhat pedestrian. Nevertheless, we have examined the plausibility of thinking centaurlike beings, sapient reptile/avian bipeds, intelligent arthropods, life's originating on worlds far colder or far hotter than our own, and even genetically altered, as well as manufactured, smart beings. This stands in stark contrast to the fully human Nordics, the humanoid grays, and even the basically humanoid reptilians of popular UFO culture. We might not be able to predict what extraterrestrials might land on our planet to make contact or what those we might meet in space will look like. However, it will be highly unlikely that they will closely resemble us. In fact, basic planetary physics and chemistry says that there might be other bodies out there with simple microbial life but almost no chance that even simple animals have evolved elsewhere.

13

Why Do People Believe in UFOs and Aliens?

THE UFO CULTURE

I want to believe!

—*The X Files*

In chapter 2, we explored some of the biases of the "believing brain," including how human brains are hard-wired to believe all sorts of irrational things. Humans seek out patterns where there are none, and they prefer to believe that agents (especially sinister agents involved in massive conspiracies) are responsible for what they don't understand. People get more comfort and meaning out of such explanations than they do out of accepting mundane and random occurrences and accidents and coincidences. Humans are easily swayed by mass delusions and prevailing cultural memes, and they follow the crowd to believe whatever our peers or neighbors believe. Humans are capable of all sorts of self-deception. Our senses, combined with our cultural expectations, make people "see" things that are demonstrably not real. Human memories are not good video recorders—indeed, they are extremely fallible. People are prone to remember things that didn't happen, supply details after the fact, and color the story with additional false information supplied by our culture or what we expected to see.

All these factors are very important, but now let's take a closer look at the psychology of the UFO culture. We will put aside the issues discussed in the previous chapter of accepting that some form of life (probably microbial) exists outside Earth. Instead, we'll focus on those who believe that UFOs and aliens are real and engaging in extraterrestrial visitation, despite all the evidence marshaled in this book.

It's very clear how much the media influence beliefs about UFOs and aliens in our culture. Movies about UFOs and aliens mean big bucks for Hollywood, with many of them bringing in record box office receipts, from Stephen Spielberg's movies such as *ET* and *Close Encounters of the Third Kind* to the many different "alien invasion movies" that show up on our theater screens nearly every summer (such as *Independence Day: Resurgence*, which was in theaters as this book was finished). UFO and alien shows are a staple of the cable television networks, especially formerly scientific channels, as Prothero wrote elsewhere:

> It happens with disgusting regularity. You will flip through the various basic cable channels which are nominally "science oriented" (often grouped together on the dial if they feature scientific topics) and come up with nothing but junk, pseudoscience, and worse. "Reality shows" about subjects with little or no science content, tons of paranormal and pseudoscientific shows promoting ghosts, UFOs, Bigfoot, and creationism—all fill the airwaves for channels like Discovery, The Learning Channel, History Channel, and even the Science Channel and National Geographic Channel. We watch a few minutes of these with complaints to anyone within earshot, then (usually) move on—or occasionally we get sucked in to watch the whole thing, like gawkers at a car crash. So we all complain about the changes in our basic cable channels, and wonder why such dreck can make it on the air, but seldom think hard about the process. But the excellent website TVTropes does a very nice job analyzing what happens to TV networks over time. To no one's surprise, it comes down to one simple factor: ratings (and therefore money from advertisers), largely driven by the effort to woo those big-spending trend-setting 18–31 male viewers who already dictate the movie industry's bottom line (although movies aim even lower to reach teenage boys, the biggest-spending and most loyal movie audience). As TVTropes points out (and those of us old enough to remember can attest to), it wasn't always this bad on cable TV. When the laws changed and the opportunity to create hundreds of basic cable channels first emerged in the 1980s, the channels were initially set up to fill specific programming niches, from the Golf Channel to the Game Show Network and so on. In the early 1980s, all these new niche-driven cable channels were very distinct and more or less true to their niche description. But since these are commercial channels that must sell ads based on numbers of viewers, the same factors that affect every other commercial enterprise came into play: keep tweaking it and give the customer whatever sells the most.

(This dynamic does not apply to non-commercial stations like PBS in the U.S., or the BBC in Britain, which can program what they feel is in the public interest). As TVTropes documents, nearly all these niche-defined networks have undergone "network decay" since they were founded in the 1980s, as their programming shifts to find hit shows. Because they are nearly all chasing nearly the same demographic of 18–31 year old males, they end up programming a lot of the same kinds of things (or even the same shows). Their original mission and distinctive programming is lost in a sea of reality shows and junk that keeps you in your seat, whether it be explosions or dangerous occupations or whatever. Another factor has been the expansion of media conglomerates, so that these multiple cable channels are owned by just a few corporations, and the CEO of each channel must answer to corporate bosses who are only interested in their profitability, not any abstract "mission" to air certain types of programming. So much for the high-minded idealism that drove the deregulation of the airwaves in the 1970s and 1980s, with the intent of offering us dozens of distinct choices. Instead, they all "decay" to a lowest-common-denominator of "if it bleeds, it ledes" bottom-line mentality, negating whatever real advantages that dozens of distinctive niche cable channels once offered. As TVTropes points out, the decisions are made by network execs worried only about their ratings and bottom lines, not any high-minded ideal like "quality television" that PBS brags so loudly about. They could (and did) notice that professional "wrestling" is popular with their 18–31 male demographic, and see no problem with programming the WWE next to a show about science.[1]

The effects of this huge cultural immersion in paranormal beliefs are apparent when you do any kind of survey or polling of what people believe. There are many different surveys out there, but the most famous[2] is the Baylor Religion Survey. This huge data set of multiple-choice questions, collected almost each year since 2005, looks not only at beliefs but also at the demographic data behind them. In the section "Paranormal Experiences," the Baylor survey found that 17% of Americans claimed to have witnessed a UFO. Another 26% agreed that UFOs are spaceships from other worlds. These beliefs are highly correlated, so that people who embrace one type of paranormal experience (such as UFOs and aliens) are very likely to also believe in Bigfoot and ghosts. In their book *Paranormal America*, Christopher Bader, Frederick Carson Mencken, and Joseph O. Baker point out that belief in some form of the paranormal is held by about two-thirds of Americans and that the average American believes at least two paranormal ideas to be true.[3] Beliefs in Bigfoot or the Loch Ness monster are also strongly correlated with acceptance of other paranormal ideas, such as UFOs, psychics, Atlantis, ghosts, and so on.

In other words, belief in the paranormal is the norm, not the exception. The paranormal *is* normal!

In an interview connected with the publication of *Paranormal America*, coauthor F. Carson Mencken said the following:

> Two-thirds of Americans have at least one paranormal belief. We argue in our book that two factors drive paranormal belief and research: enlightenment and discovery. In terms of sociological theories, we use Hirschi's social control approach in conjunction with Stark and Bainbridge's work on religious compensators to explain how discovery and enlightenment would lead the disconnected to unconventional beliefs. Those who have strong bonds to society (people in power, high status careers, structural positions of authority and responsibility) are highly conventional in their lifestyles. It is an expected (i.e. normative) pattern. One aspect of living a conventional lifestyle is to believe and practice conventional beliefs/rituals (i.e. Christianity, Judaism in the United States). When someone who is expected to live a conventional lifestyle strays from expected (i.e. normative) behavior patterns, they risk sanctions (as Nancy Reagan did with her use of astrology in the White House). Those who are less connected to society, which may represent the poor, uneducated on one end, and the hyper-educated, non-conformist types on the other end, do not risk the same sanctions by pursuing alternative belief systems. But this in and of itself is not enough to spur people to the paranormal. There has to be a reason why people will seek alternative belief systems. One theory which applies to those of lower socioeconomic status is religious compensators. Since conventional religiosity is for and run by highly conventional people and provides many empirical rewards for this group, those from lower socioeconomic status groups will not gain many spiritual or conventional rewards from participating in conventional religion. Alternative belief ystems can be empowering. The discovery of something spiritually unique (an unknown secret to the universe) that the rest of society does not have gives those from excluded groups a sense of purpose, a status as someone important. Moreover, all humans are seeking enlightenment and discovery. New information helps us to reduce risk in our lives, and to make better informed decisions. Many paranormal practices (psychics, mediums, communication with the dead, astrology, etc.) are about giving people an insight into their future. Those groups not bound to conventional religious systems are freer to explore these alternative systems in order to gain information that may help them improve their lives.

Sharps, Matthews, and Asten[4] conducted an interesting study looking at cross-correlations of belief in Bigfoot and UFOs with other aspects of personality and behavior. They found that belief in UFOs and in ghosts were highly correlated with belief in Bigfoot and other paranormal beliefs, consistent with the results of the Baylor Religion Survey already cited. Even more interesting, they found that people who believed in Bigfoot and other paranormal ideas tended to score very highly for ADHD (attention-deficit hyperactivity disorder), depression, and dissociation (the idea that "part of the person is elsewhere and not available at the present time"). People who have high scores in dissociation tend to be removed from everyday human experience and have diminished critical assessment skills, so it's not surprising that they correlate highly with belief in the paranormal. The high correlation with depression was more surprising, but the authors speculate that the belief in the paranormal opens up "potential avenues of escape from perceived difficulties" in their lives. The high correlation with ADHD might be part of our past as hunter-gatherers, when hyperfocus, rapid task-shifting, and hyperactivity were valuable to hunters. The authors speculate that "these individuals might be more attracted to a world of unknown animals and unexplored possibilities, in part as a refuge from the modern world to which specific aspects of their intellects and psyches are unsuited."

An important study by Viren Swami and others in 2011 used a test called the EBS (Extraterrestrial Belief Scale). It has 37 items that measure general beliefs about extraterrestrial life, attitudes toward the importance of scientific research on the topic, and beliefs in alien visitation, including beliefs that governments are participating in cover-ups to mask the evidence of alien visitation. The study also assesses how these beliefs correlate with magical thinking, beliefs in ESP, psychokinesis, and the belief in the afterlife. They also rank people on whether they tend to follow the flock ("sheep") or think independently ("goats"). Here are some of those 37 items that survey participants were asked to rank on a scale from 1 (strongly disagree) to 7 (strongly agree):

- The government of this country is covering up the existence of extraterrestrial life.
- Unidentified flying objects (UFOs) observed in the skies are in fact sightings of the spacecraft of intelligent extraterrestrials.
- Intelligent extraterrestrial life has visited Earth.
- The military know that extraterrestrial life does exist.
- Extraterrestrial creatures visited Earth in the distant past or at the dawn of human civilization.

· Crop circle patterns are caused by the actions of extraterrestrials
· Extraterrestrial life once existed on Mars.
· Ancient monuments, such as the pyramids of Egypt, are evidence of extraterrestrial life because they could not have been built without technical abilities beyond those of people at that time.
· Scientists know that extraterrestrial life does exist.
· Skepticism about extraterrestrial life is a psychosocial trend conditioned or perpetuated by those in power.

As reported in *Psychology Today,*

Finally, and perhaps most controversially, Swami's team asked participants to complete a scale in which they stated whether they have schizophrenic-like symptoms (hallucinations, delusions, disordered thinking, and limited social networks). They administered this "schizotypy" scale based on evidence from previous studies showing that people who have unusual perceptual experiences (such as hallucinations) are also more likely to believe in the paranormal and therefore might be more likely to hold alien visitation beliefs as well. Finally, the researchers measured basic demographics including age, gender, education, and religious orientation. After crunching the numbers, Swami and his co-authors concluded that the strongest believers in alien visitations and cover-ups were in fact more likely to have paranormal and superstitious beliefs, particularly beliefs in ESP, the afterlife, and to have had unusual perceptual experiences. The UFO believers were also likely to be high on the personality factor of "Openness to Experience," meaning that they were high on such qualities as willingness to fantasize, explore new ideas, and try out new things. Men and women were equally likely to endorse beliefs in alien visitation and cover-ups, and adults with higher education were lower in their extraterrestrial beliefs. Overall, the average endorsement of the EBS items was low (approximately a 2 on a 5-point scale) as were most scores on the Sheep-Goat and schizotypy scales. The highest score (2.5) was on the belief in afterlife scale. Nevertheless, enough people endorsed the EBS items to allow the researchers to detect these significant patterns.[5]

Another study,[6] this one conducted by Kelton Research for *National Geographic,* showed that about 36% of Americans believe that UFOs exist (that's 80 million Americans!), that about 10% believe they have seen one, and that 80% believe the government has covered up information about UFOs. A 2012 survey[7] in the United Kingdom found that more people believed in aliens (33 million) than in God (only 27 million), that 20% believe UFOs have landed on Earth, and that about 10% believe they have seen a UFO. In addition, 52% believe that evidence of UFOs has been covered up because widespread knowledge of their existence would threaten government stability, and more than 5 million Britons believe the Apollo moon landings were faked. A Public Policy Polling survey[8] found that 12 million Americans believe that interstellar lizards in people suits rule our country and that the Queen of England is a blood-drinking shape-shifting alien, as well as that about 66 million Americans believe that aliens landed in Roswell, New Mexico.

BUT ARE POLLS PROOF OF UFOS?

Despite this discouraging picture, we must take some of these polls with a grain of salt. The wording of the questions is critical, so if you ask people whether life exists in outer space, it could mean simple microbial life (which even many scientists accept as possible—see chapter 12). The question could also mean humanoid aliens, an idea that almost no scientist accepts but that the general public has been indoctrinated with for years. And many of these polls have not only poorly worded questions but also problems with sample sizes or methodology. For example, one poll[9] said that 48% of Americans thought that alien spacecraft are observing our planet—but the sample size was very small, and there were lots of other problems with the wording of its questions. Nevertheless, it got huge media coverage and attention.

Considering the evidence that we have outlined in this book against the likelihood of humanlike aliens' actually visiting Earth and UFOs' really being alien spacecraft, numbers such as this are truly frightening and discouraging. But we must not let these polls lead us into the *Argumentum ad populum* fallacy (Latin for "appeal to the people"), also known as the *Vox populi* ("voice of the people" in Latin) fallacy or "bandwagon" fallacy. This fallacy tries to convince doubters that a claim is true solely because lots of people believe it. Consider this: thanks to creationist efforts to cripple public science education, a high percentage of Americans think the Earth is only 6,000 years old and believe that humans walked with the dinosaurs. What's more, millions of people think that Elvis is still alive. At one time, people believed that smoking was healthful—because everyone did it. To go even further back, not that long ago, people believed that the Earth was flat, that the sun revolved around the Earth, and that the Earth was the center of the universe. They believed that cold weather (rather than viruses) caused colds and that leeches could cure you of all sorts of ailments. Thus "popular wisdom" is no guide to the truth, especially when we have scientific evidence showing that the less popular idea is actually true.

As John Oliver put it on his HBO show *Last Week Tonight*, while talking about the widespread doubt about climate change in America even though more than 97% of climate scientists agree that it's real and that humans are causing it,

> a poll shows that 1 out of 4 Americans are skeptical of global warming. Who gives a shit? That doesn't matter. You don't need people's opinions on a fact. You might as well have a poll asking: Which is bigger? 5 or 15? Do owls exist? Are there hats? . . . The only accurate way to report that 1 out of 4 Americans are skeptical of global warming is to say that, "A poll finds that 1 out of 4 Americans are wrong about something."[10]

However, there is lots of evidence that at least the public face of UFO-logy is beginning to die off, at least in the newspapers (though perhaps not on the Internet). Studies show that the media coverage of UFOs and aliens was greatest during the Cold War years but had died down considerably as the tensions and paranoia of the Cold War subsided. According to one study,

> the UFO phenomenon has ebbed and flowed since the late 1940s, but has recently died down from its height, something he attributes to the end of the Cold War. "I think there is evidence that it's not the same phenomenon that it once was, at least as a mass movement that people are mesmerized by," said Eghigian. "It was Cold War anxieties—but also Cold War enthusiasms and passions—that had a direct impact on this."
>
> According to Eghigian, in a sample of 25 newspapers, the number of headlines mentioning UFOs or flying saucers has fallen to less than 20 headlines each year in the post-Cold War era. Prior to the 1990s, there tended to be more than 40 headlines about the phenomena annually. "The United States is coming out of World War II, so the country is concerned about national security and still has a war-like mentality," Eghigian said. "It's a world that is rife with secrecy and preserving secrecy, so this helps inform both how governments and militaries deal with evidence."[11]

ALIEN CULTURES

Another important aspect of the psychology of UFO believers is their social organization. Groups that bring together believers in aliens and UFOs form an important social network that gives members comfort, affirmation, and reassurance that there are other people who think like they do. In many ways, it is analogous to belonging to a church congregation: your own subcommunity of friends who have similar interests and beliefs, who give you solace when you need it. Like fans of any particular topic (from NASCAR to *Twilight* to *True Blood*), UFO believers form their own "subculture" of people who believe strongly in the reality of aliens and UFOs and who spend a significant amount of their time and resources researching these topics. They have their own meetings, their own jargon, their own shared body of accepted knowledge, and their own distinctive way of looking at the world. As sociologists have long pointed out, acquiring the argot, or distinctive lingo, of a subculture is part of the process of becoming a member of the subculture, for it distinguishes insiders from outsiders, and having mastered it is a mark of acceptance. Thus these people not only casually and routinely talk about Area 51 and Roswell and Betty and Barney Hill but also use a whole series of other terms that are shorthand for the more famous cases of UFOs and aliens—and everyone within the field knows immediately what they mean.

Under the title "This Event Is Not for Skeptics," Rick Rojas of the *Los Angeles Times* reported[12] on the MUFON (Mutual UFO Network) convention in Irvine, California, in 2011. As he describes it, many of the attendees reported experiences of "alien abductions," and some think they are actually alien–human hybrids. Many of them view aliens as godlike, benign omnipotent protectors who beckon to them in the night using bright lights. Typical of them, 61-year-old Cynthia Crawford, who

sold sculptures of aliens, said there was no reason to fear contact by extraterrestrials. She said she has a spiritual connection to her alien guides who have made medical ailments disappear and once manifested a crisp $20 bill. She told others they should experience the same. "Send the light and the unconditional love, and they will come to you," she told one young man. "When you start seeing our star family—oh my God—you'll love it." Another topic discussed at the

convention was human-extraterrestrial hybrids. Crawford, who lives near the
Superstition Mountains in Arizona, said that she is one of them. The hybrids, she
said, often have high foreheads and thin faces with long, skinny noses. Crawford,
however, has a round face framed by thick blond hair. "I think I look human," she
said. She turned her head and widened her eyes. "Do you think I look human?"[13]

As the article reports, the UFO fans were particularly intent on being
taken seriously by scientists, aping scientific methods with their own
"certified field investigator" program (including a genuine MUFON
badge!) that required them to carry recording devices, Geiger counters,
and a respirator. Thus, as the "certified investigator" David MacDonald is
quoted as saying, "We all want to believe, we all want to believe bad [sic],
but you've got to look at the evidence. You've got to come at this like a
scientific researcher." Just like the Bigfooters and other cryptozoologists
that Loxton and Prothero described in their book *Abominable Science*,
the UFOlogists have a huge chip on their shoulder about the fact that
scientists do not take them seriously—but have a distorted, superficial
idea of how science is really done. According to psychotherapist Barbara
Lamb, who works with "experiencers" (people claiming alien contact),
"We do have what we consider evidence, but the scientific community
doesn't want to consider that as evidence. There's a kind of booga-booga
about ETs and UFOs." According to author and UFO "researcher" Richard Dolan, "Just below that level of snicker, snicker, is fear."

The analogy with a church or cult is not a casual one. As we saw in
chapter 6, some UFO believer groups *are* actually religious cults with
extreme beliefs and dogmas as well as the ability to manipulate their
members to do anything (such as the Heaven's Gate cult, whose members committed ritual mass suicide when they thought a spacecraft was
coming to get them). As we discussed in chapter 2, the groups who fer-
vently believe in conspiracy theories behave much like religious cults
as well.

So we must not underestimate the importance of community and a
shared subculture in the growth and reinforcement about believers in
UFOs and other paranormal topics. They are important agents that not
only propagate the numbers of true believers but also help produce all
the websites and books and other media that spread these ideas across
modern culture.

IS THE STUDY OF UFOS AND ALIENS SCIENTIFIC?

A number of investigators and organizations (such as MUFON) claim to be rigorous and scientific in their studies of phenomena associated with aliens and UFOs. But as many people have pointed out, it's not sufficient just to do "research" or to use scientific instruments and paraphernalia, then claim that you're doing science. That's only "sham science" or pseudoscience, as geologist and science blogger Sharon Hill pointed out. To investigate these paranormal phenomena in a rigorous scientific manner, you must adopt the critical attitude of scientists around the world. Sharon Hill wrote the following while describing cryptozoology (the study of monsters not proven to exist, such as Bigfoot, the Yeti or the Loch Ness monster), but it could also describe the UFO "researchers" with the appropriate substitutions:

> I actually have scientific training, a degree in a scientific field, done lab work, field work and research. However, I would NOT feel comfortable calling myself a cryptozoologist for several reasons—I'm not a trained specialist in wildlife biology or zoology and I'm not devoting a large portion of my time studying it. I've enjoyed the subject since I learned to read . . . . I want to earn the label and respect . . . .
>
> One can't get away with that in science. A "scientist" in our society is assumed, at least, to have a specialized training, a degree in a scientific field. Many hold them to an even higher standard such as currently working in the field actively doing research, producing data and publishing it. But, the education and training of a scientist is a CRITICAL piece. It requires practice to learn how to think scientifically. It takes effort to put together careful research results. And, science is a special culture that has rules. The norms of science—how you are expected to conduct your work—are strict which give science its high credibility.[14]

Hill goes on to say:

> I've researched and published on why amateur investigation groups fail to reach the high bar of science. I see these groups doing what I call "sham inquiry." It sounds sciencey, it looks sciencey and it can fool a lot of people into thinking it's scientific but there are clear reasons why it is not. "Sham inquiry" is about the process and why the results they get out of that process are inferior to scientific inquiry.
>
> The primary problem . . . is that cryptozoologists, by and large, assume that a mystery creature is out there for them to find. They begin with a bias. They are advocates. They are not testing a hypothesis but instead seeking evidence

to support their position. The answer is already in their head. (Same is true for ghost hunters and UFO investigators.) They also begin with the wrong question. Instead of "what happened?" they ask "Is it a cryptid?" They have narrowed the possible solutions immediately off the mark....

I don't admire when the basic ideals of science are ignored—good scholarship in research, quality data collection and documentation, proper publication, skepticism, and open criticism. Instead, the bulk of popular cryptozoology is a jumble of the same old poor quality evidence, a ton of hype, rampant speculation and unfounded assumptions, even conspiracy theories and, too often, paranormal explanations.

... [A]mateurs and non-scientists CAN do science. And, second, cryptozoology CAN be a science. But right now, I don't see that occurring often. It takes a lot of effort to do this and resources that the average enthusiast does not have. Too much is missing to call cryptozoology a science at this moment in time.

It doesn't have to be that way. I want the quality to improve and for the field to resolve itself into a rational examination of people's sightings and reports of strange incidents. You can't get there by following the same paths as those past cryptozoologists. I'm talking about the proper examination of claims, thorough investigation, a good cryptozoology news and information site without hype, rational literature, and well-thought out responses to criticism.... I find the critical, detailed, careful, objective evaluation of these curious crypto questions WAY more satisfying than those hyped up adventure stories or breaking new claims of evidence (that end up fizzling out).[15]

As Loxton and Prothero pointed out in a book on cryptozoology (and this also applies to studies of other paranormal phenomena, such as UFOs and aliens), certain things are incumbent on "true believers" who want scientists to take their claims seriously:

· *Rethink their fundamental assumptions:* As Sharon Hill points out, when something odd happens, the UFO believer's first thought is *Is it a UFO?* instead of the proper scientific question: *What really happened?* One of the basic rules of science is Ockham's Razor: the simplest explanation is probably the best, and there's no need to invoke additional factors unless the evidence really demands it. A real scientist doesn't try to force the data to fit his or her pre-existing expectations and biases but rather tries to look at all hypotheses that explain something and rule them out, one by one. But UFO believers first make the assumption that any odd observation or bit of data is probably due to a UFO or alien rather than the simpler assumption that the phenomenon is something natural, not paranormal, and requires no unusual explanations. As we have seen throughout this book, nearly every "observation" or piece of "evidence" has simple natural explanations, which true believers discard in favor of a more outlandish, less parsimonious explanation: "It was an alien."

· *Null hypothesis:* Related to this point is the basic concept in science that comes from statistics, known as the "null hypothesis," or the hypothesis of no significant difference. In statistics, one makes the assumption that two samples are *not* significantly different from one another (the "null hypothesis") until it can be demonstrated statistically that they *are* significantly different. Applied to the study of aliens and UFOs, the fundamental scientific assumption should be that any strange observation can be explained by natural causes; *only* if there is strong evidence to rule out simple natural causes can we jump to the less likely conclusion that some sort of strange paranormal explanation is required.

· *The unexplained is not necessarily unexplainable:* In science, there are lots of unsolved problems and unexplained mysteries at any given time. That's the way it's supposed to work when you are doing research at the cutting edge of a field. But at no time does a scientist say, "I can't explain this now, so it can never be explained by anyone." Instead, we take a "wait-and-see" attitude with the puzzling pieces of our data, and sooner or later most anomalies in science are explained. This is inherent in the label "Unidentified Flying Object." These things are merely *unidentified* at the time when they are spotted, which does *not* mean that they must be alien spacecraft. As we saw in this book, many of these originally unidentified objects have now been identified. So whenever another strange sighting occurs, the proper scientific response is to withhold judgment on it, not to jump to the conclusion that it is beyond scientific explanation.

· *Burden of proof:* In the sciences, most conventional ideas require only a bit of evidence to show that they are reasonable. But the more outlandish and extreme the idea, the greater the burden of proof. As Marcello Truzzi said, "Extraordinary claims require extraordinary proof"—or as Carl Sagan put it, "Extraordinary claims require extraordinary evidence." If the evidence is not out of the ordinary, then there must be a *lot* of it to overcome the strong evidence against the existence of UFOs and aliens. For scientists to take the highly improbable claims of the paranormal seriously, UFO believers need to provide extremely solid proof in the form of bones or carcasses, not a few inconsistent "eyewitness accounts" or inconclusive photos or videos. So far, none of their pieces of evidence has come close to meeting this burden.

· *High-quality data collection*: There is still no credible physical evidence, in the way of carcasses, crashed spacecraft, unusual relicts, or other actual samples, for any alien manifestation discussed in this book. This is the barest minimum that real scientists would demand before taking stories of the paranormal seriously. High-quality photographic evidence would be convincing as well, although every single photo and video produced so far has been inconclusive or an obvious hoax. "Eyewitness testimony" is virtually useless, because psychologists have shown that it is subject to all sorts of errors and prone to hallucinations. But instead of being honest about their data, we have seen again and again that UFO believers get their priorities mixed up and take "eyewitness evidence" far too seriously. They don't discriminate between clearly false or made-up stories, and they often exaggerate or misreport stories when they compile them. They rarely do the hard work of interviewing witnesses and double-checking their stories for consistency, something that was clear when skeptical UFO researchers such as Robert Shaeffer or the late Phillip Klass got to the bottom of the lies and myths behind many of the accounts. UFO believers falsely assume that a long list of "sightings" all describe the same phenomenon when we have shown that a careful look at the evidence of "sightings" shows that they are internally inconsistent and contradictory and seem to reflect cultural expectations, not scientific reality. This carelessness with evidence, sloppy scholarship when reporting things, and tendency to believe the most unbelievable stories without any skepticism is simply not acceptable scientific practice.

· *Proper documentation and publication:* True scientists and researchers publish normal mainstream science all the time. They publish these papers in legitimate peer-reviewed scientific journals and exhibit a degree of skepticism and caution that befits a real scientific paper. Most UFO believers, by contrast, write only for their own audience, publish in their own journals and websites, and make no real effort to write high-quality papers that could be acceptable in a mainstream science journal. They claim that scientists are biased against them, but if they only made sure their research met scientific standards, they might find that they could get published in legitimate peer-reviewed journals and be taken seriously.

· *Openness to criticism:* All legitimate scientists must get accustomed to the fact that peer review is a harsh process that bruises the ego and that can even be biased or unfair at times. But good scientists have a thick skin and a strong sense of the merits of their research as well as how to improve their methods or look for better approaches, and they persevere to get their best work published. Such peer criticism is essential, because it makes sure that only the best science eventually gets published. But when UFO believers cry "Bias!" or claim that scientists are not open-minded about their ideas, then retreat to their own comfort zones, they are doing themselves a big disservice. If they want to be taken seriously as scientists and not be regarded as pseudoscientists, they must get used to the criticism, improve and refine their work, and throw out the substandard evidence. Then they will find that the scientific community is remarkably receptive to publishing unconventional ideas—so long as they are strongly supported by evidence and meet the other standards in science.

· *Skepticism of one's own data:* Good scientists must be skeptical of their own work as well as that of others. They have to be willing to toss out data or results if they are not good enough. As Richard Feynman said, "The first principle is that you must not fool yourself—and you are the easiest person to fool." If UFO believers want to be taken seriously, they must do the hard task of weeding out the questionable sightings and inconclusive pictures and videos, be more careful in their scholarship about what eyewitnesses actually said, and try to be honest when the data aren't up to snuff. They must stop defending obviously indefensible data, which makes them look stubborn and foolish. If they did this, they would be taken more seriously and admitted to the "High Table" of legitimate scientific inquiry.

Recently, Sharon Hill wrote about the modern state of "research" in UFOlogy:

The UFO research field is having a bit of a crisis these days. Reports come in by the hundreds. There are not enough people to investigate them. Yet, decades of UFO research by private and military organizations have resulted in disappointment for those who surely thought there was something out there to reveal. Many of the historic figures of UFOlogy are aging or have passed away. Who is doing the work now? And what exactly are they doing?

The major organization remaining in the U.S. for investigating UFO sightings is MUFON, the Mutual UFO Network. MUFON is not in good shape. Their stated mission is to conduct scientific investigation of UFO reports for the benefit of mankind. But there is dispute about their ability to actually do that. The current version of MUFON, according to those observing the situation, is focused on everything except proper UFO investigation and is nowhere near scientific. Membership in the organization has fallen off and some local MUFON groups are disgruntled. Leadership upheavals over the past few years may have been distracting and overall, they are experiencing a serious case of mission creep.

MUFON consists of chapters covering each state across the country who operate somewhat independently with members paying dues to the main headquarters. They promote a scientific method. But do they actually accomplish that goal? Recent commentators say no, they do not. The focus in local MUFON chapters meetings these days is decidedly unscientific with talks on alien abduction, conspiracy theories, human-ET hybrids, hypnotic regression, and repressed memories. That's a wide range of pseudoscience in one place. It's dragging down the credibility of the entire subject as well as missing the point of improving actual UFO investigations.

A comprehensive two-part piece recently appeared online describing the changing of the guard at MUFON that is installing its fourth director since 2009. The UFO Trail blog critiques the current status the field and takes note of the rising voices in the community, some of whom wish to elevate the investigations and methods out of the realm of pseudoscience. Author of *The UFO Trail*, Jack Brewer, is critical of the current methodology, characterized it as "sham inquiry"—a label I used to describe amateur paranormal investigation and one he thought also applied in this case.

The newly named director of MUFON, Jan Harzon, states that UFOlogy is a science and intends to put a scientific face back on UFO investigations. Their latest symposium, held in Las Vegas this past July had the theme "Science, UFOs, and the Search for ET." The conference featured presentations from several science professionals (current and former) but did not provide any blockbuster information or do much to promote science.

"We hope to bridge the gap between science and UFOlogy," said Jan Harzon, state director of network's Orange County bureau. "They're one in the same." — Las Vegas Sun (19 July 2013).

Many skeptical critics would dispute the claim that UFOlogy is a science but that depends on how you wish to define "science." A general definition such as a "systematically collected body of knowledge" is not very descriptive of a subject area that contains a lot of data but few constructive hypotheses to provide a framework. The UFO sighting data is mostly witness reports and much of it is of questionable veracity or too old to be of much value any longer. The National UFO Reporting Center has a database of reported sightings but I haven't found any compelling reports about flaps or trends to make sense of the data. It could be that I'm not reading cutting-edge UFO research but if there really was a good report that solidly concluded that there was a pattern and subsequent explanation for UFO flaps, I would be interested. I would hope I'd have heard of it, at least from those who have a more in-depth knowledge. But, as with claims of proof of psychics or hauntings, we only have popular, often biased reports about particular events from individuals that have a belief to promote. Those case studies in addition to being problematic in their accuracy (since it's hard to confirm many of the events via witnesses) are not robust enough to aid in explaining the phenomenon.

The discussion coming from a small group of today's modern UFO researchers suggests that UFOlogy is on the wrong track these days. With a focus on abductions, conspiracies, and exopolitics/disclosure, the core of the field is no longer about investigating and identifying what people are reporting to have seen in the sky.

Antonio Paris runs the API Aerial Phenomenon Investigation Team, which has a somewhat different focus than MUFON. He wishes to return to the "nuts and bolts" idea of UFO investigation and get away from the conspiracy and fringe topics that so often dominate the symposiums and local MUFON talks.

I asked Antonio what sorts of tools his organization uses to do investigations. He noted that Internet sites can help identify some of the man-made objects like aircraft and satellite. MUFON also mentions these tools on their sites along with astronomical sites to identify bright celestial objects that are sometimes confusing to people viewing them on the ground. Paris is also familiar with the shape of many military aircraft and says he can typically identify them in association with military bases nearby. API has tackled about 300 or so cases but does not pursue those that look like jokes, hoaxes, or give them nothing to go on. There is no lack of UFO reports. An initial screening to determine viable cases is necessary to remove those cases not worthy of investigation or they would be overwhelmed.

MUFON trains their investigators through a manual and an exam. Paris noted to me that the test requires no specialized skills and many people could potentially pass it without even looking at the manual. The certification as a field investigator is a worthwhile effort by MUFON to standardize their methods and provide a framework for consistency of methods but it's only internal to MUFON. When each MUFON chapter operates pretty much on its own, inconsistency and regional differences creep in. Paris told me he is frustrated by the lack of sharing of information both internally and externally of MUFON noting that an object of interest can fly over a wide area. Coordination of reports that may be of the same object would be a worthwhile effort. Science is dependent on sharing information either through collaboration or peer review of findings. UFOlogy appears weak in that area having no established journal or even an online location for filing results.

Even more fundamental to UFOlogy than answering "Is it a science?" is "Can it ever be scientific?"

UFOs are uniquely difficult to investigate for several reasons. The observation is fleeting. It may not repeat. It is difficult to reproduce. If it remains airborne, it leaves no physical evidence behind, only the story of the witness. The observation is often made in the dark under conditions in which it is difficult (or impossible) to accurately judge size and distance.

The options for making a UFO into an IFO (identified) are many and various. Along with the typical reports (satellites, aircraft, flares and planets), we have more man-made things in the air now than ever: experimental balloons and aircraft, weather balloons, Chinese lanterns, drones, dirigibles, toys and deliberate hoaxes. Even people can become UFOs.

Can UFOs be scientifically investigated? If by "scientific" we mean methodical, objective observations in consideration of natural laws, logic and reason, then, yes. I think UFO investigation can be scientific but the sea change that is needed would be pretty huge for the field and I don't know if they can pull it off. As with paranormal investigators, UFO researchers tend to lean towards the believer side. That's what keeps them passionate. But it's also their undoing. A bias towards belief in a mystery or in alien craft is the first giant misstep in UFOlogy. The first step for a rejuvenated field to gain credibility is to drop the default belief that ETs are visiting earth and back up to the very basic question, "What, if anything, did this witness see?" Begin looking for real world answers instead of "proof" to support a belief in alien life. . . . .

I asked CFI fellow and Skeptical Inquirer UFO columnist Robert Sheaffer his thoughts about scientific investigation of UFOs. It's a bit tricky. Many have assumed it's possible, he says, but it turns out to be more difficult than it seems. Sheaffer has documented the several times "rapid response" teams have been attempted by UFO organizations. It was hoped that by gathering "reliable witness reports," and implementing a "rapid response team" to capture the UFO with professional filming techniques, better evidence could be put forth for the claim that something worth paying attention to was really occurring.

Rapid response teams turned out to be disappointing, says Sheaffer. Antonio Paris was part of one such team, the "STAR Team" for MUFON. Millionaire

Robert Bigelow funded the project. "MUFON has been somewhat tight-lipped about the results," Sheaffer tells me, "but they are generally conceded to be hugely disappointing." It did not give them the results they hoped for.

... MUFON still gets hundreds of cases a month and there is considerable backlog of investigations. That's a hefty workload for volunteers. There is a need to sort the wheat from the utter chaff but there still are valid means to find out what people probably saw in the sky. Most reports will have a satisfying answer if diligently investigated. But that may not happen or the eyewitness may not accept it.

Paris's API group is in contact with the new leadership at MUFON and is encouraged that a more sound approach to the field is on the horizon. This may be a new dawn for UFOlogy as the old guard dies away and the new, more centered, serious thinkers take over. UFOlogy is undergoing a transformation once again.

For now, UFOlogy attempts to sound sciencey, but it is not nearly up to the standards to be called "science." Can it be science? Only with a wholesale change in assumptions and approach. Drop the fascination with conspiracies and abductions—go back to nuts and bolts.[16]

SUMMARY

Belief in UFOs is popular in our culture, but just because lots of people believe something doesn't make it true. Moreover, the people who believe in UFOs tend to believe in lots of other paranormal activities as well. They often form a subculture of believers in the paranormal, resistant to evidence and wanting to believe no matter what. Some of these subcultures are actually formal organizations, such as MUFON. But a close examination of MUFON and other UFO groups shows that they do not use critical thinking or the scientific method but rather practice "sham science" and try to sound "sciencey" without actually thinking like a scientist must. They have a long way to go, and many clear steps that they must take, if they want to be taken seriously by real scientists.

14

The Verdict

In this book, we have gone over nearly every piece of evidence and every UFO story in as much detail as we could for a book of this length. Throughout our critical examination, we have tried to apply Ockham's Razor and other tools of the scientific method and critical thinking. Again and again we have asked whether it is really more plausible that some alien force made these things happen—or is there another simpler nonparanormal explanation? Here are some of our conclusions:

· As we saw in chapter 3, Area 51 was (and still is) a top-secret military base, but because the military tested spy planes and stealth aircraft there, not because it was hiding an alien spacecraft like the one seen in movies such as *Independence Day*. Many once classified documents and former Area 51 personnel (known as the Roadrunners, whose number included my own father, Clifford Prothero) bear witness to this.
· In addition, many of the UFO "sightings" of the 1950s, 1960s, 1970s, and 1980s have now been explained by high-altitude supersonic spy planes such as the SR-71 Blackbird and its predecessors, which were secret then and flew at higher altitudes, and much faster, than any jet plane or commercial airliner.
· As we saw in chapter 4, there's a perfectly good explanation for what happened in Roswell in 1947—but the UFO believers refuse to accept it. The entire mythology of Roswell grew up almost

30 years after the incident, and it includes no evidence from anyone who was there in 1947.
- As other chapters have demonstrated, nearly every reported UFO sighting that has been properly investigated has now been demonstrated to be a hoax, misinterpreted, or a mass delusion—or simply unclear but *not* necessarily unexplainable.
- The claims of alien abductions have all fallen apart under closer scrutiny.
- Silly things such as crop circles are well documented as hoaxes, yet some people prefer to believe that aliens from outer space came all that distance just to disturb some farmer's field—and then left with no other trace.
- The bizarre "ancient astronauts" ideas involving the pyramids, the Nazca lines, and other things are full of fakery and exaggeration, and they fail to account for the fact that ancient humans were much more intelligent and resourceful than we imagine.

SHOW ME THE EVIDENCE!

Where is the evidence? As we stated from the beginning, eyewitness accounts are useless, because people so often "see" things that aren't there or enhance or modify their "memories" of what they have seen. Photos of aliens and UFOs are legendary for being hoaxes or mistaken interpretations of much more mundane phenomena—or just bad camera work. So far, *not one* piece of reliable physical evidence has *ever* been produced—just as there is *not one* piece of reliable physical evidence for Bigfoot or the Loch Ness monster, even after decades of looking by thousands of people. If aliens have been visiting this planet, and if UFOs in the sky are real, there should be much *more* evidence now than in the past, not less.

Seth Shostak runs the SETI Institute and takes the search for extraterrestrial life seriously (and considers it a realistic scientific possibility). In a post entitled "The Trail Is Stale," he bemoaned the poor quality of the evidence for UFOs and aliens, especially the more recent evidence:

> The few alternatives to this vintage archive are contemporary photos and videos of vague lights in the sky, low-resolution and low-confidence material that isn't likely to sway many scientists. The good stuff seems to be the old stuff.

To better judge if this is really true, I trawled the web for listings of "the best UFO cases." I quickly collected nearly 100 events that were considered worthy, of which 60 were unique, in the sense of not being repeats (e.g., the Roswell incident appears on most of these lists).

I then plotted up the year in which each of these unique events took place, virtually all since 1940. And guess what? By far the majority occurred in the first half of the last 76 years.

The quality UFO evidence is getting long in the tooth.

So what's going on? Our technology for documenting alien spacecraft—if you assume they're real—is substantially better than even a few decades ago. An Apple iPhone's camera now boasts 8 megapixels, which I reckon is a hundred times as many as the 8 millimeter movie film we had in the 1960s. These fabulous cameras are in the hands of nearly two billion smartphone users world-wide. And yet the UFO photos are as blurry and muddy as ever. You'd think at least a few people could make snaps that aren't ambiguous or hoaxed. And I haven't mentioned the surveillance provided by the 1,100 active satellites in orbit above our heads.

Now, some people deflect these puzzling facts by stating that excellent evidence for cosmic visitors really exists but is kept under wraps by the government. This may be reassuring to some, but it's utterly goofy. Can anyone explain how beings from other worlds have managed to arrange their itineraries so that only governments are solidly aware of their presence?[1]

A wonderful comic strip on the Internet says it all. It shows a graph of the increasing percentage of people who now carry a cell phone camera with them at all times, yet notes that now there are *fewer* good images of UFOs and aliens than in the past, not *more*. The caption reads: "In the last few years, with very little fanfare, we've conclusively settled the question of flying saucers, lake monsters, ghosts, and Bigfoot." According-ing to some reports, even Stephen Spielberg, the man who probably influenced more UFO and ET believers than anyone with his movies *ET* and *Close Encounters of the Third Kind*, now doubts[2] the existence of aliens—because now, when everyone has a camera at all times, there is less evidence of these things, not more.

The most damning line of evidence of all comes from astronomy and biology, when we consider the enormous distances involved, the difficulty of travel and communication across those distances, and the unlikelihood that anything more than microbes could live on other plan-ets outside our solar system. Those who claim that complex life is evolv-ing on other planets fail to take into account all the unique and special events and coincidences and accidents that made it possible for complex life to evolve on Earth in the first place. Those who claim that aliens look a lot like humans are guilty of poor imaginations and extreme anthropomorphism, not taking into account the odds against the possibility that *any* other organism would evolve to look like us. Once again, people have watched too much science fiction, in which writers (who have limited imaginations and limited costume budgets) always make the aliens look pretty much like us.

IT'S JUST CULTURAL IMPRINT, NOT REALITY

Perhaps the most telling evidence that alleged UFOs and alien visitations are not real is the fact that they are highly culturally dependent—that their memes and mythology change with the culture. Little more than a century ago, there were no reports of "aliens" or "Martians" visiting Earth. If strange objects were seen up in the sky, they were attributed to angels or devils or whatever cultural meme prevailed then. In ancient times, strange objects in the sky were attributed to the gods, be they Zeus or Thor or the god of any other culture.

Reports of aliens from outer space and of spaceships do not start showing up in newspaper reports and other records until the turn of the 20th century, the same time when the first "spaceships" and "Martians" began to appear in science fiction. Jules Verne began publishing his first tales of rocket-powered journeys to the moon in 1865. The science fiction genre began to blossom by the late 1900s and took off in the early 20th century with the invention of airplanes. The first real "UFO" account on record occurred in 1897 in Texas, when a newspaper reporter from the *Dallas Morning News*, E. E. Haydon, described an encounter with a crashed spacecraft, dozens of eyewitnesses, a recovered dead Martian, and metallic wreckage (much like the 1947 Roswell account exactly 50 years later). After a detailed investigation, it all turned out to be a publicity stunt to draw tourists and their dollars.

As we discussed in chapter 4 and again in chapter 10, there were no reports of "flying saucers" before the 20th century. The numerous versions of "flying saucers" were taken from comic strips and pulp magazine covers of the 1920s and 1930s but date back as far as 1915. The first serious newspaper report of a "flying saucer" is as recent as 1947, when pilot

Kenneth Arnold reported nine boomerang-shaped objects in the sky. He described their movements as being like those of "a saucer if you skip it across the water," which a careless reporter then conflated into the idea that the objects were saucer-shaped, not boomerang-shaped—and from then on, UFO believers have been conditioned to see any object in the sky that is remotely saucer-shaped as an alien spaceship. If flying saucers were real, then why do we find no reports of them before 1900?

The rest of the imagery of aliens and UFOs comes from media as well. The notion of alien implants came from the 1953 movie *Invaders from Mars*. Alien abductions were derived from the *Buck Rogers* comic strip, *Invaders from Mars*, *This Island Earth*, and *The 27th Day*. The "dying worlds" motif is from *This Island Earth* and *The 27th Day*. The notion of interbreeding aliens with humans is from both *Village of the Damned* and *V*.

The aliens of the science fiction movies shape what people expect to see when they report their own "sightings." As we discussed in chapter 10, the robot aliens of *The Day That Earth Stood Still* in 1951 fit the mentality of the Cold War, whereas the more familiar benevolent humanoid shape with long limbs and fingers and huge eyes (the "gray aliens") is a later invention largely propagated by the movies, especially those of Steven Spielberg and *Close Encounters of the Third Kind* in 1977. Now everyone expects "aliens" to look like this, not like a metallic robot from the 1950s.

The Betty and Barney Hill case of UFO abduction (see chapter 5) has been copycatted by many others after having become a famous story in the media—even though it has been completely debunked. The 1997 Phoenix lights (chapter 2) were just flares dropped by the military, yet they have spawned a huge number of later copycat reports of strange lights in the skies.

If aliens really looked like our post-1960s conception of them, and if flying saucers were real, then why do they show up in recent accounts only right after their introduction to the culture through the media— and not before? Why are there no reports of aliens or UFOs of modern description before 1897? The answer is simple: they are cultural manifestations of things we don't understand and have forced into whatever meme is current at the time. Before the 20th century, unidentified flying things in the sky were taken to be angels or demons or gods.

As scientists and critical thinkers, we need to rely the best evidence and not be fooled by all the false leads, bad evidence, and biased thinking that have long plagued and corrupted UFOlogy. We must always keep an open mind to the possibility that such things could exist, but for the moment, the evidence is *very* strongly against the idea of advanced beings' from outer space contacting us in any way. In our analysis, the verdict is clear: case not proven—and almost certainly false.

NOTES

CHAPTER 1

1. "Did UFO Visit Vancouver Canadians Baseball Game?" *Huffington Post*, September 9, 2013, http://www.huffingtonpost.com/2013/09/09/ufo-canadian-baseball-game-photo -video_n_3895294.html.

2. Ibid.

3. Ibid.

4. Ibid.

5. Ibid.

6. Lee Spelgel, "UFO Hoax at Canada Baseball Game Exposed; H.R. MacMillan Space Centre Admits Bizarre Stunt," *Huffington Post*, last updated September 14, 2013, http:// www.huffingtonpost.com/2013/09/12/canada-baseball-game-ufo-hoax_n_3908612.html.

7. "Did UFO Visit Vancouver Canadians Baseball Game?"

8. Massimo Pigliucci, *Nonsense on Stilts: How to Tell Science from Bunk* (Chicago: University of Chicago Press, 2010).

CHAPTER 2

1. Donald Prothero, "Bigfoot DNA? It's Playing Possum!" *SkepticBlog*, July 10, 2013, http://www.skepticblog.org/2013/07/10/bigfoot-dna-bogus/.

2. Michael Shermer, *The Believing Brain* (New York: Henry Holt and Company, 2011), 5.

3. Michael Shermer, "Show Me the Body," on Michael Shermer's official website, May 2003, http://www.michaelshermer.com/2003/05/show-me-the-body/.

4. Elizabeth Loftus and Katherine Ketcham, *Witness for the Defense: The Accused, the Eyewitness, and the Expert Who Put Memory on Trial* (New York: St. Martin's, 1991); Austin Cline, "Eyewitness Testimony and Memory: Human Memory Is Unreliable and So Is Eyewitness Testimony," *About Religion*, July 27, 2015, http://atheism.about.com/od /parapsychology/a/eyewitness.htm.

5. Daniel Simons, "Selective Attention Test," *YouTube* video, March 10, 2010, http:// www.youtube.com/watch?v=vJG698U2Mvo.

6. A. Leo Levin and Harold Cramer, *Problems and Materials on Trial Advocacy* (Mineola, NY: Foundation Press, 1968), 269.

7. Elizabeth Loftus, Memory: Surprising New Insights into How We Remember and Why We Forget (Reading, MA: Addison-Wesley), 37.

8. Benjamin Radford and Joe Nickell, *Lake Monster Mysteries: Investigating the World's Most Elusive Creatures* (Lexington: University Press of Kentucky, 2006), 160–63.

9. Elsbeth Bothe, "Facing the Beltway Snipers, Profilers Were Dead Wrong," *The Baltimore Sun*, December 15, 2002, http://articles.baltimoresun.com/2002-12-15/entertainment/0212160297_1_van-zandt-serial-killers-snipers/2.

10. Craig Whitlock and John White, "Cops Slow to Link Caprice to Killings," *The Washington Post*, October 26, 2002, http://www.sfgate.com/news/article/Cops-slow-to-link-Caprice-to-killings-They-2759421.php.

11. Shermer, *The Believing Brain*, 88–89.

12. Ibid., 59.

13. Ibid.

14. "Unmasking the Face on Mars," *NASA Science*, May 24, 2001, http://science1.nasa.gov/science-news/science-at-nasa/2001/ast24may_1/.

15. Phil Plait, "Mesa Me!" *Bad Astronomy*, April 13, 2008, http://www.badastronomy.com/bitesize/marsmesa.html.

16. See the full details in Donald R. Prothero, *Reality Check: How Science Deniers Threaten Our Society* (Bloomington: Indiana University Press, 2013).

17. David W. Pfennig, William R. Harcombe, and Karin S. Pfennig, "Frequency-Dependent Batesian Mimicry," *Nature* 410, no. 6826 (2001): 323, doi:10.1038/35066628.

18. Shermer, *The Believing Brain.*, 87.

19. Ibid.

20. Ibid.

21. Michael Shermer, "Agenticity," on Michael Shermer's official website, June 2009, http://www.michaelshermer.com/2009/06/agenticity/; Tony Ortega, "The Great UFO Cover-up," *Phoenix New Times*, June 26, 1997.

22. William Saletan, "Conspiracy Theorists Aren't Really Skeptics," *Slate*, November 19, 2013, http://www.slate.com/articles/health_and_science/science/2013/11/conspiracy_theory_psychology_people_who_claim_to_know_the_truth_about_jfk.html.

23. "Democrats and Republicans Differ on Conspiracy Theory Beliefs," *Public Policy Polling*, April 2, 2013, http://www.publicpolicypolling.com/pdf/2011/PPP_Release_National_ConspiracyTheories_040213.pdf.

24. Zogby International, "Zogby America Likely Voters 8/23/07 thru 8/27/07," *911Truth.org*, September 6, 2007, http://www.911truth.org/images/ZogbyPoll2007.pdf.

25. "Democrats and Republicans Differ on Conspiracy Theory Beliefs."

26. I highly recommend watching "Conspiracies (Web Exclusive): Last Week Tonight with John Oliver (HBO)," *YouTube* video, March 27, 2016, https://www.youtube.com/watch?v=fNS4lecOaAc.

27. Tony Ortega, "The Hack and the Quack," *Phoenix New Times*, March 3, 1998.

28. See www.dailymotion.com/video/xy923j-phoenix-lights-1997 and www.ufospy.com/phoenix-lights-original-ufo-footage. See also "Phoenix Lights," *Wikipedia*, https://en.wikipedia.org/wiki/Phoenix_Lights; and Benjamin Radford, "Mysterious Phoenix Lights a UFO Hoax," *Live Science*, April 23, 2008, http://www.livescience.com/2483

-mysterious-phoenix-lights-ufo-hoax.html. Original videos of the Phoenix Lights are available at "They're Back! Phoenix Lights Caught over Arizona Again!" *WN.com*, March 31, 2016, https://wn.com/ufo_phoenix_lights_debunked_video_tape_analysed _3_13_1997.

29. Radford, "Mysterious Phoenix Lights a UFO Hoax."

CHAPTER 3

1. "Extraterrestrial Highway (Near Area 51)," official Las Vegas travel website, accessed August 22, 2016, http://www.vegas.com/attractions/near-las-vegas /extraterrestrial-highway/.

2. Andrew Maxwell, "Conspiracy Road Trip: UFOs," *YouTube* video, October 15, 2012, http://www.youtube.com/watch?v=7ByWCFX4ZQs.

3. "Area 51 and Alien Landscapes," *Skeptics Society*, October 21, 2013, http://www .skeptic.com/geology_tours/2014/Area-51-and-Alien-Landscapes/downloads/Area -51-and-Alien-Landscapes-2014.pdf.

4. Jonathan Strickland and Patrick J. Kiger, "How Area 51 Works," *HowStuffWorks Science*, July 5, 2007, http://science.howstuffworks.com/space/aliens-ufos/area-51б.htm.

5. Alexander Aciman, "Area 51 Is Alive and Unwell," *Time*, August 16, 2013, http:// newsfeed.time.com/2013/08/16/area-51-is-alive-and-unwell/.

6. Annie Jacobsen, *Area 51: An Uncensored History of America's Top Secret Military Base* (Boston: Back Bay Books, 2012).

7. "FAQ: Is It Really Called Area 51?" *Dreamland Resort*, September 9, 2010, http:// www.dreamlandresort.com/faq/faq_name.html.

8. "FAQ: What Is the TTR (Tonopah Test Range)?" Dreamland Resort, September 11, 2007, http://www.dreamlandresort.com/faq/faq_ttr.html.

9. Curtis Peebles, *Dark Eagles* (rev. ed.) (Novato, CA: Presidio Press, 1999).

10. Erik Lacitis, "Area 51 Vets Break Silence: Sorry, but No Space Aliens or UFOs," *Seattle Times Newspaper*, March 27, 2010.

11. Gregory W. Pedlow and Donald E. Welzenbach, *The Central Intelligence Agency and Overhead Reconnaissance: The U-2 and OXCART Programs, 1954–1974* (Washington DC: History Staff, Central Intelligence Agency, 1992).

12. Jacobsen, *Area 51*.

13. "Mission Helo Was Secret Stealth Black Hawk," *Military Times*, March 29, 2013, http://www.militarytimes.com/story/military/archives/2013/03/29/mission -helo-was-secret-stealth-black-hawk/78535250/

14. Maxwell, "Conspiracy Road Trip: UFOs."

15. Strickland and Kiger, "How Area 51 Works"; Tom Mahood, "The Bob Lazar Corner," *Dreamland Resort*, September 21, 2007, http://www.dreamlandresort.com/area51/lazar /index.htm.

16. "FAQ: But, What about Bob Lazar?" *Dreamland Resort*, September 20, 2007, http:// www.dreamlandresort.com/faq/faq_lazar.html.

17. Sebastian Anthony, "Element 115: How Chemists Discovered the Newest Member of the Periodic Table," *ExtremeTech*, August 30, 2013, http://www.extremetech.com/extreme /165357-element-115-how-chemists-discovered-the-newest-member-of-the-periodic-table.

18. "Customer Reviews for *Area 51: An Uncensored History of America's Top Secret Military Base*," Amazon, accessed August 22, 2016, http://www.amazon.com/Area-51 -Uncensored-Americas-Military/product-reviews/0316132942/ref=cm_cr_pr_hist_1?ie =UTF8&filterBy=addOneStar&showViewpoints=0&sortBy=bySubmissionDateDesce nding.

19. Lee Spelgel, "Area 51 Personnel Feel 'Betrayed' By Annie Jacobsen's Soviet–Nazi UFO Connection," *The Huffington Post*, December 8, 2011, http://www.huffingtonpost .com/2011/06/07/area-51-ufos-aliens-annie-jacobsen-nazi-soviet_n_869706.html.

CHAPTER 4

1. B. D. "Duke" Gildenberg, "A Roswell Requiem," *Skeptic* 10, no. 1 (2003): 60–73.

2. Ibid.

3. "Harassed Rancher who Located 'Saucer' Sorry He Told About It," *Roswell Daily Record*, July 9, 1947.

4. Richard Weaver and James McAndrew for the U.S. Air Force, *The Roswell Report: Fact vs. Fiction in the New Mexico Desert* (Washington, DC: U.S. Government Printing Office, 1995).

5. Weaver and McAndrew, *The Roswell Report*.

6. Gildenberg, "A Roswell Requiem."

7. Ibid.

8. Ibid.

9. Ibid.

10. Weaver and McAndrew, *The Roswell Report*.

11. James McAndrew for the U.S. Air Force, *The Roswell Report: Case Closed* (Washington, DC: U.S. Government Printing Office, 1997).

12. K. Pflock, *Roswell: Inconvenient Facts and the Will to Believe* (Amherst, NY: Prometheus, 2001), 222–23.

13. Kal Korff. *The Roswell UFO Crash: What They Don't Want You to Know* (NY: Prometheus, 1997), 248.

14. Charles Berlitz and William L. Moore, *The Roswell Incident* (New York: Grosset and Dunlap, 1980); Kal K. Korff, "What Really Happened at Roswell?" *Skeptical Inquirer* 21, no. 4 (1997): 24–30.

15. Joe Nickell and James McGaha, "The Roswellian Syndrome: How Some UFO Myths Develop." *The Skeptical Inquirer* 36, no. 3 (2012). http://www.csicop.org/si/show /the_roswellian_syndrome_how_some_ufo_myths_develop.

16. Gildenberg, "A Roswell Requiem."

CHAPTER 5

1. Roger Leir, "Third Phase of the Moon," *YouTube* video, March 17, 2014, www.youtube .com/watch?v=wifuvvUKbfo.

2. Joe Nickell, *Real-Life X-Files: Investigating the Paranormal* (Louisville: University Press of Kentucky, 2001), 207, 208.

3. Ibid.

4. "3 Facts About Alien Implants," *Alien Hunter*, accessed December 19, 2016, www .alienhunter.org/implants.

5. Derrel Sims, "One Kid Who Didn't Buy It," *Journal of Abduction Encounter Research* (1Q 2007), www.alienhunter.org/alien-abduction/derrels-story/.

6. Susan A. Clancy, *Abducted: How People Come to Believe They Were Kidnapped by Aliens* (Cambridge, MA: Harvard University Press, 2005), 28.

7. Jerry Decker, "Touched by an Alien," *Unexplained Mysteries*, Episode 23, aired May 7, 2004 (Santa Monica, CA: Ann Daniel Productions).

8. Colin Johnston, "The Truth about Zeta Reticuli," *Astronotes, Armagh Planetarium's Stellar Blog!*, May 9, 2013, www.armaghplanet.com/blog/the-truth-about-zeta-reticuli.html.

9. Ibid.

10. Colin Johnston, "The Truth about Betty Hill's UFO Star Map," *Astronotes, Armagh Planetarium's Stellar Blog!*, August 19, 2011, www.armaghplanet.com/blog/betty-hills-ufo -star-map-the-truth.html.

11. Ibid.

12. Ibid.

13. "Bob White Case Ufo Object UFO Hunters," *UFO Hunters*, season 3, episode 7, "UFO Relics," aired May 6, 2009 (New York: History Channel), www.metatube.com/en /videos/103730/Bob-White-case-ufo-object-UFO-Hunters/.

14. Ibid.

15. Pat Linse with Ean Harrison, "Bob White's Great UFO Artifact Mystery—Solved! Sometimes All It Takes Is Finding the Right Expert," *Skeptic*, October 12, 2011, www .skeptic.com/e-skeptic/11-10-12/.

16. "1971, The Delphos Kansas UFO Landing Ring," *UFO Casebook*, accessed December 21, 2016, ufocasebook.com/Kansas.html.

17. Ibid.

18. Phyllis A. Budinger, "New Analysis of Soil Samples from the Delphos UFO Case," *Journal of UFO Studies, New Series 8* (2003): 1–25; Keith Basterfield, "Delphos—Soil Analyses," *Unidentified Aerial Phenomena—Scientific Research*, June 21, 2013, http://ufos -scientificresearch.blogspot.com/2013/06/delphos-soil-analyses.html.

19. Geoffrey M. Gadd, "Fungal Production of Citric and Oxalic Acid: Importance in Metal Speciation, Physiology and Biogeochemical Processes," *Advances in Microbial Physiology* 41 (1999): 47–92.

20. "Lawn Mushrooms and Fairy Rings," *Missouri Botanical Garden*, accessed December 21, 2016, www.missouribotanicalgarden.org/gardens-gardening/your-garden/help-for -the-home-gardener/advice-tips-resources/pests-and-problems/diseases/fairy-rings/lawn -mushrooms.aspx.

21. F. J. Stevenson, *Humus Chemistry: Genesis, Composition, Reactions* (New York: John Wiley and Sons, 1994).

22. Ángel M. Nieves-Rivera, "About the So-Called 'UFO Rings' and Fungi," supplement, *Mycologia* 52, no. S6 (2001): S3–S6, www.escepticospr.com/Archivos/uforings.htm.

23. Koichi Takaki, Kohei Yoshida, Tatsuya Saito, Tomohiro Kusaka, Ryo Yamaguchi, Kyusuke Takahashi, and Yuichi Sakamoto, "Effects of Electrical Stimulation on Fruit Body Formation in Cultivating Mushrooms," *Microorganisms* 2 (2014): 58–72, doi:10.3390/ microorganisms2010058.

24. Michael Shermer, "How to Talk to a UFOlogist (If You Must)," *SkepticBlog*, August 25, 2009, http://www.skepticblog.org/2009/08/25/how-to-talk-to-a-ufologist/.

25. Leon Jaroff, "It Happens in the Best Circles," *Time*, September 23, 1994, http://content.time.com/time/magazine/article/0,9171,157904,00.html.

26. Tom Barry, producer and director, *The Truth behind Crop Circles* (London: Zigzag Productions, 2010).

27. Colin Andrews, "The Julia Set: UFO Crop Circle Witnesses Come Forward," *Free Republic*, October 16, 2009, www.freerepublic.com/focus/chat/2364207/posts.

28. Michael Lindemann, "LA Story," *Circle Makers*, accessed December 21, 2016, www.circlemakers.org/la.html.

29. Barry, *The Truth behind Crop Circles*.

30. John D. Alexander, *UFOs: Myths, Conspiracies and Realities* (New York: Thomas Dunne Books, 2011), 156.

31. Ibid.

32. Ridpath, Ian, "The Rendlesham UFO Witness Statements," official website of Ian Ridpath, December 2012, www.ianridpath.com/ufo/rendlesham2c.htm.

33. Ibid.

34. Ibid.

35. "The Binary Codes," *The Rendlesham Forest Incident*, accessed December 22, 2016, www.therendleshamforestincident.com/The_Binary_Codes.html.

36. Ibid.

37. Ibid.

38. Ian Ridpath, "Nick Pope's 'Investigation' of the Rendlesham Forest Radiation Readings," official website of Ian Ridpath, April 2005, www.ianridpath.com/ufo/rendlesham.4a.htm.

39. Thomas E. Bullard, *The Myth and Mystery of UFOs* (Lawrence: University Press of Kansas, 2010), 83.

40. "Transcript of Bergstrom AFB Interview of Betty Cash, Vicky & Colby Landrum," *The Computer UFO Network*, August 1981, www.cufon.org/cufon/cashlan1.htm.

41. Ibid.

42. "The Piney Woods Incident, Cash-Landrum," *UFO Casebook*, accessed December 22, 2016, www.ufocasebook.com/Pineywoods.html.

43. As quoted by Robert Sheaffer, "Between a Beer Joint and Some Kind of Highway Warning Sign: The 'Classic' Cash–Landrum Case Unravels," *Bad UFOs*, November 13, 2013, http://badufos.blogspot.com/2013/11/between-beer-joint-and-some-kind-of.html.

44. Brad Sparks, "Cash - Landrum: NOT Ionizing Radiation, " January 1999, web.archive.org/web/20070817173738/http://www.qtm.net/~geibdan/a1999/cash3.htm.

45. Ibid.

46. Sheaffer, "Between a Beer Joint and Some Kind of Highway Warning Sign."

47. Ibid.

48. April Holloway, "Initial DNA Analysis of Paracas Skull Released—with Incredible Results," *Ancient Origins*, February 5, 2014, www.ancient-origins.net/news-evolution-human-origins/initial-dna-analysis-paracas-elongated-skull-released-incredible.

49. Andy Campbell, "Bigfoot DNA Tests: Melba Ketchum's Research Results Are Bogus, Claims *Houston Chronicle* Report," *Huffington Post*, July 3, 2013, www.huffingtonpost.com/2013/07/33/bigfoot-dna-test-results-n-3541431.html.

50. Steven Novella, M.D., "The Starchild Project," *New England Skeptical Society*, February 2006.

51. "The Rhodope Skull: Evidence for the Existence of Aliens on Earth?" *The Event Chronicle*, March 9, 2015, www.theeventchronicle.com/metaphysics/galactic /the-rhodope-skull-evidence-for-the-existence-of-aliens-on-earth/.

52. "The Bolshoi Tjach Skulls, Adygea, Russia," *Soul Guidance*, accessed December 22, 2016, http://soul-guidance.com/houseofthesun/dp09/dp09bolshoitjach.html.

53. "The Rhodope Skull."

54. Anton Spangerberg, "Sealand Skull Photos Released," *Unexplained Mysteries*, September 9, 2010, www.unexplained-mysteries.com/column.php?id=189988.

55. "Sealand Skull Update," Terra Forming Terra, November 11, 2010, http://global warming-arclein.blogspot.com/2010/11/sealand-skull-update.html.

56. "Alien Scientist Find Giant Virus Called Pandoravirus," *World Of UFO*, July 24, 2013, http://worldofufo.blogspot.com/2013/07/alien-scientist-find-giant-virus-called .html.

57. Christine Dell'Amore, "Biggest Virus Yet Found, Maybe Fourth Domain of Life?" *National Geographic News*, July 19, 3013, http://news.nationalgeographic.com /news/2013/07/130718-viruses-pandoraviruses-science-biology-evolution/.

58. Bullard, *The Myth and Mystery of UFOs*, 180.

59. Sharon Begley, "The IQ Puzzle," *Newsweek*, May 6, 1996, 70–72.

60. Ibid.

61. Ellen Lloyd, "Humans' Extraterrestrial DNA: We All Carry a Coded Alien Message Left by the Star Gods," *Message to Eagle*, December 14, 2014, www.messagetoeagle.com /extraterrdnahum.php#.VsbNFykgouo.

62. "Scientists Have Found an Alien Code in Our DNA: Ancient Engineers," *Ancient Code*, April 27, 2015, http://www.ancient-code.com/scientists-have-found-an-alien -code-in-our-dna-ancient-engineers/.

CHAPTER 6

1. Daniel Loxton, "Space Brothers from Venus?" *Junior Skeptic, Skeptic Magazine* 20, no. 4 (2015): 68.

2. James W. Moseley, "Some New Facts about 'Flying Saucers Have Landed,'" *Saucer News* 1 (October 1957).

3. George Hunt Williamson," *Obscurantist*, May 31, 2013, http://obscurantist.com /oma/Williamson-george-hunt/.

4. Loxton, "Space Brothers from Venus?" 66.

5. Ibid., 67.

6. Mike Stackpole, "Dr. Stranges Lives Up to His Name," *The Phoenix Skeptics News* 1, no. 6 (May/June 1988), http://www.discord.org/~lippard/Arizona_Skeptic/PSNv1n6 -May-Jun-1988.pdf.

7. Ibid.

8. Philip Klass, "Walton Abduction Cover-up Revealed," *UFO Investigator*, June 1976.

9. Ibid.

10. Derek Bartholomaus, "Dinosaur Photo Deconstruction," *Billy Meier UFO Case*, July 2011, http://billymeierufocase.com/dinosaurphotodeconstruction.html.

11. Derek Bartholomaus, "Asket and Nera Photo Deconstruction," *Billy Meier UFO Case*, accessed December 27, 2016, http://billymeierufocase.com/asketdeconstruction.html.

12. Derek Bartholomaus, "World War III Prophecy Deconstruction," *Billy Meier UFO Case,* January 2009, http://billymeierufocase.com/wwiiideconstruction.html.

13. Billy Meier, "Henoch Prophecies," *They Fly,* February 28, 1987, http://www.theyfly.com/henoch-prophecies.

14. Ibid.

15. Alan Friswell, "Faking It," *Fortean Times,* June 2005.

16. Derek Bartholomaus, "Wedding Cake UFO Deconstruction," *Billy Meier UFO Case,* accessed December 27, 2016, http://billymeierufocase.com/wcufodeconstruction.html.

17. Kal K. Korff, *Spaceships of the Pleiades: The Billy Meier Story* (Amherst, NY: Prometheus Press, 1995), 275.

18. Ibid., 276.

19. Ibid., 278.

20. Ibid., 284, 285, 293.

21. Ibid., 46.

22. Robert Todd Carroll, "Raëlian," *The Skeptic's Dictionary,* updated October 27, 2015, http://www.skepdic.com/raelian.html.

23. "Carbon Dating," UK.Rael.org, accessed December 30, 2016, http://uk.rael.org/page.php?11.

24. As quoted by Adam Lee, "Strange and Curious Sects: Raelianism," *Patheos,* July 2009, http://www.patheos.com/blogs/daylightatheism/2009/07/strange-and-curious-sects-vi/.

25. "About Jean Sendy and Books Comparison with Those of Claude Vorilhon Rael," *Raelian.com,* accessed December 30, 2016, http://raelian.com/en/jean_sendy.php.

26. *Today Tonight,* "Space Aliens Are at the Very Root of the Belief System of Scientology," February 4, 2009.

27. Russel Miller, *Bare-faced Messiah: True Story of L. Ron Hubbard* (London: Michael Joseph, 1987), 107.

28. L. Sprague De Camp, letter of August 26, 1946, as quoted in George Pendle, *Strange Angel: The Otherworldly Life of Rocket Scientist John Whiteside Parsons* (New York: Harcourt Publishers, 2006), 271.

29. Claudia Puig, "Scientology Doc 'Going Clear' Shocks Sundance Filmgoers," *USA Today,* January 26, 2015, http://www.usatoday.com/story/life/movies/2015/01/26/controversial-scientology-documentary/22337995/.

30. John Horn, "Alex Gibney on *Going Clear,* His Scientology Documentary That's the Talk of Sundance," *Vulture,* January 30, 2015, http://www.vulture.com/2015/01/alex-gibney-on-his-new-scientology-documentary.html.

31. John Nocera, "Scientology's Chilling Effect," The New York Times, February 24, 2015, http://www.nytimes.com/2015/02/24/opinion/joe-nocera-scientologys-chilling-effect.html.

32. Bilge Ebiri, "HBO's Scientology Exposé *Going Clear* Is Jaw-Dropping," *Vulture,* March 13, 2015, http://www.vulture.com/2015/01/sundance-going-clear-is-a-jaw-dropping-expos.html.

33. Kira Bindrim, "HBO's 'Going Clear' Is a Perfect Scientology Primer," *Newsweek,* March 25, 2015, http://www.newsweek.com/hbos-going-clear-perfect-scientology-primer-316567.

34. Joanna Robinson, "The 6 Most Disturbing Moments from HBO's Scientology Documentary *Going Clear,*" *Vanity Fair,* March 30, 2015, http://www.vanityfair.com/hollywood/2015/03/going-clear-disturbing-moments.

35. Jim Lippard, "The Decline and (Probable) Fall of the Scientology Empire!" *Skeptic Magazine* 17, no. 1 (2011), http://www.skeptic.com/reading_room /the-decline-and-probable-fall-of-the-scientology-empire/.

36. Douglas Frantz, "Scientology's Puzzling Journey from Tax Rebel to Tax Exempt," *The New York Times*, March 9, 1997, http://www.nytimes.com/1997/03/09/us/scientology -s-puzzling-journey-from-tax-rebel-to-tax-exempt.html.

37. Tony Ortega, "Scientology Targeted South Park's Parker and Stone in Investigation," *The Village Voice*, October 23, 2011, http://blogs.villagevoice.com/runninscared /2011/10/scientology_tar.php.

38. Kira Bindrim, "This Is Why HBO's 'Going Clear' Has the Church of Scientology Worried," *Newsweek*, March 25, 2015, http://www.rawstory.com/rs/2015/03/this -is-why-hbos-going-clear-has-the-church-of-scientology-worried/.

39. Lippard, "The Decline and (Probable) Fall of the Scientology Empire!"

40. Lawrence Wright, "The Apostate: Paul Haggis vs. the Church of Scientology," *The New Yorker*, February 14, 2011, http://www.newyorker.com/magazine/2011/02/14 /the-apostate-lawrence-wright.

41. Ibid.

42. Stephen Barrett, MD, "The Origin and Current Status of Christian Science," *Quackwatch*, updated March 16, 2016, http://www.quackwatch.org/01QuackeryRelatedTopics /cs2.html.

43. Stanton T. Friedman, "The Bob Lazar Fraud," Stanton Friedman's official website, updated January 2011, http://www.stantonfriedman.com/index.php?ptp=articles &fdt=2011.01.07.

CHAPTER 7

1. Brian Dunning, "Skeptoid #124: Betty and Barney Hill—the Original UFO Abduction," *Skeptoid* (podcast), October 21, 2008, https://skeptoid.com/episodes/4124/.

2. Ibid.

3. Decker, "Touched by an Alien."

4. Teresa of Avila, *The Life of St. Teresa of Jesus of the Order of Our Lady of Carmel*, ch. XXIX, item 17, http://www.ccel.org/ccel/teresa/life.viii.xxx.html.

5. Decker, "Touched by an Alien."

6. "MUFON Sighting Report," *Mutual UFO Network*, accessed December 30, 2016, https://mufoncms.com/cgi-bin/report_handler.pl.

7. Jims Space Agency, "REAL ALIENS CONTACT !!! - Alien Message Proof Evidence *****," *YouTube* video, September 16, 2011, https://www.youtube.com/watch?v =xUFNZUxBYp8.

8. Daniel L. Schacter, *Searching for Memory: The Brain, the Mind and the Past* (New York: Basic Books, 1996), 256.

9. Elizabeth Loftus, "How Reliable Is Your Memory?" TED Talks, June 2013, http://www.ted.com/talks/elizabeth_loftus_the_fiction_of_memory.

10. K. Zirpolo and D. Nathan, "I'm Sorry: A Long-Delayed Apology from One of the Accusers in the Notorious McMartin Pre-School Molestation Case," *Los Angeles Times Magazine*, October 30, 2005.

11. Clancy, *Abducted*, 59.

12. "The Allagash Abduction," *UFO Casebook*, accessed December 30, 2016, http: //www.ufocasebook.com/Allagash.html.

13. Clancy, *Abducted*, 56.

14. "Reptilians and Human Abductions," *Arcturi*, accessed December 30, 2016, http: //www.arcturi.com/ReptilianArchives/ReptiliansAndAbductions.html.

15. Ibid.

16. Donald Worley, "Alien Abduction Rape by Reptilians," *Abduct.com*, accessed December 30, 2016, http://www.abduct.com/worley/worley68.php.

17. Ibid.

18. Ibid.

19. "Nordic Aliens," *UFO–Alien Database*, updated August 23, 2015, http://ufoalienda tabase.wikia.com/wiki/Nordic_Aliens.

20. "David Icke and the Reptilian Agenda," *Arcturi*, accessed December 30, 2016, http: //www.arcturi.com/ReptilianArchives/DavidIcke.html.

21. "The World Order and the Reptilians," *Arcturi*, accessed December 30, 2016, http: //www.arcturi.com/ReptilianArchives/ReptiliansAndTheRothchilds.html.

22. Ibid.

23. Bobby Brewer, "Seven Things You Should Know about UFOs," *Christian Research Journal* 25, no. 2 (2002), http://www.equip.org/article/seven-things-you -should-know-about-ufos/.

24. Flavius Josephus (translated by William Whiston), *Antiquities of the Jews* (Peabody, MA: Hendrickson Publishers, 1987), book 1, chapter 3, item 1.

25. Richard Abanes, *End-Times Visions* (New York: Four Walls Eight Windows, 1998), 141, 142.

26. Chuck Missler, "The Return of the Nephilim," *Personal Update*, September 1997.

27. James Hough, "What Is the Catholic Church's View on Aliens?" *Answers*, accessed December 30, 2016, http://www.answers.com/Q/What_is_the_Catholic _Church's_view_on_aliens?/.

28. Anthony Enns, "Alien Abduction: A Return to the Medieval," *Iowa Journal of Cultural Studies* 1 (1999).

29. Enns, "Alien Abduction."

CHAPTER 8

1. "Hans Schindler Bellamy," *Atlantipedia*, updated April 17, 2016, http://www.atlan tipedia.com/doku.php?id=hans_schindler_bellamy.

2. "The Baghdad Battery," *World Mysteries*, accessed December 30, 2016, http://old .world-mysteries.com/sar_11.htm.

3. Ibid.

4. Arran Frood, "Riddle of 'Baghdad's Batteries,'" *BBC News*, February 27, 2003.

5. Aaron Sakulich, "The 'Baghdad Battery,'" *The Iron Skeptic*, accessed December 30, 2016, http://www.theironskeptic.com/articles/battery/battery.htm.

6. Ibid.

7. Ibid.

8. Lenny Flank, "The Baghdad Battery: An Update," *The Daily Kos*, February 10, 2015, http://www.dailykos.com/story/2015/2/10/1361589/-The-Baghdad-Battery-An-Update.

9. R. Balasubramaniam, "On the Growth Kinetics of the Protective Passive Film of the Delhi Iron Pillar," *Current Science* 82, no. 11 (2002), http://tejas.serc.iisc.ernet.in/~currsci /jun102002/1357.pdf.

10. Steven Dutch, "The Piri Reis Map," official website of Steve Dutch, updated June 2, 2010, http://www.uwgb.edu/dutchs/PSEUDOSC/PiriRies.HTM.

11. Nickell, *Real-Life X-Files*, 5.

12. Ibid., 7.

13. Erich von Daniken, *Chariots of the Gods?* (New York: G. P. Putnam's Sons, 1968).

14. David Childress, "Evidence for Ancient Atomic Warfare," *Nexus Magazine* 7, no. 5 (November/December 2000), http://www.bibliotecapleyades.net/ancientatomicwar /esp_ancient_atomic_01.htm.

15. "Tap O' Noth," *Canmore*, accessed January 2, 2017, https://canmore.org.uk/site /17169/tap-o-noth.

16. "Archaeology Answers on Ancient Indus River Valley Civilizations," *Before Us*, accessed January 4, 2017, http://www.beforeus.com/indusa.htm.

17. "Mohenjo-daro," RationalWiki, updated December 27, 2016, http://rationalwiki .org/wiki/Mohenjo-daro.

18. "Archaeology Answers on Ancient Indus River Valley Civilizations."

19. Chris Morton and Ceri Louise Thomas, *The Mystery of the Crystal Skulls* (Rochester, VT: Inner Traditions Bear and Company, 2002).

20. Richard A. Lovett and Scot Hoffman, "Crystal Skulls Fuel Controversy, Fascination," *National Geographic*, accessed January 2, 2017, http://science.nationalgeographic .com/science/archaeology/crystal-skulls/.

21. Jason Colavito, "Review of Ancient Aliens S04E06: The Mysteries of Puma Punku," official website of Jason Colavito, March 17, 2012, http://www.jasoncolavito.com/blog /review-of-ancient-aliens-s04e06-the-mysteries-of-puma-punku.

22. Kevin Burns, "Puma Punku," *Ancient Aliens*, season 4, episode 6, aired March 16, 2012 (Los Angeles, CA: Prometheus Entertainment).

23. University of Kansas, "Mysterious Stone Spheres in Costa Rica Investigated," *Science Daily*, March 23, 2010.

24. David Hatcher Childress, *Ancient Technology in Central America* (Kempton, IL: Adventures Unlimited Media, 2009).

25. von Daniken, *Chariots of the Gods?*, 77.

26. Chris White, with commentary by Dr. Michael Heiser, "Vimanas," *Ancient Aliens Debunked*, accessed January 2, 2017, http://ancientaliensdebunked.com/references-and -transcripts/vimanas/#sthash.hsS6cc8w.dpuf/.

27. von Daniken, *Chariots of the Gods?*, 100, 101.

28. Moshe Greenberg, "Ezekiel 1–20," in *Anchor Bible*, vol. 22, edited by William Foxwell Allbright and David Noel Freedman (Garden City, NY: Doubleday & Co. Inc., 1983), 58.

29. Rawles, "The Abydos-Hieroglyph Does NOT Depict a Helicopter."

30. Burns, "Alien Power Plants."

31. Ibid.

32. Quiltingmamma, "Photo: 'Before and after Cleaning,'" *Trip Advisor*, June 18, 2011, www.tripadvisor.com/LocationPhotoDirectLink-g1598532-d472016-i32011744-Temple _of_Hathor_at_Dendera-Qena_Qena_Governorate_Nile_River_Valley.html.

33. Burns, "Alien Power Plants."

34. von Daniken, *Chariots of the Gods?* 122, 123.

35. Diego Cuoghi, "The Art of Imagining UFOs," *Skeptic Magazine* 11, no. 1 (2004): 45, http://www.diegocuoghi.com/pdf/SKEPTIC_CUOGHI_The_Art_Of_Imagining_UFOs.pdf.

36. Sitchin, Zecharia. *The 12th Planet* (New York: Stein and Day, 1976), 13.

37. Will Hart, The *Genesis Race* (Rochester, VT: Inner Traditions Bear & Company, 2003), 65.

38. Sitchin, *12th Planet*, 53.

39. Georges Roux, *Ancient Iraq* (London: Allen and Unwin, 1992), 48.

40. "The Mysterious Origins of the World's First City Builders," *Mysterious Universe*, 2014, mysteriousuniverse.org/2014/the-mysterious-origins-of-the-worlds-first-city-builders/.

41. Arkadiusz Soltysiak. "Physical Anthropology and the 'Sumerian Problem.'" *Studies in Historical Anthropology* 4, no. 2004 (2006): 145–58.

42. J. A. Black, G. Cunningham, J. Ebeling, E. Flückiger-Hawker, E. Robson, J. Taylor, and G. Zólyomi, eds. and trans., *The Lament for Ur* from *The Electronic Text Corpus of Sumerian Literature*, Oxford, 2006.

43. Ibid.

44. Ibid.

45. Michael S. Heiser, "The Myth of a 12th Planet: 'Nibiru' According to the Cuneiform Sources," http://www.sitchiniswrong.com/nibirunew.pdf.

46. J. A. Black, G. Cunningham, J. Ebeling, E. Flückiger-Hawker, E. Robson, J. Taylor, and G. Zólyomi, eds. and trans., *Hymn to Utu* from *The Electronic Text Corpus of Sumerian Literature*, Oxford, 2003.

47. Michael Heiser, "The Myth of a 12th Planet: A Brief Analysis of Cylinder Seal VA 243," www.michaelsheiser.com/VA243seal.pdf.

48. Walter E. A. Van Beek, "Dogon Restudied: A Field Evaluation of the Work of Marcel Griaule," *Current Anthropology* 32, no. 2 (April 1991).

49. J. M. Bonnet-Bidaud and E. Pantin. "ADONIS High Contrast Infrared Imaging of Sirius B," *Astronomy and Astrophysics* 484, no. 2 (2008): 651–655.

50. Laurie R. Godfrey, ed., *Scientists Confront Creationism* (New York: W.W. Norton, 1983), 200–2.

51. Francis Crick, *Life Itself: Its Origin and Nature* (New York: Simon and Schuster, 1981), 85.

52. Fred Hoyle and Chandra Wickramasinghe, *Cosmic Life Force: The Power of Life across the Universe* (New York: Paragon House, 1990), 600.

53. Ibid., 60.

54. Ibid., 138.

55. Ibid., 144.

56. Michael Drosnin. *The Bible Code II: The Countdown* (New York: Viking), 143, 144.

57. Francis Crick, "Seeding the Universe," *Science Digest* 89, November 1981, 153.

58. Timothy Ferris. "Playboy Interview: Erich von Daniken," *Playboy Magazine*, August 1974.

CHAPTER 9

1. "The World Teacher Is Now Here," *Share International*, accessed January 30, 2017, http://shareinternational.org.au/.

2. "Key Beliefs," *The Aetherius Society*, accessed June 2, 2017, http://www.aetherius.org/key-beliefs/.

3. "Unarius" is an acronym for Universal Articulate Interdimensional Understanding of Science.

4. "Key Beliefs."

5. "An Astral Visit To Venus," *Unariun Wisdom*, accessed January 2, 2017, http://www .unariunwisdom.com/an-astral-visit-to-venus/

6. Ernest L. Norman, "The Truth about Mars," *FireDocs*, updated 1967, http://www .firedocs.com/remoteviewing/mars/mars.html.

7. Ibid.

8. Robert Stone, *Farewell, Good Brothers* (Los Angeles: Robert Stone Productions, 1992).

9. "Atlantis, Lemuria & Maldek," *The Aetherius Society*, accessed June 2, 2017, http: //www.aetherius.org/the-mother-earth/atlantis-lemuria-maldek/.

10. Ibid.

11. "Paper 57: The Origin of Urantia," in *The Urantia Book*, 658, http://www.urantia .org/urantia-book-standardized/paper-57-origin-urantia.

12. "Atlantis, Lemuria & Maldek."

13. Ibid.

14. Ibid.

15. Physics 4 You, "Chronicles of Atlantis," *Unarius Wisdom*, accessed January 2, 2017, http://web.archive.org/web/20150614113150/http://www.unariuswisdom.com /chronicles-of-atlantis/.

16. Maurizio Martinelli, "George Hunt Williamson–Michael D'Obrenovic: The Herald of the Encounter among Father and Sons," *duepassinelmistero.com*, September 2009, http://www.duepassinelmistero.com/williamson.htm.

17. *His New Word*, www.hisnewword.org.

18. "The Sound Bath," *Integratron*, accessed January 4, 2017, http://integratron.com /sound-bath/.

19. Frank E. Stranges, "A Holy Stranger in the Pentagonal Lodge," in *Stranger at the Pentagon* (New Brunswick, NJ: Inner Light Publications, 1967), http://www.theeventchronicle .com/editors-pick/valiant-thor-a-venusian-at-the-pentagon-rev-frank-e-stranges/.

20. Andreas Gruenschloss, "Fiat Lux," official website of Andreas Gruenschloss, September 2001, http://wwwuser.gwdg.de/~agruens/UFO/fiatlux.html.

21. Louis Farrakhan, "The Divine Destruction of America: Can She Avert It?" *The Final Call*, June 9, 1996.

22. Becky Schlikerman, "Farrakhan: Revolution Imminent in U.S." *Chicago Tribune*, February 27, 2011.

23. Ramon Watkins, "The Black Jew Hebrew Israelite Study Guide," *Prophet Yahweh UFOs*, October 2011, http://prophetyahwehufos.blogspot.com/2011/10/black-jew -hebrew-israelite-study-guide.html.

24. *Introduction to UFOs in the Bible*, accessed January 30, 2017, https://video.search. yahoo.com/yhs/search;_ylt=A86.J7oStYNY32UAWyAnnIlQ?p=Prophet+Yahweh +on+UFOs+in+the+Bible&fr=yhs-mozilla-001&fr2=piv-web&hspart=mozilla&hsimp =yhs-001#id=1&vid=ea01f3418834c8251be7e87c64a0d660&action=view.

25. Ryan J. Cook, "Nuwaubians," *Center for AnthroUfology*, updated December 31, 2005, http://web.archive.org/web/20151009082634/http://www.anthroufo.info/un-nuwab .html.

26. "Core Beliefs," *Nuwaubian Nation*, October 24, 2014, http://nuwaubianinfo.weebly .com/information/core-beliefs.

27. Ibid.

28. Willy Ley, "Pseudoscience in Naziland," *Astounding Science Fiction*, May 1947.

29. Louis Pauwels and Jacques Bergier, translation by Rollo Myers, *The Morning of the Magicians* (New York: Stein and Day, 1963), 147.

30. Juan Graña, "Maria Orsitsch," *Supercomentarios*, May 14, 2014, http://juan-jackblog .blogspot.com/2014/05/maria-orsitsch-eglish-version.html.

31. Nicholas Goodrick-Clarke, *Black Sun: Aryan Cults, Esoteric Nazism and the Politics of Identity* (New York: New York University Press, 2002), 169, 170.

32. Ibid., 164.

33. Jean-François Mayer, translation by Elijah Siegler, "'Our Terrestrial Journey Is Coming to an End': The Last Voyage of the Solar Temple," *Nova Religio* 2, no. 2 (April 1999), http://english.religion.info/2003/06/12/our-terrestrial-journey-is-coming -to-an-end-the-last-voyage-of-the-solar-temple/.

34. Ibid.

35. Ibid.

36. Ryan J. Cook, "Heaven's Gate," *Center for AnthroUfology*, updated December 31, 2005, http://web.archive.org/web/20130731093726/http://www.anthroufo.info/un -hgate.html.

37. *Heaven's Gate*, http://www.heavensgate.com/.

38. Leon Festinger, Henry W. Riecken, and Stanley Schachter, *When Prophecy Fails* (Minneapolis: University of Minnesota Press, 1956), 208.

39. *Heaven's Gate Cult Documentary, History TV*, accessed January 30, 2017, https: //video.search.yahoo.com/yhs/search;_ylt=AwrTccUCwoNYtsgAajcnnIlQ?p=Terrie +Nettles+Heaven%27s+Gate&fr=yhs-mozilla-001&fr2=piv-web&hspart=mozilla &hsimp=yhs-001#id=3&vid=d7a369c71ab457bc188448ad1022b842&action=view

CHAPTER 10

1. Teresa of Avila, *The Life of St. Teresa*, ch. XXIX, item 17.

2. Martin Kottmeyer, "Entirely Unpredisposed: The Cultural Background of UFO Abduction Reports," *Magonia Magazine*, January 1990, http://www.debunker.com/texts /unpredis.html.

3. Kenneth Arnold, "Transcript of Ed Murrow–Kenneth Arnold Telephone Conversation," *CUFOS Associate Newsletter*, February/March 1984: 3, http://www.project1947 .com/fig/kamurrow.htm.

4. Robert Sheaffer, "The First 'Flying Saucer' Sighting: Kenneth Arnold, Mt. Rainier, Washington—June 24, 1947," *Debunker*, updated August 5, 2016, http://www.debunker .com/arnold.html.

5. Martin Kottmeyer, "The Saucer Error," *The REALL News* 1, no. 4 (May 1993), http: //www.debunker.com/texts/SaucerError.html.

6. See Prison Planet Forum, http://forum.prisonplanet.com/index.php?topic =23030.240.

7. Armando Simon, "Pulp Fiction UFOs," *Skeptic Magazine* 16, no. 4 (2011): 19.

8. H. G. Wells, *The First Men in the Moon* (1901; New York: Berkley Highland Books, 1970), 73.

9. Aaron Sakulich, "A Media History of Gray Aliens," *The Triangle*, May 20, 2005, http://www.theironskeptic.com/articles/gray/gray_history.htm.

10. "Grey Aliens Agenda: Human Hybrid Integration Program," *Arcturi*, accessed January 4, 2017, http://www.arcturi.com/GreyArchives/HumanHybridIntegrationProgram .html.

11. Bullard, *The Myth and Mystery of UFOs*, 271.

12. Ibid., 198, 199.

13. "Reptilian Alien Reports," *MUFON Forum*, last updated October 27, 2015, http: //mutualufonetwork.proboards.com/thread/221/reptilian-alien-reports/.

14. David Icke, *The Biggest Secret* (Scottsdale, AZ: Bridge of Love Publications, 1999), 127.

15. Ibid., 288.

16. Ibid.

17. Ibid., 243.

18. Ibid., 246, 247.

19. Olga Oksman, "Conspiracy Craze: Why 12 Million Americans Believe Alien Lizards Rule Us," *The Guardian*, April 7, 2016.

20. Simon, "Pulp Fiction UFOs," 18.

21. "The Smallest Spy Drone in the World—Pic and Video," *Gadgetonics*, July 5, 2012, http://www.gadgetonics.com/2012/07/smallest-spy-drone-in-world-pic-and.html.

22. John W. Whitehead, "Roaches, Mosquitoes and Birds: The Coming Micro-Drone," *The Huffington Post*, April 17, 2013, http://www.huffingtonpost.com/john-w-whitehead /micro-drones_b_3084965.html.

CHAPTER 11

1. Melissa Chan, "'UFO Clouds' Form in Cape Town, Spark New Fears of Alien Invasion," *New York Daily News*, updated November 10, 2015, http://www.nydailynews.com /news/world/ufo-clouds-form-cape-town-spark-new-fears-aliens-article-1.2429217.

2. Puzuzu Ba'al, "UFO Clouds," *Alien Sky Ships*, accessed January 4, 2017, http://www .alienskyships.com/UFO_clouds.html.

3. Michael Roesch, "Harmonic Convergence," Mount Shasta Companion, College of the Siskiyous, accessed January 4, 2017, http://www.siskiyous.edu/shasta/fol/har/index.htm.

4. Puzuzu Ba'al, "UFO Clouds."

5. Christine Shearer, Mick West, Ken Caldeira, and Steven J. Davis, "Quantifying Expert Consensus against the Existence of a Secret, Large-Scale Atmospheric Spraying Program," *Environmental Research Letters* 11, no. 8 (2016), http://iopscience.iop.org /article/10.1088/1748-9326/11/8/084011.

6. Kyle Hill, "A Million Poisoning Planes," *The Committee for Skeptical Inquiry*, August 14, 2013, http://www.csicop.org/specialarticles/show/a_million_poisoning_planes.

7. Mick West, "A Brief History of 'Chemtrails,'" *Contrail Science*, May 11, 2007, http: //contrailscience.com/a-brief-history-of-chemtrails/.

8. "Chemtrails," *RationalWiki*, updated December 27, 2016, http://rationalwiki.org /wiki/Chemtrails.

9. Mick West, "WWII Contrails," *Contrail Science*, August 16, 2008, http://contrailscience .com/wwii-contrails/.

10. West, "Debunked."

11. Mick West, "Germans Admit They Used Düppel!" *Contrail Science*, March 14, 2008, http://contrailscience.com/germans-admit-they-used-duppel/.

12. Ibid.

13. Mick West, "Kucinich, Chemtrails and HR 2977 – The 'Space Preservation Act,'" *Contrail Science*, May 19, 207, http://contrailscience.com/kucinich-chemtrails-and-hr-2977/.

14. Mick West, "How to Debunk Chemtrails," *Contrail Science*, June 5, 2011, http://contrailscience.com/how-to-debunk-chemtrails/.

15. Hill, "A Million Poisoning Planes."

CHAPTER 12

1. Elizabeth Howell, "How Many Stars Are in the Milky Way?" *Space.com*, May 21, 2014, http://www.space.com/25959-how-many-stars-are-in-the-milky-way.html.

2. Nell Greenfieldboyce, "Galaxy Quest: Just How Many Earth-Like Planets Are Out There?" *NPR*, November 5, 2013, http://www.npr.org/2013/11/05/242991030/galaxy-quest-just-how-many-earth-like-planets-are-out-there.

3. Robert L. Griffith, Jason T. Wright, Jessica Maldonado, Matthew S. Povich, Steinn Sigurdsson, and Brendan Mullen, "The G Infrared Search for Extraterrestrial Civilizations with Large Energy Supplies III: The Reddest External Sources in WISE," *Astrophysical Journal Supplement Series* 247, no. 2 (April 2015), https://arxiv.org/pdf/1504.03418v2.pdf.

4. "SETI FAQ: How Long Have Astronomers Been Looking for Extraterrestrial Signals?" SETI, accessed January 5, 2017, http://www.seti.org/faq#obs19.

5. George Gaylord Simpson, "On the Nonprevalence of Humanoids," *Science* 143: 769–75.

6. D. T. Anderson, Sidnie M. Manton, and J. P. Harding, "Studies on the Onychophora VIII: The Relationship between the Embryos and the Oviduct in the Viviparous Placental Onychophorans *Epiperipatus trinidadensis* Bouvier and *Macroperipatus torquatus* (kennel) from Trinidad," *Philosophical Transactions of the Royal Society B, Biological Sciences*, July 20, 1972.

CHAPTER 13

1. Donald Prothero, "Science TV 'Network Decay,'" *SkepticBlog*, January 25, 2012, http://www.skepticblog.org/2012/01/25/science-tv-sell-out/.

2. Christopher Bader, Kevin Dougherty, Paul Froese, Byron Johnson, F. Carson Mencken, Jerry Z. Park, and Rodney Stark, "American Piety in the 21st Century: New Insights to the Depth and Complexity of Religion in the US," Baylor University, September 2006, http://www.baylor.edu/content/services/document.php/33304.pdf.

3. Ibid.

4. M. J. Sharps, J. Matthews, and J. Asten, "Cognition and Belief in Paranormal Phenomena: Gestalt/Feature-Intensive Processing Theory and Tendencies toward ADHD, Depression and Dissociation," *The Journal of Psychology* 140, no. 6 (2006): 579–90.

5. Susan Krause Whitbourne, "What Psychology Can Tell You About Your Extraterrestrial Beliefs," *Psychology Today*, June 12, 2012, https://www.psychologytoday.com/blog

/fulfillment-any-age/201206/what-psychology-can-tell-you-about-your-extraterrestrial-beliefs.

6. Alon Harish, "UFOs Exist, Say 36 Percent in National Geographic Survey," *ABC News*, June 27, 2012, http://abcnews.go.com/Technology/ufos-exist-americans-national-geographic-survey/story?id=16661311.

7. Lee Spelgel, "More Believe in Space Aliens Than in God According to U.K. Survey," *The Huffington Post*, updated October 18, 2012, http://www.huffingtonpost.com/2012/10/15/alien-believers-outnumber-god_n_1968259.html.

8. "In U.S., 12 Million Americans Believe the World Is Run by Lizards from Outerspace," *The Intellectualist*, April 8, 2016, http://theintellectualist.co/in-u-s-12-million-americans-believe-the-world-is-run-by-a-group-of-lizards-from-outerspace/.

9. idoubtit, "Questionable Poll Touts UFO/ET Belief," *Doubtful News*, September 12, 2013, http://doubtfulnews.com/2013/09/questionable-poll-touts-ufoet-belief/.

10. "Climate Change Debate: Last Week Tonight with John Oliver (HBO)," *YouTube* video, May 11, 2014, https://www.youtube.com/watch?v=cjuGCJJUGsg.

11. "Mutual Mistrust May Have Added a Few X-files to the UFO Era," *ScienceDaily*, February 16, 2016, https://www.sciencedaily.com/releases/2016/02/160216181704.htm.

12. Rick Rojas, "UFO Investigators' Convention Emphasizes Scientific Methods," *Los Angeles Times*, July 31, 2011, http://articles.latimes.com/2011/jul/31/local/la-me-ufo-20110731.

13. Ibid.

14. Sharon Hill, "Want to Shed the Pseudoscience Label? Try Harder," *Doubtful Blog*, May 23, 2011, http://idoubtit.wordpress.com/2011/05/23/want-to-shed-the-pseudoscience-label-try-harder/.

15. Ibid.

16. Sharon Hill, "UFO Research Is Up in the Air: Can It Be Scientific?" *The Committee for Skeptical Inquiry*, August 28, 2016, http://www.csicop.org/specialarticles/show/ufo_research_is_up_in_the_air.

CHAPTER 14

1. Seth Shostak, "UFOs: The Trail Is Stale," *The Huffington Post*, updated September 1, 2016, http://www.huffingtonpost.com/seth-shostak/ufos-the-trail-is-stale_b_8071452.html.

2. "Steven Spielberg Surprised by Drop in UFO Sightings," *Contact Music*, June 28, 2005, http://www.contactmusic.com/steven-spielberg/news/spielberg-confused-by-decrease-in-ufo-sightings.

BIBLIOGRAPHY

Abanes, Richard. *End-Times Visions*. New York: Four Walls Eight Windows, 1998.

"About Jean Sendy and Books Comparison with Those of Claude Vorilhon Rael." *Raelian. com*, accessed December 30, 2016. http://raelian.com/en/jean_sendy.php.

Aciman, Alexander. "Area 51 Is Alive and Unwell." *Time*, August 16, 2013. http://newsfeed .time.com/2013/08/16/area-51-is-alive-and-unwell/.

Alexander, John D. *UFOs: Myths, Conspiracies and Realities*. New York: Thomas Dunne Books, 2011.

"Alien Scientist Find Giant Virus Called Pandoravirus." *World Of UFO*, July 24, 2013. http://worldofufo.blogspot.com/2013/07/alien-scientist-find-giant-virus-called .html.

"The Allagash Abduction." *UFO Casebook*, accessed December 30, 2016. http://www .ufocasebook.com/Allagash.html.

Anderson, D. T., Sidnie M. Manton, and J. P. Harding. "Studies on the Onychophora VIII: The Relationship between the Embryos and the Oviduct in the Viviparous Placental Onychophorans *Epiperipatus trinidadensis* Bouvier and *Macroperipatus torquatus* (kennel) from Trinidad." *Philosophical Transactions of the Royal Society B, Biological Sciences*, July 20, 1972.

Andrews, Colin. "The Julia Set: UFO Crop Circle Witnesses Come Forward." *Free Republic*, October 16, 2009. www.freerepublic.com/focus/chat/2364207/posts.

Anthony, Sebastian. "Element 115: How Chemists Discovered the Newest Member of the Periodic Table." *ExtremeTech*, August 30, 2013. http://www.extremetech.com /extreme/165357-element-115-how-chemists-discovered-the-newest-member-of -the-periodic-table.

"Archaeology Answers on Ancient Indus River Valley Civilizations." *Before Us*, accessed January 4, 2017. http://www.beforeus.com/indusa.htm.

"Area 51 and Alien Landscapes." *Skeptics Society*, October 21, 2013. http://www.skeptic .com/geology_tours/2014/Area-51-and-Alien-Landscapes/downloads/Area -51-and-Alien-Landscapes-2014.pdf.

Arnold, Jack (director; screenplay by Harry Essex and Arthur Ross). *Creature from the Black Lagoon*. Universal International, 1954.

Arnold, Kenneth. "Transcript of Ed Murrow–Kenneth Arnold Telephone Conversation." *CUFOS Associate Newsletter*, February/March 1984, p. 3. http://www.project1947 .com/fig/kamurrow.htm.

Asher, William (director; screenplay by John Mantley). *The 27th Day*. Culver City, CA: Columbia Pictures, 1957.

Asimov, Isaac. *I, Robot*. New York: Gnome Press, 1950.

Asimov, Isaac. "Not as We Know It." In *View From A Height*. Garden City, NY: Doubleday, 1963.

"Atlantis, Lemuria & Maldek." *The Aetherius Society*, accessed June 2, 2017. http://www .aetherius.org/the-mother-earth/atlantis-lemuria-maldek/.

Bader, Christopher, Kevin Dougherty, Paul Froese, Byron Johnson, F. Carson Mencken, Jerry Z. Park, and Rodney Stark. "American Piety in the 21st Century: New Insights to the Depth and Complexity of Religion in the US." Baylor University, September 2006. http://www.baylor.edu/content/services/document.php/33304.pdf.

"The Baghdad Battery." *World Mysteries*, accessed December 30, 2016. http://old.world -mysteries.com/sar_11.htm.

Balasubramaniam, R. "On the Growth Kinetics of the Protective Passive Film of the Delhi Iron Pillar." *Current Science* 82, no. 11 (2002): 1357–65. http://tejas.serc.iisc.ernet .in/~currsci/jun102002/1357.pdf.

Bark, Deborah (producer). *Dark Fellowships: The Nazi Cult, the Vril Society*. Discovery Channel, 2008.

Barrett, Stephen, MD. "The Origin and Current Status of Christian Science." *Quackwatch*, updated March 16, 2016. http://www.quackwatch.org/01QuackeryRelatedTopics /cs2.html.

Barry, Kevin (writer and producer). *The Unexplained: Heaven's Gate and Other UFO Cults*. Towers Productions for A&E Network, 1997.

Barry, Tom, producer and director. *The Truth behind Crop Circles*. London: Zigzag Productions, 2010.

Bartholomaus, Derek. "Asket and Nera Photo Deconstruction." *Billy Meier UFO Case*, accessed December 27, 2016. http://billymeierufocase.com/asketdeconstruction .html.

———. "Dinosaur Photo Deconstruction." *Billy Meier UFO Case*, July 2011. http://bil lymeierufocase.com/dinosaurphotodeconstruction.html.

———. "Wedding Cake UFO Deconstruction." *Billy Meier UFO Case*, accessed December 27, 2016. http://billymeierufocase.com/wcufodeconstruction.html.

———. "World War III Prophecy Deconstruction." *Billy Meier UFO Case*, January 2009. http://billymeierufocase.com/wwiiideconstruction.html.

Basterfield, Keith. "Delphos—Soil Analyses." *Unidentified Aerial Phenomena—Scientific Research*, June 21, 2013. http://ufos-scientificresearch.blogspot.com/2013/06 /delphos-soil-analyses.html.

Begley, Sharon. "The IQ Puzzle." *Newsweek*, May 6, 1996, pp. 70–72.

Berlitz, Charles, and William L. Moore. *The Roswell Incident*. New York: Grosset and Dunlap, 1980.

"The Binary Codes." *The Rendlesham Forest Incident*, accessed December 22, 2016. www .therendleshamforestincident.com/The_Binary_Codes.html.

Bindrim, Kira. "HBO's 'Going Clear' Is a Perfect Scientology Primer." *Newsweek*, March 25, 2015. http://www.newsweek.com/hbos-going-clear-perfect-scientology-primer-316567.

———. "This Is Why HBO's 'Going Clear' Has the Church of Scientology Worried." *Newsweek*, March 25, 2015. http://www.rawstory.com/rs/2015/03 /this-is-why-hbos-going-clear-has-the-church-of-scientology-worried/.

Black, J. A., G. Cunningham, J. Ebeling, E. Flückiger-Hawker, E. Robson, J. Taylor, and G. Zólyomi (editors and translators). *Hymn to Utu* from *The Electronic Text Corpus of Sumerian Literature*. Oxford: Oxford University, 2003.

———. *The Lament for Ur* from *The Electronic Text Corpus of Sumerian Literature*. Oxford: Oxford University, 2006.

Black, J. A., Cunningham, G., Ebeling, J., Flückiger-Hawker, E., Robson, E., Taylor, J., and Zólyomi, G., ed. and trans. *Letter from Sin-iddinam to the god Utu about the distress of Larsa* from *The Electronic Text Corpus of Sumerian Literature*. Oxford: Oxford University, 1998

Blish, James. *The Seedling Stars*. New York: Gnome Press, 1957.

"Bob White Case UFO Object UFO Hunters." *UFO Hunters*. Season 3, episode 7, "UFO Relics," aired May 6, 2009. New York: History Channel. www.metatube.com/en /videos/103730/Bob-White-case-ufo-object-UFO-Hunters/.

"The Bolshoi Tjach Skulls, Adygea, Russia." *Soul Guidance*, accessed December 22, 2016. http://soul-guidance.com/houseofthesun/dp09/dp09bolshoitjach. html.

Bonnet-Bidaud, J. M., and E. Pantin. "ADONIS High Contrast Infrared Imaging of Sirius B." *Astronomy and Astrophysics* no. 484, 2 (2008): 651–55.

Bothe, Elsbeth. "Facing the Beltway Snipers, Profilers Were Dead Wrong," *The Baltimore Sun*, December 15, 2002. http://articles.baltimoresun.com/2002-12-15 /entertainment/0212160297_1_van-zandt-serial-killers-snipers/2.

Brahm, John (director; written by Joseph Stefano and Lou Morheim). "The Bellero Shield." *The Outer Limit*. Season 1, episode 20. Aired February 10, 1964.

Brewer, Bobby. "Seven Things You Should Know about UFOs." *Christian Research Journal* 25, no. 2 (2002). http://www.equip.org/article/seven-things-you-should -know-about-ufos/.

Brin, David. *The Uplift War*. New York: Bantam Spectra, 1987.

Budinger, Phyllis A. "New Analysis of Soil Samples from the Delphos UFO Case." *Journal of UFO Studies*, 8 (2003): 1–25.

Bullard, Thomas E. *The Myth and Mystery of UFOs*. Lawrence: University Press of Kansas, 2010.

Burns, Kevin (executive producer). "Alien Power Plants." *Ancient Aliens*. Season 4, episode 13. Aired January 4, 2013. Los Angeles, CA: Prometheus Entertainment.

———. "Puma Punku." *Ancient Aliens*. Season 4, episode 6. Aired March 16, 2012. Los Angeles, CA: Prometheus Entertainment.

Campbell, Andy. "Bigfoot DNA Tests: Melba Ketchum's Research Results Are Bogus, Claims *Houston Chronicle* Report." Huffington Post, July 3, 2013. www.huffington post.com/2013/07/33/bigfoot-dna-test-results-n-3541431.html.

Capek, Karel. *R.U.R. (Rossum's Universal Robots)*. Translated by Paul Selver. London: Oxford University Press, 1923.

"Carbon Dating." UK.Rael.org, accessed December 30, 2016. http://uk.rael.org/page .php?11.

Carroll, Robert Todd. "Raëlian." *The Skeptic's Dictionary*, updated October 27, 2015. http: //www.skepdic.com/raelian.html.

Chan, Melissa. "'UFO Clouds' Form in Cape Town, Spark New Fears of Alien Invasion." *New York Daily News*, updated November 10, 2015. http://www.nydailynews.com/news /world/ufo-clouds-form-cape-town-spark-new-fears-aliens-article-1.2429217.

Chase, Rachel. "Calm Down: The Paracas Skulls Are Not from Alien Beings." *Living in Peru*, February 13, 2014, http://www.peruthisweek.com/blogs-calm-down -the-paracas-skulls-are-not-from-alien-beings-102258.

"Chemtrails." *RationalWiki*, updated December 27, 2016. http://rationalwiki.org/wiki /Chemtrails.

Childress, David. "Evidence for Ancient Atomic Warfare." *Nexus Magazine* 7, no. 5 (Nov./ Dec. 2000). http://www.bibliotecapleyades.net/ancientatomicwar/esp_ancient _atomic_01.htm.

Childress, David Hatcher (writer/director). *Ancient Technology in Central America*. Kempton, IL: Adventures Unlimited Media, 2009.

Clancy, Susan A. *Abducted: How People Come to Believe They Were Kidnapped by Aliens*. Cambridge, MA: Harvard University Press, 2005.

Clarke, Arthur C. *Childhood's End*. New York: Ballantine Books, 1953.

"Climate Change Debate: Last Week Tonight with John Oliver (HBO)." *YouTube* video, May 11, 2014. https://www.youtube.com/watch?v=cjuGCJJUGsg.

Cline, Austin. "Eyewitness Testimony and Memory: Human Memory Is Unreliable and So Is Eyewitness Testimony." *About Religion*, July 27, 2015. http://atheism.about .com/od/parapsychology/a/eyewitness.htm.

Colavito, Jason. *The Cult of Alien Gods: H. P. Lovecraft and Extraterrestrial Pop Culture*. Amherst, NY: Prometheus, 2005.

———. "Review of Ancient Aliens S04E06: The Mysteries of Puma Punku." *JasonColavito.com*, March 17, 2012. http://www.jasoncolavito.com/blog /review-of-ancient-aliens-s04e06-the-mysteries-of-puma-punku.

"Conspiracies (Web Exclusive): Last Week Tonight with John Oliver (HBO)." *YouTube* video, March 27, 2016. https://www.youtube.com/watch?v=fNS4lecOaAc.

Cook, Ryan J. "Heaven's Gate." *Center for AnthroUfology*, updated December 31, 2005. http://web.archive.org/web/20130731093726/http://www.anthroufo.info /un-hgate.html.

———. "Nuwaubians." *Center for AnthroUfology*, updated December 31, 2005. http://web.archive.org/web/20151009082634/http://www.anthroufo.info /un-nuwab.html.

"Core Beliefs." *Nuwaubian Nation*, October 24, 2014. http://nuwaubianinfo.weebly.com /information/core-beliefs.

Crème, Benjamin. "The World Teacher Is Now Here." *Share International*, 2015, http: //shareinternational.org.au/.

Crick, Francis. *Life Itself: Its Origin and Nature*. New York: Simon and Schuster, 1981.

———. "Seeding the Universe." *Science Digest* 89 (November 1981).

Cuoghi, Diego. "The Art of Imagining UFOs." *Skeptic Magazine* 11, no. 1 (2004): 43–53. http://www.diegocuoghi.com/pdf/SKEPTIC_CUOGHI_The_Art_Of _Imagining_UFOs.pdf.

"Customer Reviews for *Area 51: An Uncensored History of America's Top Secret Military Base*." *Amazon*, accessed August 22, 2016. http://www.amazon.com/Area-51 -Uncensored-Americas-Military/product-reviews/0316132942/ref=cm_cr_pr_ hist_1?ie=UTF8&filterBy=addOneStar&showViewpoints=0&sortBy=bySubm issionDateDescending.

Dales, George F. "The Mythical Massacre at Mohenjo-Daro" *Expedition* (spring 1964): 36–43.

"David Icke and the Reptilian Agenda." *Arcturi*, accessed December 30, 2016. http://www
    .arcturi.com/ReptilianArchives/DavidIcke.html.
Decker, Jerry. "Taken: The Abduction Phenomenon." *Unexplained Mysteries*. Episode 9.
    Aired November 14, 2003. Santa Monica, CA: Ann Daniel Productions.
————— "Touched by an Alien." *Unexplained Mysteries*. Episode 23. Aired May 7, 2004.
    Santa Monica, CA: Ann Daniel Productions.
Dell'Amore, Christine. "Biggest Virus Yet Found, Maybe Fourth Domain of Life?"
    National Geographic News, July 19, 3013. http://news.nationalgeographic.com
    /news/2013/07/130718-viruses-pandoraviruses-science-biology-evolution/.
"Democrats and Republicans Differ on Conspiracy Theory Beliefs." *Public Policy Poll-
    ing*, April 2, 2013. http://www.publicpolicypolling.com/pdf/2011/PPP_Release
    _National_ConspiracyTheories_040213.pdf.
Dick, Philip K. *Do Androids Dream of Electric Sheep?* Garden City, NY: Doubleday, 1968.
"Did UFO Visit Vancouver Canadians Baseball Game?" *Huffington Post*, September 9,
    2013. http://www.huffingtonpost.com/2013/09/09/ufo-canadian-baseball-game
    -photo-video_n_3895294.html.
Drosnin, Michael. *The Bible Code II: The Countdown*. New York: Viking, 2002.
Dunning, Brian. "Skeptoid #124: Betty and Barney Hill—the Original UFO Abduction."
    Skeptoid (podcast), October 21, 2008. https://skeptoid.com/episodes/4124/.
Dutch, Steven. "The Piri Reis Map." Updated June 2, 2010. http://www.uwgb.edu/dutchs
    /PSEUDOSC/PiriRies.HTM.
Ebiri, Bilge. "HBO's Scientology Exposé *Going Clear* Is Jaw-Dropping." *Vulture*, March 13,
    2015. http://www.vulture.com/2015/01/sundance-going-clear-is-a-jaw-dropping
    -expos.html.
Emmerich, Roland (director; screenplay by Roland Emmerich and Dean Devlin). *Inde-
    pendence Day*. Century City, CA: 20th Century Fox, 1996.
Enns, Anthony. "Alien Abduction: A Return to the Medieval." *Iowa Journal of Cultural
    Studies* 1 (1999).
"Extraterrestrial Highway (Near Area 51)." *Vegas.com*, accessed August 22, 2016. http:
    //www.vegas.com/attractions/near-las-vegas/extraterrestrial-highway/.
"FAQ: But, What about Bob Lazar?" *Dreamland Resort*, September 20, 2007. http://www
    .dreamlandresort.com/faq/faq_lazar.html.
"FAQ: Is It Really Called Area 51?" *Dreamland Resort*, September 9, 2010. http://www
    .dreamlandresort.com/faq/faq_name.html.
"FAQ: What Is the TTR (Tonopah Test Range)?" *Dreamland Resort*, September 11, 2007.
    http://www.dreamlandresort.com/faq/faq_ttr.html.
Farrakhan, Louis. "The Divine Destruction of America: Can She Avert It?" *The Final Call*,
    June 9, 1996.
Ferris, Timothy. "Playboy Interview: Erich von Daniken." *Playboy Magazine*, August
    1974.
Festinger, Leon, Henry W. Riecken, and Stanley Schachter. *When Prophecy Fails*. Min-
    neapolis: University of Minnesota Press, 1956.
Flank, Lenny. "The Baghdad Battery: An Update." *The Daily Kos*, February 10, 2015. http:
    //www.dailykos.com/story/2015/2/10/1361589/-The-Baghdad-Battery-An
    -Update.
Flavius Josephus *ca.* CE 90 (translated by William Whiston). *Antiquities of the Jews*. Pea-
    body, MA: Hendrickson Publishers, 1987.

Frantz, Douglas. "Scientology's Puzzling Journey from Tax Rebel to Tax Exempt." *The New York Times*, March 9, 1997. http://www.nytimes.com/1997/03/09/us/scientology-s-puzzling-journey-from-tax-rebel-to-tax-exempt.html.

Frazier, Kendrick, Barry Karr, and Joe Nickell, eds. *The UFO Invasion: The Roswell Incident, Alien Abductions, and Government Cover-Ups.* Amherst, NY: Prometheus, 1997.

Friedman, Stanton T. "The Bob Lazar Fraud." *StantonFriedman.com*, updated January 2011. http://www.stantonfriedman.com/index.php?ptp=articles&fdt=2011.01.07.

Friedman, Stanton T., and Don Berliner. *Crash at Corona: The U.S. Military Retrieval and Cover-Up of a UFO.* New York: Paragon House, 1992.

Friswell, Alan. "Faking It." *Fortean Times*, June 2005.

Frood, Arran. "Riddle of 'Baghdad's Batteries.'" *BBC News*, February 27, 2003.

Gadd, Geoffrey M. "Fungal Production of Citric and Oxalic Acid: Importance in Metal Speciation, Physiology and Biogeochemical Processes." *Advances in Microbial Physiology* 41 (1999): 47–92.

"George Hunt Williamson." *Obscurantist*, May 31, 2013. http://obscurantist.com/oma/Williamson-george-hunt/.

Gildenberg, B. D. "A Roswell Requiem." *Skeptic* 10, no. 1 (2003): 60–73.

Gish, Duane T. *Evolution? The Fossils Say No!* Green Forest, AR: New Leaf, 1978.

Godfrey, Laurie R., ed. *Scientists Confront Creationism.* New York: W. W. Norton, 1983.

Goodrick-Clarke, Nicholas. *Black Sun: Aryan Cults, Esoteric Nazism and the Politics of Identity.* New York: New York University Press, 2002.

Graña, Juan. "Maria Orsitsch." *Supercomentarios*, May 14, 2014. http://juan-jackblog.blogspot.com/2014/05/maria-orsitsch-eglish-version.html.

Greenberg, Moshe (commentator). "Ezekiel 1–20." In *Anchor Bible*, vol. 22, edited by William Foxwell Allbright and David Noel Freedman. Garden City, NY: Doubleday & Co. Inc., 1983.

Greenfieldboyce, Nell. "Galaxy Quest: Just How Many Earth-Like Planets Are Out There?" *NPR*, November 5, 2013. http://www.npr.org/2013/11/05/242991030/galaxy-quest-just-how-many-earth-like-planets-are-out-there.

"Grey Aliens Agenda: Human Hybrid Integration Program." *Arcturi*, accessed January 4, 2017. http://www.arcturi.com/GreyArchives/HumanHybridIntegrationProgram.html.

Griffith, Robert L., Jason T. Wright, Jessica Maldonado, Matthew S. Povich, Steinn Sigurdsson, and Brendan Mullen. "The G Infrared Search for Extraterrestrial Civilizations with Large Energy Supplies III: The Reddest External Sources in WISE." *Astrophysical Journal Supplement Series* 247, no. 2 (April 2015). https://arxiv.org/pdf/1504.03418v2.pdf.

Gruenschloss, Andreas. "Fiat Lux." September 2001. http://wwwuser.gwdg.de/~agruens/UFO/fiatlux.html.

"Hans Schindler Bellamy." *Atlantipedia*, updated April 17, 2016. http://www.atlantipedia.com/doku.php?id=hans_schindler_bellamy.

Harish, Alon. "UFOs Exist, Say 36 Percent in National Geographic Survey." *ABC News*, June 27, 2012. http://abcnews.go.com/Technology/ufos-exist-americans-national-geographic-survey/story?id=16661311.

Hart, Will. *The Genesis Race.* Rochester, VT: Inner Traditions Bear & Company, 2003.

Head, Tom. "The Mysterious Origins of the World's First City-Builders." *Mysterious Universe*, June 27, 2014. http://mysteriousuniverse.org/2014/06/the-mysterious-origins-of-the-worlds-first-city-builders/.

Heaven's Gate. http://www.heavensgate.com/.

Heiser, Michael S. "The Myth of a 12th Planet: A Brief Analysis of Cylinder Seal VA 243." *MichaelSHeiser.com*, December 5, 2003. http://www.michaelsheiser.com /VA243seal.pdf.

———. "The Myth of a 12th Planet: 'Nibiru' according to the Cuneiform Sources." *Sitchin Is Wrong*, updated August 7, 2010. http://www.sitchiniswrong.com /nibirunew.pdf.

Hesseman, Michael. *Space Brother Contactees*. Dusseldorf: 2000 Film Productions, 1996.

Hill, Kyle. "A Million Poisoning Planes." *The Committee for Skeptical Inquiry*, August 14, 2013. http://www.csicop.org/specialarticles/show/a_million_poisoning_planes.

Hill, Sharon. "UFO Research Is Up in the Air: Can It Be Scientific?" *The Committee for Skeptical Inquiry*, August 28, 2016. http://www.csicop.org/specialarticles/show /ufo_research_is_up_in_the_air.

———. "Want to Shed the Pseudoscience Label? Try Harder." *Doubtful Blog*, May 23, 2011. http://idoubtit.wordpress.com/2011/05/23/want-to-shed-the-pseudoscience -label-try-harder/.

His New Word. www.hisnewword.org.

Holloway, April. "Initial DNA Analysis of Paracas Skull Released—with Incredible Results." *Ancient Origins*, February 5, 2014. www.ancient-origins.net/news-evolution -human-origins/initial-dna-analysis-paracas-elongated-skull-released-incredible.

Horn, John. "Alex Gibney on *Going Clear*, His Scientology Documentary That's the Talk of Sundance." *Vulture*, January 30, 2015. http://www.vulture.com/2015/01/alex -gibney-on-his-new-scientology-documentary.html.

Hough, James. "What Is the Catholic Church's View on Aliens?" *Answers*, accessed December 30, 2016. http://www.answers.com/Q/What_is_the_Catholic _Church's_view_on_aliens?/.

Howell, Elizabeth. "How Many Stars Are in the Milky Way?" *Space.com*, May 21, 2014. http://www.space.com/25959-how-many-stars-are-in-the-milky-way.html.

Howells, William W. *Mankind in the Making*. Garden City, NY: Doubleday, 1959.

Hoyle, Fred, and Chandra Wickramasinghe. *Cosmic Life Force: The Power of Life across the Universe*. New York: Paragon House, 1990.

Icke, David. *The Biggest Secret*. Scottsdale, AZ: Bridge of Love Publications, 1999.

idoubtit. "Questionable Poll Touts UFO/ET Belief." *Doubtful News*, September 12, 2013. http://doubtfulnews.com/2013/09/questionable-poll-touts-ufoet-belief/.

"In U.S., 12 Million Americans Believe the World Is Run by Lizards from Outerspace." *The Intellectualist*, April 8, 2016. http://theintellectualist.co/in-u-s-12-million -americans-believe-the-world-is-run-by-a-group-of-lizards-from-outerspace/.

Jacobsen, Annie. *Area 51: An Uncensored History of America's Top Secret Military Base*. Boston: Back Bay Books, 2012.

Jaroff, Leon. "It Happens in the Best Circles." *Time*, September 23, 1994. http://content .time.com/time/magazine/article/0,9171,157904,00.html.

Jims Space Agency. "REAL ALIENS CONTACT !!! - Alien Message Proof Evidence *****." *YouTube* video, September 16, 2011. https://www.youtube.com /watch?v=xUFNZUxBYp8.

Johnson, Kenneth (director). *V*. TV miniseries. Burbank, CA: Warner Brothers Home Entertainment, 1983.

Johnston, Colin. "The Truth about Betty Hill's UFO Star Map." *Astronotes, Armagh Planetarium's Stellar Blog!*, August 19, 2011. www.armaghplanet.com/blog/betty -hills-ufo-star-map-the-truth.html.

———. "The Truth about Zeta Reticuli," *Astronotes, Armagh Planetarium's Stellar Blog!*, May 9, 2013. www.armaghplanet.com/blog/the-truth-about-zeta-reticuli.html.

Kaufman, Marc. "Most Earthlike Planets Found Yet: A 'Breakthrough.'" *National Geographic News*, April 18, 2013.

"Key Beliefs." *The Aetherius Society*, accessed June 2, 2017. http://www.aetherius.org /key-beliefs/.

Klass, Philip. "Walton Abduction Cover-up Revealed." *UFO Investigator*, June 1976.

Klass, P. J. *The Real Roswell Crashed Saucer Coverup*. New York: Prometheus, 1997.

Korff, Kal. *The Roswell UFO Crash: What They Don't Want You to Know*. Amherst, NY: Prometheus, 1997.

Korff, Kal K. *Spaceships of the Pleiades: The Billy Meier Story*. Amherst, NY: Prometheus, 1995.

———. "What Really Happened at Roswell?" *Skeptical Inquirer* 21, no. 4 (1997): 24–30.

Kottmeyer, Martin. "Entirely Unpredisposed: The Cultural Background of UFO Abduction Reports." *Magonia Magazine*, January 1990. http://www.debunker.com /texts/unpredis.html.

———. "The Saucer Error." *The REALL News* 1, no. 4 (May 1993). http://www.debunker .com/texts/SaucerError.html.

Krofft, Sid and Marty (producers). *Land of the Lost*. NBC and CBS, 1974–1976.

Lacitis, Erik. "Area 51 Vets Break Silence: Sorry, but No Space Aliens or UFOs." *Seattle Times Newspaper*, March 27, 2010.

"Lawn Mushrooms and Fairy Rings." *Missouri Botanical Garden*, accessed December 21, 2016. www.missouribotanicalgarden.org/gardens-gardening/your-garden /help-for-the-home-gardener/advice-tips-resources/pests-and-problems /diseases/fairy-rings/lawn-mushrooms.aspx.

Lee, Adam. "Strange and Curious Sects: Raelianism." *Patheos*, July 2009. http://www .patheos.com/blogs/daylightatheism/2009/07/strange-and-curious-sects-vi/.

Lee, Adrien. "Solved: The Mystery of the Mary Celeste." *UCL News*, May 20, 2006. https: //www.ucl.ac.uk/news/news-articles/inthenews/itn060522.

Leir, Roger. "Third Phase of the Moon." *YouTube* video, March 17, 2014. www.youtube .com/watch?v=wifuvvUKbfo.

Levin, A. Leo, and Harold Kramer. *Trial Advocacy Problems and Materials*. Mineola, NY: Foundation Press, 1968.

Ley, Willy. "Pseudoscience in Naziland." *Astounding Science Fiction*, May 1947.

Lindemann, Michael. "LA Story." *Circle Makers*, accessed December 21, 2016. www.circle-makers.org/la.html.

Linse, Pat, with Ean Harrison. "Bob White's Great UFO Artifact Mystery—Solved! Sometimes All It Takes Is Finding the Right Expert." *Skeptic*, October 12, 2011. www .skeptic.com/e-skeptic/11-10-12/.

Lippard, Jim. "The Decline and (Probable) Fall of the Scientology Empire!" *Skeptic Magazine* 17, no. 1 (2011). http://www.skeptic.com/reading_room/the-decline -and-probable-fall-of-the-scientology-empire/.

Lloyd, Ellen. "Humans' Extraterrestrial DNA: We All Carry a Coded Alien Message Left by the Star Gods." *Message to Eagle*, December 14, 2014. www.messagetoeagle .com/extraterrdnahum.php#.VsbNFykgouo.

Loftus, Elizabeth. "How Reliable Is Your Memory?" TED Talks, June 2013. http://www .ted.com/talks/elizabeth_loftus_the_fiction_of_memory.

Loftus, Elizabeth, and Katherine Ketcham. *Witness for the Defense: The Accused, the Eyewitness, and the Expert Who Put Memory on Trial.* New York: St. Martin's Press, 1991.

Lovett, Richard A., and Scot Hoffman. "Crystal Skulls Fuel Controversy, Fascination." *National Geographic,* accessed January 2, 2017. http://science.nationalgeographic.com/science/archaeology/crystal-skulls/.

Loxton, Daniel. "Space Brothers from Venus?" *Junior Skeptic, Skeptic Magazine* 20, no. 4 (2015).

Mahood, Tom. "The Bob Lazar Corner." *Dreamland Resort,* September 21, 2007. http://www.dreamlandresort.com/area51/lazar/index.htm.

Martinelli, Maurizio. "George Hunt Williamson–Michael D'Obrenovic: The Herald of the Encounter among Father and Sons." duepassinelmistero.com, September 2009. http://www.duepassinelmistero.com/williamson.htm.

Maxwell, Andrew. "Conspiracy Road Trip: UFOs." *YouTube* video, October 15, 2012. http://www.youtube.com/watch?v=7ByWCFX4ZQs.

Mayer, Jean-François (translation by Elijah Siegler). "'Our Terrestrial Journey Is Coming to an End': The Last Voyage of the Solar Temple." *Nova Religio* 2, no. 2 (April 1999). http://english.religion.info/2003/06/12/our-terrestrial-journey-is-coming-to-an-end-the-last-voyage-of-the-solar-temple/.

McAndrew, James; U.S. Air Force. *The Roswell Report: Case Closed.* Washington, DC: U.S. Government Printing Office, 1997.

Meier, Billy. "Henoch Prophecies." *They Fly,* February 28, 1987. http://www.theyfly.com/henoch-prophecies.

Menzies, William Cameron (director). *Invaders From Mars.* Century City, CA: Twentieth Century Fox, 1953.

Milius, John (director; screenplay by John Milius and Oliver Stone). *Conan the Barbarian.* Universal Pictures and 20th Century Fox, 1982.

Miller, Russel. *Bare-faced Messiah: True Story of L. Ron Hubbard.* London: Michael Joseph, 1987.

Missler, Chuck. "The Return of the Nephilim." *Personal Update,* September 1997.

"Mohenjo-daro." *RationalWiki,* updated December 27, 2016. http://rationalwiki.org/wiki/Mohenjo-daro.

Morton, Chris, and Ceri Louise Thomas. *The Mystery of the Crystal Skulls.* Rochester, VT: Inner Traditions Bear and Company, 2002.

Moseley, James W. "Some New Facts about 'Flying Saucers Have Landed,'" *Saucer News* 1 (October 1957): 2–12.

"MUFON Sighting Report." *Mutual UFO Network,* accessed December 30, 2016. https://mufoncms.com/cgi-bin/report_handler.pl.

"Mutual Mistrust May Have Added a Few X-files to the UFO Era." *ScienceDaily,* February 16, 2016. https://www.sciencedaily.com/releases/2016/02/160216181704.htm.

Newman, Joseph M., and Jack Arnold (directors; screenplay by Raymond F. Jones, Frank Coen and Edward G. O'Callaghan). *This Island Earth.* Universal International, 1955.

Nickell, Joe. *Entities: Angels, Spirits, Demons, and Other Alien Beings.* Amherst, NY: Prometheus, 1995.

Nickell, Joe. *Real-Life X-Files: Investigating the Paranormal.* Louisville: University Press of Kentucky, 2001.

Nickell, Joe. "Return to Roswell." *Skeptical Inquirer* 33, no. 1 (January/February): 10–12.

Nickell, Joe, and James McGaha. "The Roswellian Syndrome: How Some UFO Myths Develop." *Skeptical Inquirer* 36, no. 3 (2012). http://www.csicop.org/si/show /the_roswellian_syndrome_how_some_ufo_myths_develop

Nieves-Rivera, Ángel M. "About the So-Called 'UFO Rings' and Fungi." Supplement, *Mycologia* 52, no. S6 (2001): S3–S6. www.escepticospr.com/Archivos/uforings .htm.

"1971, The Delphos Kansas UFO Landing Ring." *UFO Casebook*, accessed December 21, 2016. ufocasebook.com/Kansas.html.

Nocera, John. "Scientology's Chilling Effect." *The New York Times*, February 24, 2015. http://www.nytimes.com/2015/02/24/opinion/joe-nocera-scientologys-chilling -effect.html.

"Nordic Aliens." *UFO–Alien Database*, updated August 23, 2015. http://ufoaliendatabase .wikia.com/wiki/Nordic_Aliens.

Norman, Ernest L. "The Truth about Mars." *FireDocs*, updated 1967. http://www.firedocs .com/remoteviewing/mars/mars.html.

Novella, Steven, M.D. "The Starchild Project." *New England Skeptical Society*, February 2006.

Oksman, Olga. "Conspiracy Craze: Why 12 Million Americans Believe Alien Lizards Rule Us." *The Guardian*, April 7, 2016.

O'Neil, Gerald. *The High Frontier: Human Colonies in Space.* New York: Morrow, 1976.

Ortega, Tony. "The Great UFO Cover-up." *Phoenix New Times*, June 26, 1997.

———, Tony. "The Hack and the Quack." *Phoenix New Times*, March 3, 1998.

———. "Scientology Targeted South Park's Parker and Stone in Investigation." *The Village Voice*, October 23, 2011. http://blogs.villagevoice.com/runninscared/2011/10 /scientology_tar.php.

"Paper 57: The Origin of Urantia." In *The Urantia Book*. http://www.urantia.org /urantia-book-standardized/paper-57-origin-urantia.

Pauwels, Louis, and Jacques Bergier (translation by Rollo Myers). *The Morning of the Magicians.* New York: Stein and Day, 1963.

Pedlow, Gregory W., and Donald E. Welzenbach. *The Central Intelligence Agency and Overhead Reconnaissance: The U-2 and OXCART Programs, 1954–1974.* Washington DC: History Staff, Central Intelligence Agency, 1992.

Peebles, Curtis. *Dark Eagles* (rev. ed.). Novato, CA: Presidio Press, 1999.

Pendle, George. *Strange Angel: The Otherworldly Life of Rocket Scientist John Whiteside Parsons.* New York: Harcourt Publishers, 2006.

Pevney, Joseph (director; screenplay by Gene L. Coon and Frederick Brown). "Arena." *Star Trek.* Season 1, Episode 18, aired January 19, 1967.

Pfennig, David W., William R. Harcombe, and Karin S. Pfennig. "Frequency-Dependent Batesian Mimicry," *Nature* 410, no. 6826 (2001): 323. doi:10.1038/35066628.

Pflock, K. *Roswell: Inconvenient Facts and the Will to Believe.* Amherst, NY: Prometheus, 2001.

"Phoenix Lights." *Wikipedia.* https://en.wikipedia.org/wiki/Phoenix_Lights.

Physics 4 You. "Chronicles of Atlantis." *Unarius Wisdom*, accessed January 2, 2017. http://web.archive.org/web/20150614113150/http://www.unariuswisdom.com /chronicles-of-atlantis/.

Pigliucci, Massimo. *Nonsense on Stilts: How to Tell Science from Bunk.* Chicago: University of Chicago Press, 2010.

"The Piney Woods Incident, Cash-Landrum." *UFO Casebook*, accessed December 22, 2016. www.ufocasebook.com/Pineywoods.html.

Plait, Phil. "Mesa Me!" *Bad Astronomy*, April 13, 2008. http://www.badastronomy.com /bitesize/marsmesa.html.

Prothero, Donald R. *Reality Check: How Science Deniers Threaten Our Society*. Bloomington: Indiana University Press, 2013.

———. "Science TV 'Network Decay.'" *SkepticBlog*, January 25, 2012. http://www.skep ticblog.org/2012/01/25/science-tv-sell-out/.

Prothero, Donald, Brian Dunning, Mark Edward, Daniel Loxton, Steven Novella, and Michael Shermer. *SkepticBlog*. http://www.skepticblog.org/.

Puig, Claudia. "Scientology Doc 'Going Clear' Shocks Sundance Filmgoers." *USA Today*, January 26, 2015. http://www.usatoday.com/story/life/movies/2015/01/26 /controversial-scientology-documentary/22337995/.

Puzuzu Ba'al. "UFO Clouds." *Alien Sky Ships*, accessed January 4, 2017. http://www.alien skyships.com/UFO_clouds.html.

Quiltingmamma. "Photo: 'Before and after Cleaning.'" *Trip Advisor*, June 18, 2011. www .tripadvisor.com/LocationPhotoDirectLink-g1598532-d472016-i32011744- Temple_of_Hathor_at_Dendera-Qena_Qena_Governorate_Nile_River_ Valley.html.

Radford, Benjamin. "Mysterious Phoenix Lights a UFO Hoax." *Live Science*, April 23, 2008. http://www.livescience.com/2483-mysterious-phoenix-lights-ufo-hoax.html.

Radford, Benjamin, and Joe Nickell. *Lake Monster Mysteries: Investigating the World's Most Elusive Creatures*. Lexington: University Press of Kentucky, 2006.

Rael. "Message from the Designers." *The Raelian Movement*. www.rael.org.

Raine, Lee, and Kathryn Zickuhr. "American Views on Mobile Etiquette." Pew Research Center, August 26, 2015. http://www.pewinternet.org/2015/08/26 /americans-views-on-mobile-etiquette/

Randle, K., and D. Schmitt. *The Truth about the UFO Crash at Roswell*. New York, Avon, 1994.

Rawles, Bruce. "The Abydos-Hieroglyph Does NOT Depict a Helicopter." *Abduction and UFO Research Association*, 1996. http://www.members.tripod.com/A_U_R_A /abydos.html.

Reitman, Janet. "Inside Scientology: Unlocking the Complex Code of America's Most Mysterious Religion." *Rolling Stone*, February 23, 2006.

"Reptilian Alien Reports." *MUFON Forum*, updated October 27, 2015. http://mutualufon etwork.proboards.com/thread/221/reptilian-alien-reports/.

"Reptilians and Human Abductions." *Arcturi*, accessed December 30, 2016. http://www .arcturi.com/ReptilianArchives/ReptiliansAndAbductions.html.

"The Rhodope Skull: Evidence for the Existence of Aliens on Earth?" *The Event Chronicle*, March 9, 2015. www.theeventchronicle.com/metaphysics/galactic /the-rhodope-skull-evidence-for-the-existence-of-aliens-on-earth/.

Ridpath, Ian. "Nick Pope's 'Investigation' of the Rendlesham Forest Radiation Readings." IanRidpath.com, April 2005. www.ianridpath.com/ufo/rendlesham.4a.htm.

———. "The Rendlesham UFO Witness Statements." *IanRidpath.com*, December 2012. www.ianridpath.com/ufo/rendlesham2c.htm.

Robinson, Joanna. "The 6 Most Disturbing Moments from HBO's Scientology Documentary *Going Clear*." *Vanity Fair*, March 30, 2015. http://www.vanityfair.com /hollywood/2015/03/going-clear-disturbing-moments.

Roesch, Michael. "Harmonic Convergence." *Mount Shasta Companion,* College of the Siskiyous, accessed January 4, 2017. http://www.siskiyous.edu/shasta/fol/har/index.htm.

Rojas, Rick. "UFO Investigators' Convention Emphasizes Scientific Methods." *Los Angeles Times,* July 31, 2011. http://articles.latimes.com/2011/jul/31/local/la-me-ufo-20110731.

Roux, Georges. *Ancient Iraq.* London: Allen and Unwin, 1992.

Sagan, Carl. *The Demon Haunted World.* New York: Random House and Ballantine, 1996.

Sakulich, Aaron. "The 'Baghdad Battery.'" *The Iron Skeptic,* accessed December 30, 2016. http://www.theironskeptic.com/articles/battery/battery.htm.

———. "A Media History of Gray Aliens." *The Triangle,* May 20, 2005. http://www.theironskeptic.com/articles/gray/gray_history.htm.

Saler, Benson, Charles A. Ziegler, and Charles B. Moore. *UFO Crash at Roswell: The Genesis of a Modern Myth.* Old Saybrook, CT: Konecky & Konecky, 1997.

Saletan, William. "Conspiracy Theorists Aren't Really Skeptics," *Slate,* November 19, 2013. http://www.slate.com/articles/health_and_science/science/2013/11/conspiracy_theory_psychology_people_who_claim_to_know_the_truth_about_jfk.html.

Schacter, Daniel L. *Searching for Memory: The Brain, the Mind and the Past.* New York: Basic Books, 1996.

Schlikerman, Becky. "Farrakhan: Revolution Imminent in U.S." *Chicago Tribune,* February 27, 2011.

"Scientists Have Found an Alien Code in Our DNA: Ancient Engineers." *Ancient Code,* April 27, 2015. http://www.ancient-code.com/scientists-have-found-an-alien-code-in-our-dna-ancient-engineers/.

Scott, Ridley (director; screenplay by Hampton Fancher and David Peoples). *Blade Runner.* Burbank, CA: Warner Brothers, 1982.

"SETI FAQ: How Long Have Astronomers Been Looking for Extraterrestrial Signals?" *SETI,* accessed January 5, 2017. http://www.seti.org/faq#obs19.

Sharps, M. J., J. Matthews, and J. Asten. "Cognition and Belief in Paranormal Phenomena: Gestalt/Feature-Intensive Processing Theory and Tendencies toward ADHD, Depression and Dissociation." *The Journal of Psychology* 140, no. 6 (2006): 579–90.

shCherbaka, Vladimir I., and Maxim A. Makukov. "The "Wow! Signal" of the Terrestrial Genetic Code." *Icarus* 224, no. 1 (May 2013): 228–42.

Sheaffer, Robert. "Between a Beer Joint and Some Kind of Highway Warning Sign: The 'Classic' Cash–Landrum Case Unravels." *Bad UFOs,* November 13, 2013. http://badufos.blogspot.com/2013/11/between-beer-joint-and-some-kind-of.html.

———. "The First 'Flying Saucer' Sighting: Kenneth Arnold, Mt. Rainier, Washington—June 24, 1947." *Debunker,* updated August 5, 2016. http://www.debunker.com/arnold.html.

Shearer, Christine, Mick West, Ken Caldeira, and Steven J. Davis. "Quantifying Expert Consensus against the Existence of a Secret, Large-Scale Atmospheric Spraying Program." *Environmental Research Letters* 11, no. 8 (2016). http://iopscience.iop.org/article/10.1088/1748-9326/11/8/084011.

Shermer, Michael. 1997. *Why People Believe in Weird Things.* New York: Holt.

———"Agenticity." MichaelShermer.com, June 2009. http://www.michaelshermer.com/2009/06/agenticity/.

———. *The Believing Brain*. New York: Holt, 2011.

———. "How to Talk to a UFOlogist (If You Must)." SkepticBlog, August 25, 2009. http://www.skepticblog.org/2009/08/25/how-to-talk-to-a-ufologist/.

———. "Patternicity: Finding Meaningful Patterns in Meaningless Noise." *Scientific American*, December 1, 2008. https://www.scientificamerican.com/article/patternicity-finding-meaningful-patterns/.

———. "Show Me the Body." *MichaelShermer.com*, May 2003. http://www.michaelshermer.com/2003/05/show-me-the-body/.

Shermer Michael. *Why People Believe Weird Things*. New York: Holt, 1997.

Shostak, Seth. "UFOs: The Trail Is Stale." *The Huffington Post*, updated September 1, 2016. http://www.huffingtonpost.com/seth-shostak/ufos-the-trail-is-stale_b_8071452.html.

Simon, Armando. "Pulp Fiction UFOs." *Skeptic Magazine* 16, no. 4 (2011).

Simons, Daniel. "Selective Attention Test." *YouTube* video, March 10, 2010. http://www.youtube.com/watch?v=vJG698U2Mvo.

Simpson, George Gaylord. "On the Nonprevalence of Humanoids." *Science* 143: 769–75.

Sims, Derrel. "One Kid Who Didn't Buy It." *Journal of Abduction Encounter Research* (1Q 2007). www.alienhunter.org/alien-abduction/derrels-story/.

Sitchin, Zecharia. *The 12th Planet*. New York: Stein and Day, 1976.

"The Smallest Spy Drone in the World—Pic and Video." *Gadgetonics*, July 5, 2012. http://www.gadgetonics.com/2012/07/smallest-spy-drone-in-world-pic-and.html.

Smith, Cordwainer. *The Ballad of Lost C'mell*. New York: Galaxy, 1962.

Soltysiak, Arkadiusz. "Physical Anthropology and the 'Sumerian Problem.'" *Studies in Historical Anthropology* 4, no. 2004 (2006): 145–58. http://www.antropologia.uw.edu.pl/SHA/sha-04-07.pdf.

"The Sound Bath." *Integratron*, accessed January 4, 2017. http://integratron.com/sound-bath/.

Spangerberg, Anton. "Sealand Skull Photos Released." *Unexplained Mysteries*, September 9, 2010. www.unexplained-mysteries.com/column.php?id=189988.

Spelgel, Lee. "Area 51 Personnel Feel 'Betrayed' By Annie Jacobsen's Soviet–Nazi UFO Connection." *The Huffington Post*, December 8, 2011. http://www.huffingtonpost.com/2011/06/07/area-51-ufos-aliens-annie-jacobsen-nazi-soviet_n_869706.html.

———. "More Believe in Space Aliens Than in God According to U.K. Survey." *The Huffington Post*, updated October 18, 2012. http://www.huffingtonpost.com/2012/10/15/alien-believers-outnumber-god_n_1968259.html.

———. "UFO Hoax at Canada Baseball Game Exposed; H.R. MacMillan Space Centre Admits Bizarre Stunt." *The Huffington Post*, updated September 14, 2013. http://www.huffingtonpost.com/2013/09/12/canada-baseball-game-ufo-hoax_n_3908612.html.

Stackpole, Mike. "Dr. Stranges Lives Up to His Name." *The Phoenix Skeptics News* 1, no. 6 (May/June 1988). http://www.discord.org/~lippard/Arizona_Skeptic/PSNv1n6-May-Jun-1988.pdf.

Stevenson, F. J. *Humus Chemistry: Genesis, Composition, Reactions*. New York: John Wiley and Sons, 1994.

"Steven Spielberg Surprised by Drop in UFO Sightings." *Contact Music*, June 28, 2005. http://www.contactmusic.com/steven-spielberg/news/spielberg-confused-by-decrease-in-ufo-sightings.

Stone, Robert (producer and director). *Farewell, Good Brothers*. Los Angeles: Robert Stone Productions, 1992.

Stranges, Frank E. "A Holy Stranger in the Pentagonal Lodge." In *Stranger at the Pentagon*. New Brunswick, NJ: Inner Light Publications, 1967. http://www.theeventchronicle .com/editors-pick/valiant-thor-a-venusian-at-the-pentagon-rev-frank-e-stranges/.

Strickland, Jonathan, and Patrick J. Kiger. "How Area 51 Works." *HowStuffWorks Science*, July 5, 2007. http://science.howstuffworks.com/space/aliens-ufos/area-51 6.htm.

Strong, James. *Strong's Exhaustive Concordance of the Bible*. Gordonville, TN: Dugan, 1890.

Sturluson, Snorri. *Heimskringla: The Norse King Sagas*. Translated by Samuel Laing. London and Toronto: J. M. Dent & Sons Ltd.; New York: E. P. Dutton & Co., 1930.

Takaki, Koichi, Kohei Yoshida, Tatsuya Saito, Tomohiro Kusaka, Ryo Yamaguchi, Kyusuke Takahashi, and Yuichi Sakamoto. "Effects of Electrical Stimulation on Fruit Body Formation in Cultivating Mushrooms." *Microorganisms* 2 (2014): 58–72. doi:10.3390/microorganisms2010058.

"Tap O' Noth." *Canmore*, accessed January 2, 2017. https://canmore.org.uk/site/17169 /tap-o-noth.

Teresa of Avila *ca.* 1580. *The Life of St. Teresa of Jesus of the Order of Our Lady of Carmel*. Ch. XXIX. http://www.ccel.org/ccel/teresa/life.viii.xxx.html.

"They're Back! Phoenix Lights Caught over Arizona Again!" *WN.com*, March 31, 2016. https: //wn.com/ufo_phoenix_lights_debunked_video_tape_analysed_3_13_1997.

Thomas, Dave. 1995. "The Roswell Incident and Project Mogul: Scientist Participant Supports Direct Links." *Skeptical Inquirer* 19, no. 4 (July/August): 15–18.

"3 Facts About Alien Implants." *Alien Hunter*, accessed December 19, 2016. www.alien hunter.org/implants.

Today Tonight. "Space Aliens Are at the Very Root of the Belief System of Scientology." February 4, 2009.

"Transcript of Bergstrom AFB Interview of Betty Cash, Vicky & Colby Landrum." *The Computer UFO Network*, August 1981. www.cufon.org/cufon/cashlani.htm.

University of Kansas. "Mysterious Stone Spheres in Costa Rica Investigated." *Science Daily*, March 23, 2010.

"Unmasking the Face on Mars." *NASA Science*, May 24, 2001. http://science1.nasa.gov /science-news/science-at-nasa/2001/ast24may_1/.

Urantia Foundation. *The Urantia Book*. Chicago: Urantia Foundation, 1955.

Van Beek, Walter E. A. "Dogon Restudied: A Field Evaluation of the Work of Marcel Griaule." *Current Anthropology* 32, no. 2 (April 1991). http://www.michaelsheiser .com/PaleoBabble/Grialle%20Sirius%20Dogon.pdf.

von Daniken, Erich. *Chariots of the Gods?* New York: G. P. Putnam's Sons, 1968.

Watkins, Ramon. "The Black Jew Hebrew Israelite Study Guide." *Prophet Yahweh UFOs*, October 2011. http://prophetyahwehufos.blogspot.com/2011/10/black-jew -hebrew-israelite-study-guide.html.

Weaver, Richard, and James McAndrew; U.S. Air Force. *The Roswell Report: Fact vs. Fiction in the New Mexico Desert*. Washington, DC: U.S. Government Printing Office, 1995.

Wells, H. G. *The First Men in the Moon*. 1901. New York: Berkley Highland Books, 1970.

———. *The Island of Dr. Moreau*. London: Heinemann, Stone & Kimball, 1896.

West, Mick. "A Brief History of 'Chemtrails.'" *Contrail Science*, May 11, 2007. http://con trailscience.com/a-brief-history-of-chemtrails/.

———. "Debunked: What in the World Are They Spraying?" *Contrail Science*, October 26, 2010. http://contrailscience.com/what-in-the-world-are-they-spraying/.

———. "Fake, Hoax, Chemtrail Videos." *Contrail Science*, July 24, 2010. http://contrailscience.com/fake-hoax-chemtrail-videos/.

———. "Germans Admit They Used Düppel!" *Contrail Science*, March 14, 2008. http://contrailscience.com/germans-admit-they-used-duppel/.

———. "How to Debunk Chemtrails." *Contrail Science*, June 5, 2011. http://contrailscience.com/how-to-debunk-chemtrails/.

———. "Kucinich, Chemtrails and HR 2977–The 'Space Preservation Act.'" *Contrail Science*, May 19, 207. http://contrailscience.com/kucinich-chemtrails-and-hr-2977/.

———. "WWII Contrails." *Contrail Science*, August 16, 2008. http://contrailscience.com/wwii-contrails/.

Whitbourne, Susan Krause. "What Psychology Can Tell You About Your Extraterrestrial Beliefs," *Psychology Today*, June 12, 2012, https://www.psychologytoday.com/blog/fulfillment-any-age/201206/what-psychology-can-tell-you-about-your-extraterrestrial-beliefs.

White, Chris (producer), with commentary by Dr. Michael Heiser. "Vimanas." *Ancient Aliens Debunked*, accessed January 2, 2017. http://ancientaliensdebunked.com/references-and-transcripts/vimanas/#sthash.hsS6cc8w.dpuf/.

———. *Ancient Aliens Debunked*. http://ancientaliensdebunked.com/reference-and-transcripts/puma-punku/.

Whitehead, John W. "Roaches, Mosquitoes and Birds: The Coming Micro-Drone." *The Huffington Post*, April 17, 2013. http://www.huffingtonpost.com/john-w-whitehead/micro-drones_b_3084965.html.

Whitlock, Craig, and John White. "Cops Slow to Link Caprice to Killings." *The Washington Post*, October 26, 2002. http://www.sfgate.com/news/article/Cops-slow-to-link-Caprice-to-killings-They-2759421.php.

Williams, David. "Mary Celeste Hit by Seaquake!" *Deaf Whale Society*, accessed January 20, 2017. www.deafwhale.com/mary-celeste/

Wise, Robert (director; screenplay by Edmund H. North). *The Day the Earth Stood Still*. Century City, CA: 20th Century Fox, 1951.

"The World Order and the Reptilians." *Arcturi*, accessed December 30, 2016. http://www.arcturi.com/ReptilianArchives/ReptiliansAndTheRothchilds.html.

Worley, Donald. "Alien Abduction Rape by Reptilians." *Abduct.com*, accessed December 30, 2016. http://www.abduct.com/worley/worley68.php.

Wright, Lawrence. "The Apostate: Paul Haggis vs. the Church of Scientology." *The New Yorker*, February 14, 2011. http://www.newyorker.com/magazine/2011/02/14/the-apostate-lawrence-wright.

Wyndham, John. *The Day of the Triffids*. Garden City, NY: Doubleday, 1951.

Zirpolo, K., and D. Nathan. "I'm Sorry: A Long-Delayed Apology from One of the Accusers in the Notorious McMartin Pre-School Molestation Case." *Los Angeles Times Magazine*, October 30, 2005.

Zogby International. "Zogby America Likely Voters 8/23/07 thru 8/27/07." *911Truth.org*, September 6, 2007. http://www.911truth.org/images/ZogbyPoll2007.pdf.

INDEX

ABOUT THE AUTHORS

Donald R. Prothero taught college geology and paleontology for 40 years at Caltech, Columbia, Cal Poly Pomona, and Occidental, Knox, Vassar, Glendale, Mt. San Antonio, and Pierce Colleges. He is the author of more than 40 books (including six leading geology textbooks and several trade books), and more than 300 scientific papers, mostly on the evolution of fossil mammals (especially rhinos, camels, and horses) and on using the earth's magnetic field changes to date fossil-bearing strata.

Though Timothy D. Callahan was trained as an artist and worked for more than 20 years in the animation industry, throughout his life he has had an active interest in the sciences. He has also maintained a strong interest in the study of mythology and religious belief and is the religion editor for *Skeptic Magazine*. He and his wife and fellow artist, Bonnie, live in the foothills of Altadena, California.